History of Modern Physics and Astronomy   *Volume* *14*

A Biography of
George Ellery Hale

Explorer of the Universe

HELEN WRIGHT

History of Modern Physics and Astronomy

History of Modern Physics and Astronomy

TITLES IN SERIES

History of Modern Physics and Astronomy

INTRODUCTORY NOTE

With the publication of the English translation of *Les Méthodes nouvelles de la Mécanique céleste* in 1992, The History of Modern Physics 1800–1950 Series title changed to History of Modern Physics and Astronomy, reflecting the inclusion of books on the history of astronomy. The History of Modern Physics and Astronomy Series is an evolution and continuation of the original series and incorporates the twelve volumes published in the original series.

History of Modern Physics and Astronomy retains the original aim of bringing to modern readers a variety of important works related to the history of physics since 1800 that are not readily available elsewhere. In addition, works specifically about the history of astronomy are included both in relation to the internal development of astronomy and to the physics of the period in which they were written, emphasizing the close bonds between physics and astronomy.

The books in the Series are all noteworthy additions to the literature of the history of physics and astronomy. They have been selected for their merit, distinction, and uniqueness. Each book in the Series is prefaced by an introductory essay by a scholar who places the work in its historical context, thus making it more valuable as a reference work.

We believe that these books will be of interest not only to the advanced scholar in the history of physics and astronomy but also to a much broader, less specialized group of readers who may wish to understand major scientific disciplines that have become central forces in society and an integral part of our twentieth-century culture. Taken in its entirety, the Series will bring to the reader a comprehensive picture of these major disciplines not readily achieved in any one work. Taken individually, the works selected will surely be enjoyed and valued in themselves.

History of Modern Physics and Astronomy   *Volume* 14

A Biography of
George Ellery Hale

Explorer of the Universe

HELEN WRIGHT

American Institute of Physics   **New York**

©1994 by American Institute of Physics
All rights reserved.
Printed in the United States of America.

AIP Press
American Institute of Physics
500 Sunnyside Boulevard
Woodbury, NY 11797-2999

Library of Congress Cataloging-in-Publication Data
Wright, Helen, 1944–
Explorer of the universe: a biography of George Ellery Hale / Helen Wright.
     p. cm. -- (History of modern physics and astronomy ; v. 14)
   Originally published: New York : Dutton, 1966. With new introd.
   Includes bibliographic references and index.
   ISBN 1-56396-249-7
   1. Hale, George Ellery, 1868–1938. 2. Astrophysics--United States--History.
3. Astrophysics--United States--Biography.
I. Title. II. Series.
QB460.72.H35W75 1994   94-1965
523.01' 092--dc20   CIP
[B]
10 9 8 7 6 5 4 3 2 1

Grateful acknowledgement is made to the following for permission to reprint copyright material:

Carnegie Institution of Washington for excerpts from *Ten Years' Work of a Mountain Observatory,* by George Ellery Hale.

Griffith Observatory of the City of Los Angeles for excerpts from article on George Ellery Hale in September, 1947, issue of *The Griffith Observer.*

J. B. Lippincott Company for excerpts from "Prefatory Note" and "Prologue, The Observatory," form "Watchers of the Sky" in *Collected Poems* by Alfred Noyes (Copyright 1906, 1934 by Alfred Noyes. Published by J. B. Lippincott Company).

The Macmillan Company for excerpt from *Astronomy of the Twentieth Century,* by Otto Struve and Velta Zebergs. Copyright © 1962, The Macmillan Company.

Nature for excerpt from article by Sir James Jeans in its March 19, 1938, issue.

Charles Scribner's Sons for excerpts from the following books by George Ellery Hale: *The Depths of the Universe, The New Heavens, Signals from the Stars,* and *Beyond the Milky Way*; for illustration from *Signals from the Stars* (Copyright 1931 Charles Scribner's Sons, renewal copyright © 1959 Evelina S. C. Hale); and for excerpts from *Pioneer to the Past* by Charles Breasted.

Sky and Telescope for diagram of spectroheliograph from its May, 1960, issue.

The Carnegie Observatories and the Huntington Library for photos credited to Mount Wilson and Palomar Observatories.

Introduction to the REF
Reprint Edition

George Ellery Hale was a key founder of astrophysics, as important to this field as Rutherford, Millikan, Lawrence, and van de Graaff were to the beginnings of modern particle physics. A prominent example, but only one of many, is the creation and direction of the Mount Wilson Observatory of the Carnegie Institution of Washington by Hale beginning in 1904. This observatory, combining solar and stellar astronomy, laid much of the foundations of modern astrophysics, and eventually observational cosmology. Mount Wilson is to astronomy what Rutherford's Cavendish laboratory was to physics.

It seems then hardly credible that many young astronomers, beginning their education just after the second world war in the early 1950's, scarcely knew of Hale's name and influence. There were various reasons for this, principal of which was Hale's lack of self advertisement, combined with the very few numbers of professional astronomers practicing in pre-war years. It is through such a group of older astronomers that his reputation would normally have spread. Furthermore, contrary to the situation surrounding the beginnings of modern physics, there have been few histories written about the origins of astrophysics.

In the year of Hale's death at the age of 70 in 1938, the total number of astronomers in the world at the research level was less than 600 as judged by the membership of the International Astronomical Union. By comparison, the number of research physicists worldwide was perhaps 30 times that number. In addition, physics was much better known as a science immediately after World War II because of its prominence in so much of the war effort, ushering in the high tech era that has followed.

The first publication of this biography of Hale by Helen Wright in 1966 changed much of the obscurity that had descended in the middle third of this century around this central figure in world astrophysics. From the historical record of letters, conversations with Mount Wilson astronomers, friends, associates and colleagues in other fields, and from the archives of the multitude of organizations associated with Hale, Helen Wright shows that Hale was greatly revered and highly honored in his lifetime. For example, in a poll taken among scientists in 1914, Hale at the age of only 45 was considered to be one of the twelve foremost living men of science. His discoveries in solar physics and his leadership in founding the Yerkes and the Mount Wilson observatories had, even at that time, become legend. The scientific discoveries and the founding of these two great observatories, and later the greatest of them all at Palomar, are several of the central themes of this book.

The reader will also find here facts too little known but of overriding importance in the history of astrophysics. A few examples from the book show the flavor of Hale's life and the electric atmosphere in which work progressed in the transcendental experience of the Mount Wilson work. The hyperenergetic ambience of that atmosphere gave a sense of great events in the making.

Helen Wright could see and appreciate part of the uniqueness of the enterprise by her connection with the observatory through her father, who was head of the Carnegie moon project using the Hooker 100-inch telescope, then the largest in the world. Fred Wright, Home Secretary of the National Academy of Sciences, and an international figure in geophysics himself, naturally became part of Hale's circle in American science policy. Helen, herself a person of letters in scientific matters, was better placed than any other person to write of Hale and his accomplishments, having known many of the staff of the Mount Wilson Observatory from her childhood. Her 10-year long labor on this biography is the splendid result.

Hale's driving goal was to understand stellar evolution: how the stars are formed and what stages they pass through as they live out their lives. He was convinced that understanding the sun would lead to an understanding of other stars. This passionate act of faith was carried through most of his own personal research on the sun.

His work toward this goal and the rewards it brought started early. His invention and perfection of the spectroheliograph when he was but 22 permitted the daily mapping of such features as prominences at the limb of the sun and their projections on the disk as flocculi. This feat (following on the earlier, crude method of Charles A. Young, and the later, disputed, claim of priority by Henri Deslandres) was accomplished by means of his brilliant invention of placing a second spectrograph slit on any particular spectral line after the initial spectral dispersion. Letting the sun drift across the focal plane of the first slit then built up an image of the sun in the light of the particular chemical element corresponding to the chosen spectral line. The invention earned him the Gold Medal of the Royal Astronomical Society in 1904, at the age of 36. The results with the new instrument in Hale's hands were so spectacular that it was said to be a "striking achievement, savoring of the miraculous," an accolade given by no less a scientist than R. G. Aitken. Hale's insistence of applying the

methods of physics to astronomy was the theme of his scientific life. The spectroheliograph was his miracle instrument.

This and many other discoveries are discussed in this book. To anticipate the flavor of the era and the charged excitement of the scientific work, it is sufficient to sketch one of the most striking feats of logic and discovery in science, making clear why others have said that Hale's contributions defined an era.

The reason for the relative darkness of sunspots compared with the solar disk was a puzzle in early solar research. Visual observations with a spectroscope by Charles Young, Norman Lockyer, Alfred Fowler and others had shown that the spectral lines in the spots differed in intensity, number, and character from the spectrum of the disk. Using the high-dispersion horizontal Snow solar spectral telescope on Mount Wilson, Hale and Walter Adams in 1906 obtained excellent photographic spectra of sunspots. The data put the spectral differences between sunspots and the disk on a quantitative basis.

Then Hale, Adams, and part of the observatory staff began laboratory experiments both on Mount Wilson and in Pasadena in two newly constructed spectroscopic laboratories. They used arc, spark, and, later, highly accurate temperature-controlled electric furnace methods to excite the spectral lines of different excitation potentials of many elements. From this it became clear that *differing temperature* was the reason for the spectral differences. All of the observed differences between the solar surface and the sunspots could be explained in great detail in this way alone.

But this was only the beginning of the story. Hale, in a characteristic show of genius, saw even finer detail in the spot spectra of the doubling of certain lines compared with disk spectra. By looking at the polarization properties of the light he found characteristic Zeeman splittings and polarizations in 1908, discovering, thereby, the existence of strong magnetic fields in sunspots.

Although a central discovery, this was not the most important result of the sunspot program. During the course of their work, Hale and Adams had noted that the sunspot spectra closely resemble the stellar spectra of K and early M type stars rather than the G-type disk spectrum of the sun. Armed with their discovery that the differences between the spot spectra and the disk were due to lower temperature, they arrived at the Alexandrian-like conclusion that the entire sequence in the spectral classification of stars, known since the time of Father P. A. Secchi, *was itself a temperature sequence*. This, of course, is now recognized as one of the foremost facts in stellar astrophysics, but known then directly by these powerful Baconian methods already in 1906, long before the Saha theory of ionization approached the problem from the theoretical side in 1920.

Again the story does not stop there. Armed with these facts, and with the results of new data from the Mount Wilson physics laboratories, the effect of pressure and density on the intensity of spectral lines could be determined by the presence or absence of Fraunhofer lines of different excitation potentials *at the same temperature*. From this, Adams and Arnold Kohlschütter could begin in 1914 to identify *pressure effects* as the explanation of the spectral differences in the sharp and the broad lines in stars of the same temperature as a *luminosity* effect. They could calibrate the effect, thereby obtaining the intrinsic luminosity of a star *from its spectrum alone*. By this work in the mid 1910s, the method of spectroscopic parallaxes was invented with its enormous consequences for studies of galactic structure. The work culminated only in the 1930s, leading to a partial solution to Hale's cherished problem of stellar evolution, here again based largely on work at Mount Wilson on the Hertzsprung–Russell diagram that was determined from the spectroscopic parallaxes. The massive observational program was done by Adams, Alfred Joy, Milton Humason, and Ada Brayton, and was published in 1936. But its complete connection to stellar evolution, *per se*, had to wait another 15 years, until the nature of the globular cluster HR diagrams could be fully understood in the 1950s.

The wonderous connection between sunspot spectra, the distance of stars, and the structure of the Galaxy was traced many years later by Walter Adams, second director of the Mount Wilson Observatory, in a stunning article on the history of those early days where a sense of the transcendent accomplishment was present in the activities of the unique staff. In his article in *Publications of the Astronomical Society of the Pacific* (Vol 51, p. 133, 1939) Adams concludes, in a way so often emphasized by Hale, that the application of physics to astronomy through knowledge of the sun is the path to stellar evolution and beyond. Adams wrote:

"In this brief outline of a relatively simple research we have passed from a sunspot to the interior of the atom, thence to the atmosphere of the sun and stars, and so to the problems of stellar constitution and development. That the behavior of certain lines in the spectrum of a sunspot should have any bearing upon the determination of the distance of a star seems at first almost inconceivable, but the successive steps in the relationship are logical and by no means complicated. The investigation forms an excellent illustration of the innumerable interconnections of the problems of physical science, of the simplicity and directness with which they may be approached, and the effectiveness of bringing to their solution varying interests and a diversity of points of view."

This was Hale's thesis throughout his life, in science and also in his many activities undertaken from a strong sense of civic responsibility. These too are discussed by Wright, and we must marvel that Hale would undertake so much. He entered the scene before there was much American science. He left when much of the structure was in place, a single example being the National Academy of Sciences and its National Research Council arm. This one organization permitted America to be poised to enter World War II with this structure in place. Hale was instrumental in the development of this structure.

But these activities were simultaneous with his leading role in the founding of the three largest Observatories in the world in their time at Yerkes, Mount Wilson, and Palomar; the founding of the modern California Institute of Technology; the founding of the Huntington Library and Art Gallery; the development, largely from Hale's vision, of Pasadena as a cultural center with its library, its civic auditorium, and its city hall; with a continuation of his scientific research; and with the founding with James Keler of *The Astrophysical Journal* in 1899. It was said later by Otto Struve, then director of the Yerkes Observatory and head of the Astronomy Department of the University of Chicago, that the Journal was the finest of Hale's achievements, and that if a decision had to be made of keeping open Yerkes or the Journal, it would have to be the Journal. Of course, Struve was both editor of the Journal and director of Yerkes at that time.

It is clear from these unusual accomplishments that Hale was not a simple man. As this biography describes, the facts surrounding his life show him to have been one of the most complex human beings of our time. Hale was a combination of opposites. He was basically shy, yet rose in each appropriate occasion to be the most charming of people. From the reservoir of his great personal charm he could create an atmosphere within which he used his unsurpassed ability to raise large amounts of money for his visionary projects. He moved in the social circles that permitted this great talent of persuasion full freedom.

Yet he badly needed solitude. His mind would run away with him on all subjects of interest to him. His passion for each subject often became debilitating. He once wrote "no archaeologist, whether Young of Champollion deciphering the Rosetta Stone, or Rawlinson copying the cuneiform inscription on the cliff of Behistun was ever faced with a more fascinating problem than that of the hieroglyphic lines of sunspot spectra." He suffered from what he himself called the "divine discontent." He was often tense—described as constantly dissatisfied—constantly racing from one problem to the next. Each solution lead to the next step along the road of understanding stellar evolution, or art, or history of civilization, or archaeology, or whatever else drove him to hypersensitivity at the moment.

The solution to his primary goal of explaining stellar evolution was not to come in his lifetime. Like all pioneers, he was before his time. Yet without his putting in place the great instruments on Mount Wilson and on Palomar Mountain, that solution would have been delayed, and perhaps would not have been reached even now.

This divine discontent led many times to what would now be called deep depression, and then to almost complete physical breakdowns. His high-strung nervous temperament, with his strong New England reserve and his abnormal sense of duty and accomplishment, all combined to form his personality, which few of the staff members of the institutes he founded could fully understand. However, upon meeting and conversing with Hale about their own ambitions, hopes, and plans, almost all left his presence feeling that they had met greatness and inspiration.

Yet Hale suffered fools poorly. Furthermore, he had strong feelings about practical people vs. visionaries in the completion of goals once set. One such instance described in this biography concerns the work habits of a man called Reese (even his first name is lost) at Yerkes, never heard from since. Another, much more serious case is that of George Ritchey, Hale's chief optician and a genius in his own right, so much so that he and Hale crossed swords over the completion of the Hooker 100-inch reflector, which eventually became the greatest telescope in the world of its time and, arguably, the most important astronomical instrument ever built.

The account given in this biography of the relations between Hale and Ritchey is, correctly I believe, sympathetic to Hale. Other recent accounts describe the other side. Yet we must all ask what we would have done in similar circumstances if the sure success of a major project was in the balance over a decision to be made in mid-stream of employing an unproven technique that would have changed the specifications of the main goal. Later, in 1931, Hale had to make a similar decision about the 200-inch mirror, pitting the General Electric fused-silica mirror (an uncertain experiment) against the Corning pyrex (a better certainty). Everyone agrees that Hale made the correct decision there also.

Ironic as it is, the Ritchey design of a wide corrected field (called the Ritchey–Chretien configuration) was redesigned and improved first by Ben Gascoigne in Australia by adding a lens, and then by Ira S. Bowen and Arthur Vaughan of the Mount Wilson and Palomar Observatories. This latter design was subsequently used for the Carnegie Observatories' wide-field 2.5 meter survey du Pont reflector at Las Campanas, Chile. The Las Campanas Observatory, in turn, has become the successor to Palomar in the long string of children and grandchildren of Hale's observatories, this time developed by Horace W. Babcock, Ira Bowen's successor as director of the Mount Wilson and Palomar Observatories. Bowen himself was the successor director to Walter Adams of the then combined Mount Wilson and Palomar observatories, later named the Hale observatories, and operated jointly by Hale's Carnegie Institution of Washington and Hale's California Institute of Technology.

Ritchey's disloyalty over the Hooker 100-inch telescope in 1918 lead Hale to act in a most uncharacteristic way by dismissing Ritchey from the Carnegie staff. Yet who among us can understand the circumstances well enough to pass judgement from this distance? Strong personalities drive events in ways unknown from extant public records, which, in any case are always incomplete at the necessary level from which one generation can comment with justice on another.

It only remains to say that, as Helen Wright's definitive biography shows, we will wait long for another like George Ellery Hale, tortured soul that he was, yet blessed with the divine discontent. Without him, the history of astrophysics as we know it today would be a far different story.

Allan Sandage
Staff Member
The Observatories of the
Carnegie Institution of Washington
Pasadena, CA

To Walter Adams and Harold Babcock,
who helped Hale explore the universe
and have helped me follow his trail

CONTENTS

8 CONTENTS

ILLUSTRATIONS

Halftones

Photographs for which sources are not indicated in the captions are reproduced by permission of the Hale family.

Diagrams

Map

"THE SEARCH for truth is, as it always has been, the noblest expression of the human spirit. This great new window to the stars will bring us into touch with those outposts of time and space which have beckoned from immemorial ages. It will bring into fresh focus the mystery of the universe, its order, its beauty, its power."

It was the 3rd of June, 1948, and Dr. Raymond Fosdick, President of the Rockefeller Foundation, was speaking at the dedication of the great 200-inch telescope on Palomar Mountain, California, to the famous astronomer George Ellery Hale. He was addressing the hundreds of noted scientists who had come from all parts of the country, even from abroad, to do honor to the astronomer who had made this unique instrument possible.

Twenty years earlier, in 1928, Hale had launched this project. He knew he might not live to see it finished, but he knew, too, that he was building for a future in which this "adventure into the unknown" would be carried on by his colleagues and by the astronomers of coming generations who shared his belief in the need for a larger tool with which the frontiers of the universe could be pushed farther and farther back.

George Hale died in 1938, ten years before the dedication on Palomar Mountain. He would have been amazed if he could have returned on this June day in 1948 to listen to the resolution read by Dr. Lee DuBridge, President of the California Institute of Technology.

"The Board of Trustees of the California Institute of Technology hereby resolve that the 200-inch telescope of the Palomar Mountain Observatory shall hereafter be known as

THE HALE TELESCOPE

"By this action the Board of Trustees seeks to recognize the great achievements of Dr. George Ellery Hale (1868–1938), who served as Director of the Mount Wilson Observatory from 1904 to 1923, who served as a member of the Board of Trustees of the California Institute from 1907 to 1938, who originated the bold conception of the 200-inch telescope and whose brilliant leadership made possible its design and construction. As this great instrument probes the secrets of the universe, it is fitting that it should stand also in memory of the great scientist and the great leader who contributed so brilliantly to the science of astronomy and who served so ably his community and his nation."

All those in the audience who had known Hale, and many who had not but who had been influenced by his work and by his spirit, shared a profound sense of sorrow that he had not lived to see the completion of his dream.

As this biography goes to press, eighteen years after the dedication of the Hale Telescope, Allan Sandage, of the Mount Wilson and Palomar Observatories, who has made some of the most spectacular discoveries with that telescope, writes in a letter to the author: "The longer I work in astronomy the more overpowering is the conviction that we owe *all* to Hale and his dreams and positive actions to put those dreams into glass and steel. Where would world astronomy be today if Hale had not been an 'empire builder'?"

The 200-inch telescope was only the last and greatest of the four great telescopes that Hale planned and built in his lifelong effort to gain more light and to penetrate more deeply the most remote regions of the universe. From his first small observatory at Kenwood, with its 12-inch refractor, he went on to plan and build the Yerkes Observatory, with its 40-inch refractor, still the largest in the world; and the Mount Wilson Observatory, with its great 60-inch and 100-inch reflectors. In all these he carried out his ideal of an observatory that is, in fact, a physical laboratory, and thereby became the dominant figure in the astrophysical revolution.

As the noted astronomer Dr. Theodore Dunham put it: "More than any other single individual Hale was responsible for the rise of modern astrophysics. He early became convinced that observational astronomy and experimental physics have so much in common that results of far-reaching importance for astronomy must surely follow the introduction of physical methods in the study of the sun and stars. His rare genius was devoted to translating this conviction into action."

To many of the younger astronomers at the Palomar dedication, Dunham's statement might have come as a surprise. They were aware of Hale's contribution to the building of great telescopes. Yet it is likely that many of them had no clear conception of his far-reaching contribution to the growth of astrophysics. Once a revolution has been accomplished, it is often difficult to realize a time before that revolution has taken place.

It is a curious fact of our history that men and women famous in their day are largely forgotten soon after their death. Their contributions may have been great, yet they are often remembered by only a few historians. This seems to have been particularly true in science, where the student is so busy learning current facts that he learns little of the background of his field, and has no notion of the origin of many of the ideas and concepts he takes for granted.

Over twenty-five years have passed since Hale's death, and something of this fate has befallen his name. Today, though many people have heard of the Hale Telescope, they know little of its builder, whose influence on the development of astronomy and of science in general has probably been as great as that of any scientist ever born in America.

He built the great telescopes that made possible the achievements of countless other astronomers. He invented instruments that revolutionized the study of that "typical star" our sun. For he recognized early in life that by a close study of the sun's characteristics, a clearer picture of the nature of other stars would be gained, and that this in turn might lead to an understanding of the complex problem of stellar evolution. In the solar field, Hale made his first notable contribution when he was only twenty years old, through his invention and successful development of the spectroheliograph. Nineteen years later he was to make an equally outstanding contribution in his discovery of the magnetic nature of sunspots.

The great astronomer Otto Struve, formerly Director of the

Yerkes and McDonald Observatories, writing in *Astronomy of the Twentieth Century* with Velta Zebergs, notes: "While many astronomers made important contributions to solar research, Hale is the 'New Era's' preeminent figure. Any discussion of his work must emphasize his role in the development of the Yerkes, Mount Wilson, and Palomar Observatories. His research progressed along with his improvement of solar telescopes and auxiliary telescopes, and Hale's most valuable contribution to astronomy may well be his unceasing effort to provide the best equipment—including the world's largest refracting and reflecting telescopes—for solar and stellar studies.

"Hale combined good training in physics and mathematics with extraordinary inventiveness in developing new instrumentation. The naming of the 200-in. telescope of the Palomar Observatory 'the Hale telescope' is a tribute to those unusual qualifications. Few astronomers of any generation have been equally competent as instrumentalists and as observers and theoreticians."

Still, Hale's contributions to astronomy, great as they were, were merely a part of his contribution to the growth of science, not only in America but also throughout the world. In the following pages we shall read of these contributions: to the revival of the National Academy of Sciences and the founding of the National Research Council, the International Astronomical Union and the organization that is now called the International Council of Scientific Unions. But we shall also read of his role in the advancement of scientific education and the development of the humanities, in the re-creation of Throop Institute into the California Institute of Technology, and the establishment of the Henry E. Huntington Library and Art Gallery.

In the realm of the humanities, as Dr. Walter Adams, his successor as Director of the Mount Wilson Observatory, writes: "His sensitive mind, like a delicate musical instrument of many strings, responded to every contact with nature or the touch of poetry or music or art. Trained as a physicist and engineer, he was also a humanist and classicist with a profound appreciation of the contributions of the older civilizations to the beauty and value of life. The rare combination of an extraordinarily clear and analytical mind with a far-reaching and yet controlled imagination explains better than anything else the extent of his influence upon the development of science during his years of activity and the permanence of his contributions to the intellectual life of the period."

This book is the account of the manifold and far-reaching

achievements of George Ellery Hale. Yet it is the account also of the very human and magnetic individual whose enthusiasm and faith made them all possible. As the noted British astronomer Sir James Jeans wrote:

"We knew he was a great figure in science, but felt that he could have been equally great at almost anything else. For Nature had not only endowed him with those qualities that make for success in science—a powerful and acute intellect, a reflective mind, imagination, patience and perseverance—but also in ample measure with qualities which make for success in other walks of life—a capacity for forming rapid and accurate judgments of men, of situations, and of plans of action; a habit of looking to the future, and thinking always in terms of improvements and extensions; a driving power which was given no rest until it had brought his plans and schemes to fruition; eagerness, enthusiasm, and above all a sympathetic personality of great charm."

A word must now be said about the arrangement of the material in this book. In recounting the story of Hale's diverse activities, I first considered telling it chronologically. But Hale worked on so many different tasks simultaneously that the result was only confusing. Therefore it has seemed wiser (except for his early years) to tell separately the story of the founding of the observatories and organizations in which he played a leading role. Throughout, as the main thread, runs the story of his astronomical research.

It was soon after Hale's death in 1938 that the National Academy of Sciences decided that a full-scale biography of George Hale should be written, and a committee was appointed. In 1947, Walter Adams, the Director of the Mount Wilson Observatory, wrote and asked me to take on the task. It has proved to be a long, and highly rewarding, job. It has led to every part of the country, even to England, and to interviews with countless scientists. Yet the writing of the book would have been impossible without the constant help and encouragement of Dr. Adams and Dr. Harold Babcock. Their vivid memories of Hale and the pioneer days on Mount Wilson form one of the most exciting parts of the biography. For her friendship and for permission to quote from Dr. Adams' papers, I should also like to thank Mrs. Walter Adams.

I am also most grateful to all members of the Hale family who have helped me in innumerable ways—especially to Evelina Conklin

Hale, his wife; to William Hale, his son; and to his daughter, Margaret Hale Scherer, whose patience and understanding through the years and the unexpected difficulties of writing this book have been invaluable. I should also like to thank Paul Scherer for his help and for the use of material from the unpublished autobiography of his father, James A. B. Scherer. Other members of the family who have contributed to the biography include Hale's sister, the late Martha Hale Harts; George Hale of Chicago, his nephew; Robert Hale, his cousin; and Margaret Brackenridge Hale, wife of William E. Hale, the son of the astronomer. Aside from family correspondence, the most prolific source is that between Hale and his close friend Harry M. Goodwin, who kept every letter Hale wrote to him. I am most grateful to him for his keen historical sense and for his generosity in letting me use this invaluable record.

One of the greatest joys of writing a biography is the making of new friends. Some of these, like Alicia Mosgrove of San Francisco, were also Hale's friends. Her memories of him have been highlights in the writing of this book. In other cases, notably that of the former director of the Lick Observatory, C. Donald Shane and his wife, Mary Shane, a lasting friendship was formed when I went to Lick to delve into the archives. I am especially grateful to all these friends for their continuing help over the years and for their reading and criticism of the manuscript. For such reading, I should also like to thank Allan Sandage of the Mount Wilson and Palomar Observatories, who encouraged me with his praise; Horace Babcock, now Director of the Mount Wilson and Palomar Observatories; Helen Dodson Prince of the McMath-Hulbert Observatory of the University of Michigan, a leading authority on the sun; and Margaret Harwood of the Maria Mitchell Observatory of Nantucket, who knew Hale; also Helen Lockwood of Vassar College and Marguerite Munson, who have offered valuable literary criticism. My thanks go also to Ira S. Bowen, former director of the Mount Wilson and Palomar Observatories, for his interest in the biography, and to his wife, Mary Bowen, for reading the manuscript in its embryonic state. Indeed, so many people have contributed in so many ways to this biography that it is impossible to thank them all. But I should like especially to mention the great help given to me by my friend Samuel Rapport.

To do the research for a biography of this kind is virtually impossible without financial help, unless one is independently wealthy. For such help I should like to thank Mrs. George Hale, who gave the first grant toward the biography and has approved the manuscript. This grant was amplified by the American Philosophical Society. Later, thanks especially to the help and interest of Margaret Carnegie Miller, the daughter of Andrew Carnegie, the Carnegie Corporation gave me a three-year grant. A final grant was made by the California Institute of Technology, with the support and approval of its President, Lee A. DuBridge. To all these people and organizations, I am most grateful. Without their help and the continuing faith and support of my father and mother, Fred and Kathleen Wright, this book could never have been written. My hope is that the final product may justify to some degree their confidence.

There seems little doubt that the realization of Hale's enormous contribution to the development of astrophysics and to the advancement of science will grow with time—particularly in an age when the release of atomic energy and the miracle of space exploration were born, and in a world that is becoming increasingly oriented toward science.

H. W.

Little Sagastaweka Island
October, 1965

INTRODUCTION

BY DR. IRA S. BOWEN
*Former Director, Mount Wilson
and Palomar Observatories*

GEORGE ELLERY HALE started his scientific career in the late 1880's. Up to this time, America had been a pioneer nation with most of its energies absorbed in opening up a new continent. Very few facilities were available for scientific research or other intellectual activities. Even advanced training in scientific subjects was limited, and it was the ambition of every young scientist to complete his training and receive his doctorate at some European university. America waited until 1907 to receive its first Nobel prize in science, and only four had been awarded to scientists from this country before 1930.

Before carrying out his astronomical studies, Hale was forced to raise the necessary funds and to design and supervise the construction of the instruments needed. Nevertheless, he made many contributions of far-reaching importance to his own field of solar astronomy. These included the invention of the spectroheliograph and its use for the first systematic study of the solar flocculi and prominences. He also discovered the magnetic field of sunspots, which was the first observation of a magnetic field in any astronomical body. This was followed by a detailed study of these fields that brought to light the laws for the polarity of these spots and for the reversal of the polarities after each sunspot cycle. These contributions were recognized by the award to Hale of most of the important American and European medals and honors for which an astronomer was eligible.

Of far greater influence on the future of American science than these discoveries were Hale's visions of great scientific institutions in this country and the skill with which he found the men and the funds to make them a reality. Starting at home with his own institution, the University of Chicago, he envisioned a department of astronomy with observational equipment second to none. Long and skillful negotiations persuaded Yerkes, the streetcar magnate, to provide funds for the construction of the Yerkes Observatory in nearby Wisconsin. After nearly seventy years of successful operation, its 40-inch telescope is still the largest refracting telescope.

Dreaming of even more powerful equipment, Hale then convinced the newly formed Carnegie Institution of Washington of the unusual opportunities for observational astronomy on mountain sites under the nearly cloudless skies of Southern California. The first fifteen years of the institution's Mount Wilson Observatory under Hale's directorship saw the construction of two very advanced tower telescopes for solar studies, and of the 60-inch and 100-inch telescopes for the investigation of more distant objects. For many years the 100-inch was the largest telescope in existence, and it was used by Adams, Hubble, and many others to make a long series of revolutionary discoveries that have changed our whole concept of the size, structure, and evolution of the universe.

Asked, soon after coming to California, to join the Board of Trustees of the nearby Throop Polytechnic Institute, Hale consented on condition that every effort would be made to develop it into a research institution of the first rank. His persuasive powers were used to bring such top scientists as A. A. Noyes, R. A. Millikan, and T. H. Morgan to the staff of the institution, which then took the name of the California Institute of Technology. In spite of its small size, its enrollment never exceeding 1,500 students, the institute has amply fulfilled Hale's dream for it. For example, eleven of its staff members or students trained in its laboratories and classrooms have already received the Nobel prize.

Later Hale planned and found the funds for the construction of the institute's 200-inch Palomar Telescope, which after thirty years finally forced the 100-inch on Mount Wilson to take second place among the world's large telescopes.

Hale's influence, however, was not limited to the California scene. In 1902 he was elected to the National Academy of Sciences and almost immediately undertook a study of the operation of the academies of the more progressive European countries. On the basis of his forceful recommendations, the National Academy was soon enlarged and reorganized to give it a much larger role in American science.

With the coming of the First World War, the United States found itself seriously handicapped by the shortage of scientific personnel and of manufacturing facilities for precision equipment needed for the defense program. Hale took the lead in the organization of the National Research Council as an operating body under the auspices of the National Academy. Under Hale's chairmanship, the council guided the scientific war effort and the programs for developing facilities to relieve the shortages. After the war, the council continued to coordinate many of the scientific activities of the country.

Through these various projects, George Ellery Hale probably did more than any other one man to awaken interest and find support for a sound and effective development of science in this country.

Moreover, Hale's interests extended far beyond the boundaries of science. As a member of the Pasadena Planning Commission, he exerted a powerful influence in obtaining for the city of Pasadena a magnificent group of coordinated municipal buildings, including a city hall, library, and civic auditorium. The fact that the Huntington Library, in nearby San Marino, is one of the great research institutions for the study of English history and literature owes much to Hale's persuasive presentation to Mr. Huntington of his ideas for such an institution.

Never a robust individual, Hale often found it necessary to operate through correspondence rather than through the rough-and-tumble of committee discussion. As a result, there remains an unusually complete documentation of his efforts and of those of the scientists and philanthropists who collaborated with him. Helen Wright has spent many years in studying this material and in organizing it to tell the story of one of the pioneers to whom America owes much for its present position in science—with all that this now means for American industry and our present high standard of living.

Because of the far-reaching role played by Hale, the exceptional completeness of his records, and the thoroughness with which Miss Wright has studied and interpreted them, this biography should long stand as a key chapter in the history of a crucial period in the development of science in this country.

EXPLORER OF THE UNIVERSE

CHAPTER ONE

A Chicago Childhood

✳ 1868-1886

April 8th, 1884

"INFUSORIA, BRANCHIPODA, ROTIFERA, FORMANIFERA, RHISOPODA, DIOTOMACEA, PORIFERA, PROTOZOA AND DIOMIDICAE."

By the light of a kerosene lamp young George Ellery Hale was recording his observations of the microscopic life he had discovered in a drop of green slime from the ditches of Kenwood, a Chicago suburb. He was writing in the hope of convincing his father, William Hale, that he had seen all he could through his little microscope and must have a larger one. If he could write a paper showing all he had learned with his first microscope, the coveted instrument might be forthcoming.[1]

Finally the paper, illustrated with pencil drawings, was finished. Apparently William Hale was satisfied, and the order for a Beck binocular microscope went off to England. To the impatient George, the time seemed endless, but at last the beautiful instrument arrived. With it he set out to explore further the wonders of the microscopic world. "These," he was to write fifty years later in his autobiographical notes, "were real adventures, as exciting as those I have had so often in later years. For I had made the discovery that simple instruments suffice to reveal new and wonderful worlds, hidden from the unaided eye. Here was the origin of a life of research."

This discovery would lead George Hale far from the microscopic world, into the most distant reaches of the universe.

When George Ellery Hale wrote his paper on infusoria, fifteen years had passed since his birth on June 29, 1868, to William Ellery Hale and Mary Scranton Browne at 236 North La Salle Street in the heart of Chicago. Here, after their marriage on Christmas Day, 1862, during the dark days of the Civil War, William Hale had struggled to establish himself as the Chicago representative of the Rock River Company of Beloit, while his wife had sewed for the sick and wounded with the Sanitary Commission. In a boarding-house on Indiana Avenue, a son, Edward Gardner, was born in 1863, only to die after seven months. Two years later came Caroline Scranton; after two weeks she too died, in a frightful spasm, the re-sult, the doctor said, of brain congestion. When, two years after-ward, George was born, his mother was sure that he too would die. Under the shadow of that concern, George's sister, Martha, was born five years later, and his brother, William B. Hale, seven years later. Despite their mother's fears, they all survived and lived to grow up.[2]

Of the house on North La Salle Street where he was born, George remembered little. With all the other houses in the center of the city, it burned to the ground in the great Chicago Fire of 1871. Fortu-nately, in the spring of 1870 the Hale family had moved to the large frame house that William Hale had built at the southwest corner of Drexel Boulevard and Forty-seventh Street, in the suburb of Hyde Park. From that rural outpost they watched Chicago continue the stupendous growth that had started after the Civil War. In its haste to surpass every other city in the West, it sprouted jerry-built houses with showy marble fronts and thin, weak walls. "Chicago," the *Tribune* warned, "is a city of everlasting shingles, sham, veneer, stucco and putty." [3]

But it warned in vain. Shortly after nine o'clock on Sunday night, October 8, 1871, the disastrous event occurred that "left the first per-manent impression" on George's three-year-old mind. "It was one of those curious flashes which strangely linger long after the vast throng of less tenacious illuminations have sunk into obscurity." [4]

Afterward he heard his father describe the desolation downtown where the "fireproof buildings" still smoldered. The morning after the fire had died down enough to allow passage, William Hale met Lyman J. Gage of the First National Bank at the corner of North State and East Washington streets, where the Hale building, fin-ished only the year before, had once stood. Now it too lay in ashes. Much later, George learned how Mr. Gage, who held the mortgage on their new house as well as on the Hale building, extended credit to his father to rebuild. He discovered how William Hale, with the boundless energy and tenacity his son would inherit, founded the hydraulic-elevator business that would make possible the tall buildings of the new Chicago. As the steel structures went up, the Hale Elevator business grew until it extended far beyond Chicago's boundaries, even to Europe.[5]

But of all these things, George at the age of three understood nothing. His world was centered in the square frame house on Drexel Boulevard. Here in autumn, when the clumps of low scrub oak flamed with color, he first felt the lingering sorrow that autumn would always bring. Only with the first signs of spring, as he wandered through the surrounding fields of tall tufted grasses, would his spirits revive. "I am no poet," he would write, "but I understand the thrill expressed in one of the letters of Keats, where he described the billowing waves in such a field, swept by the winds of spring. . . . I have detested all my life the gloomy rigors of winter, and rejoiced in the advent of spring." [6]

When it was too cold for him to play outside, there were other rewards. His earliest memories were of the upstairs bedroom where his mother, a semi-invalid with thin lips, firm chin, and brown eyes deeply set in her gaunt face, spent most of her time. Here he loved to sit by the hour, listening while she read aloud from Mother Goose:

> "He shall have an apple,
> He shall have a plum.
> He shall have a pony
> To ride upon the moon."

From Nellie Fay, the Irish cook and maid of all work, he learned songs of a different and more melancholy sort. One night while his father and mother listened, amazed, he sang verse after verse of a song he had just learned. Afterward his mother told proudly of this and other accomplishments.[7] When later he heard tales of his precocity he never quite believed them. One such story told of his making an electric bell when he was just beginning to walk; [8] another, that when he was barely able to talk he gazed at the sun through a prism and stated that some day he would find out what it was made of.[9]

While these and other legends gathered around him, Chicago continued to expand at a fabulous rate. But the country was uneasy. In

1872, when Ulysses S. Grant defeated Horace Greeley, many predicted disaster. Their prophecy was realized in 1873 when panic swept the country, and plunged it into one of the worst depressions in history. William Hale was no exception, and the family was forced to move to a small rented house at 96 Drexel (later numbered 3989) next to the northeast corner of Fortieth Street, a mile north of the old house, closer to the center of Chicago. Like others in the neighborhood, it was half of a double house, faced with gray stone blocks, and with a flight of gray stone steps that led to the front door. Here George Hale was destined to spend the formative years of his life. Here, while other children played hide-and-seek, he concocted endless schemes for exploring macroscopic as well as microscopic realms—objects that ranged all the way from the "eye of a fly" to the miracle of Saturn's rings. Impelled by an insatiable curiosity, and stirred by the "divine discontent" that would run through his entire life, his plans became bigger, his dreams of finer instruments for studying the natural world became greater.

To his father and mother, he scarcely seemed to have begun one project before another was on the way. "George always wanted things yesterday," his father said. And George Hale himself later acknowledged, "I was impatient to make rapid progress." While in time he learned to temper his impatience, these traits remained two of his outstanding characteristics. From his mother he had inherited a high-strung, nervous temperament, and often, as she watched him, she feared that, with his intensity and precocity, he would burn himself out early. Yet she soon learned that once he had set his heart on a particular object, there was little anyone could do to change it.

As time passed, his small back bedroom, which he shared with his brother Will and which he called his shop, bulged with books and bugs, rocks and the trilobites he had picked up along the Chicago breakwater. It overflowed with equipment that ranged from the box of tools his father had given him to the first handsaw, which in turn had led to a foot-power scroll saw and to his first small lathe. It included his microscope and collecting equipment—small bottles, nets, and spoons.

Soon, however, he decided he must have a "laboratory" of his own. After months of urging, his mother succumbed and gave him the upstairs hall room where she kept her dresses. Doubtless she had watched his growing scientific interest with pride. Once she wrote to

her mother: "Georgie spends all the time he can get in his shop.
. . . He seems to have quite a taste for mechanics and we hope he
will learn a good deal from his 'tinkering.'" [10]

George himself was, of course, overjoyed by the prospect of the
new shop, where he "could carry on physical and chemical experiments" and where, too, Martha and Will, infected by his boundless
enthusiasm, would continue to share his adventures. "Here," he
wrote, "each of us had a seat and an 'outfit' consisting of Bunsen
burner, batteries, galvanometers, and other devices, most of them
made by ourselves. Here we learned to draw and bend glass tubing,
to mount and photograph microscopic objects and to make gases by
various means. We poured hydrochloric acid on zinc and lit the
evolved hydrogen as it issued from the slender tube. We mixed potassium chlorate and manganese oxide and collected the oxygen set
free from the heated mass. We decomposed water into hydrogen
and oxygen by the aid of a simple battery, and reunited the gases by
a spark." Only too often "our delights were enhanced by frightening
but delicious explosions, the sound of which sometimes pierced the
carefully closed door and reached the floor below." [11]

When they moved into this shop, George was about thirteen. He
had grown from a round-faced, curly-haired, blue-eyed child into a
slender but still not very tall boy with a long oval face that accentuated his unusually high forehead. To Will, seven years his junior, he
was the object of wholehearted admiration. "He always took us in
and let us work with him. Perhaps we were slaves, but we liked it.
The point was that we could be with George and hear him discuss
his plans for work, and help in small details; for we always had the
belief that with him we were going to do things that were mysterious."

For a short time the "little room" sufficed. Soon, however, it too
was bursting its seams, and George began to dream of a shop in the
back yard. But how was he going to build it? He puzzled over the
solution for some time. It finally came one dark and dismal Saturday
in winter when the icy blasts off Lake Michigan swirled around the
house. "Heavy clouds hung over Chicago, and the stinging sleet,
borne by a northeast gale, rattled against the window pane. But
though I felt its chill, my mind was far away." *The Boy Engineers*
held him completely in its thrall. Suddenly he came across the plan
for a magnificent shop the boys had built. Deciding to follow their
example, he dashed over to his drawing table to make scale draw-

ings for a shop ten by fifteen feet in size. The next thing Martha and Will knew, they were all on their way to the nearest lumberyard, pulling a small cart.

At the lumberyard they bought some two-by-fours, soaking wet, and covered with snow and ice. They loaded the cart, and started back through the deepening drifts. At the first street crossing they encountered a step to the high wooden sidewalk. Though they pushed and pulled, they could not get their heavy load up the step. Then, as George apparently forgot, but Will remembered all his life, "there arose practically the only real dispute among us that I can remember." Martha and Will deserted, leaving George to wrestle with the cart alone. Some time later, he came struggling home, dragging his load. Then, immediately, he began rigging up a block and tackle to hoist the raw material up to the narrow window in the attic. After this came the job of mortising and fitting the soaked boards. (He could not wait for them to dry.) By the Ides of March, they were ready to be lowered into the yard; there the walls were planked and the shingling of the roof was begun. For months the work went on. The windows and the old stove with its stovepipe were bought secondhand. They built new workbenches, set up the lathe and scroll saw, and decorated the walls with tools. "To crown all," George writes, "we set up a steam engine of an eighth of a horse power that I had built." This was fed by a secondhand boiler. When they finally got the contraption going, steam hissed out of innumerable small leaks, and the entire building shook. The effect was terrifying. As the steam pressure mounted "toward forbidden heights . . . the roar of the exhaust, mingled with the vibration of the speeding engine, brought joy to the excited engineers." At this point, however, their mother arrived, took one look at the hissing demon, and cried out in horror. Yet for years the "demon" continued to run the lathe without mishap.

At last, on a cold and snowy night, the day came for the grand opening of the shop. While the wind howled through the cracks, George and Martha and Will served oyster soup and ice cream to a shivering circle of family and friends. "I do not recall," says Will, "that there were any speeches." [12]

So ended what would prove to be the first of many dedications in which George Hale would play a leading role.

If he could have spent all his time in this way, George would have been entirely happy. Unfortunately, he had to spend much of it in school. He was to write: "I never enjoyed the confinements and the fixed duties of school life. Born a free lance, with a thirst for personal adventure, I preferred to work at tasks of my own selection." [13]

His schooldays began in the Oakland Public School, where the books they read were so dull and the teachers so dreary he never could remember anything learned there. It all ended abruptly when he came home with typhoid. When, despite his mother's fears, he recovered, she decided he must never again return to that school, where, she was sure, he had caught the germ. Instead she decided to send him to the Allen Academy, the famous Chicago school run by Ira W. Allen.

Throughout his childhood George was delicate, suffering from stomach trouble, backache, and fainting spells. These troubles were doubtless accentuated by his mother's constant worrying. Yet, if he suffered from her overconcern, George was grateful for her "love and physical care." Years later he was to write: "Nervously organized as she was, her courage was of a high order, especially in any emergency. I remember how she immediately set my arm after I had fallen out of a cherry tree at my grandmother's house in Madison, Connecticut, where we spent our summers, and no boy ever owed more to a mother's care in the many illnesses which I was unfortunate enough to experience."

He was also grateful for the love of literature she instilled in him at an early age. Later he recalled his "delight in Homer, aroused in childhood when she read aloud with a friend the *Iliad* and the *Odyssey*." Through her he was led to read *Robinson Crusoe* in its original form, *Don Quixote*, *Grimm's Fairy Tales* and countless other classics of prose and poetry that "helped greatly to arouse my imagination and prepare me for scientific research." [14] This interest was further stimulated by Dr. E. F. Williams, the Congregational minister who with his wife lived with the Hales for several years and subscribed to many of the English journals.

Still, if Mary Hale's fear for her son's body and mind were great, her concern for his soul was greater. This concern, he learned in time, had its origin in her New England childhood. She was born in Hinsdale, New Hampshire, to Mary Prudden Scranton and Dr. Gardner Browne, a Congregational clergyman who later became a homeopathic physician and an ardent believer in phrenology. When she was small, her father and mother were divorced—an unheard-of thing in those days, especially for a minister. In time both were remarried—her father to Ada Merrill, her mother to Philemon Scranton of Madison, Connecticut. Meanwhile, the little Mary was sent to live with a grandfather she had never seen (in fact, only an adopted grandfather, since he, in turn, had adopted her own mother). This stern individual, Erastus Scranton, also a Congregational minister, lived in Burlington, Connecticut. There Mary Browne spent her childhood in the shadow of a Calvinistic discipline, dominated by an avenging God, which would have saddened the gayest child and which inevitably darkened her outlook on life and colored the religious upbringing of her own children.[15]

When he started off to the Allen Academy in the heart of Chicago, George was twelve. From the time he entered the four-story red-brick building, which stood between Michigan and Wabash avenues on Twenty-third Street, he entered another world. The quiet of the suburbs was exchanged for the city's turmoil, the familiar gray-stone fronts of Drexel Boulevard for the showy marble palaces along Michigan Boulevard. Suddenly he felt grown up. Mr. Allen, a tall and imposing gentleman who was "strict and vigorous," was a Harvard graduate, had studied in Germany, and had taught physics and astronomy at Antioch. Some of George's schoolmates remembered him as a "big strapping man like an ox," wielding a horsewhip. They considered him the devil incarnate.[16] To George, in whom he discovered an embryonic interest in astronomy, his approach was mercifully different. "Astronomy," he told his pupil one day, "offers a glorious field of work."

Soon, too, Mr. Allen observed that George was different from anyone he had ever taught. His ability to concentrate was greater, his absorption more marked. Here he saw the spark that would in time ignite a brilliant flame. One day he asked George to become the "unofficial curator" of the "philosophical instruments"—an air pump, an electric machine, some Leyden jars, a few test tubes and a Bunsen burner. He also served as the professor's assistant when experiments were performed before the whole student body.

Nor was Mr. Allen the only one who was impressed. George's classmates, too, recognized unusual qualities in the quiet, rather shy boy. One of these, Burton Holmes, recalled: "When I first saw him, sitting across the aisle in Miss Haskins' class, I thought he was the best looking boy in the world. I admired and envied his fine blond, curly hair, his noble head, his 'brow of Jove,' and his manly gentle ways." [17]

Here their studies included geometry and chemistry. While he was not fond of mathematics, George liked chemistry, especially the experiments. Once his mother wrote to her stepmother, "Georgie has gone from Geometry to Chemistry and related accounts of experiments, saying 'just hear this' and 'did you know that' until now he has gone to try one of the experiments for himself to see if it can be." [18]

Still, if he enjoyed the academy, he continued to find his greatest pleasure in outside things—such as the conjuring which, with the help of Burton Holmes, led to magic shows performed before an admiring audience of family and friends. Uncle George, their father's gay and cheerful brother, who lived with them, led the guffaws. Years later, Holmes, recalling these shows, remarked, "George was the more clever of the two. I was the better showman." Then he added: "I regarded George as a genius and looked forward to our triumphant career on the stage as the Boy Magicians, Hale and Holmes. But George became a serious student. I bought a camera. The triumphant tour was forgotten." [19]

On other days in spring and fall, after school, George, an active member of the Owl Bicycle Club, and his friends would bicycle out to Ranch Ten, the cabin they had built on the lake, to paddle the canoe he had made and to pursue adventures inspired by Harry Castlemon's *Frank Among the Rancheros* and *Frank at Don Carlos Rancho,* or by the Indians in Cooper's novels. One winter they decided they needed an iceboat. As usual, once the idea was born, George worked ahead pell-mell, forgetful of everything else. On a glistening December day the job was done. In the attic at 96 Drexel stood a beautiful iceboat, complete with runners and mast, sail and rigging. Only then did the classic question arise, "How are we going to get it out of the attic?" [20] The stairs were too narrow; the window and the scuttle in the roof were too small. So, his brother Will notes, "The ice boat remained in the attic until we left it there when we moved away."

Still, to the Hale children, these Chicago days could never compare with summer days in Madison, Connecticut, a Puritan village, hardly changed from Revolutionary times, where their mother's own mother lived. "The mysteries of Christmas had hardly been revealed before we began straining our gaze toward June," and the familiar treasures of their "summer domain"—the wood house, the corn house, headquarters for a magnificent fleet of toy yachts; the underground icehouse, the ancient barns, the pear and apple orchard.

Built in 1739, their grandmother's house was redolent with odors that had seeped into the wood, odors George remembered vividly all his life. The cellar retained a pungent smell of cider, roast oysters, and huckleberry pie. There was indeed something mysterious about the entire place, which appealed to the same vein that caused him to delight in ghost stories. Out of a remote and shadowy past came its builder, a bloodthirsty pirate and slave dealer who, it was rumored, had walled up two slaves in the house and left them to die. "To this day their groans may be heard coming from the traditional place of their immurement." [21]

In these surroundings, with the fascination of sea and shore added to the joys of life along the Hammonasset River, George was happy. "Daily swims in Long Island Sound, building and racing model yachts, hours at the old shipyard, and my initiation into fly-fishing for trout are among the vivid memories of these idyllic summers." [22] Here for a time each summer he could forget the absorbing activities of his Chicago life. Yet, with the coming of autumn he would return eagerly to his shop, which in time assumed the more impressive title of "laboratory," and eventually, as a result of his growing interest in astronomy, evolved into an "observatory."

Exactly when he first became interested in astronomy, he was never sure. Perhaps it began with his reading of the copy of Olmsted's *Astronomy* his mother had used at the Hartford Female Seminary. More likely, however, it began with Jules Verne's *From the Earth to the Moon*. Dr. Williams gave the novel to Uncle George in 1873, and young George must have read it when he was still quite small. "The gigantic telescope on a peak of the Rocky Mountains, into which 'the unfortunate man disappeared' while watching the projectile on its way to the moon, especially struck my fancy." [23] Long afterward he had reason to recall that telescope.

As a small boy, too, he had often noticed a small boxlike structure in the back yard of a house at the corner of Vincennes Avenue. He had been told, "A queer man lives nights in that cheese box and tells fortunes by the stars." This "wizard" turned out to be the astronomer Sherburne Wesley Burnham, who spent his days as a court reporter and his nights observing double stars. Soon after failing in an attempt to make a telescope with a single large lens, George decided to call on Burnham. He was fourteen, and approached the noted astronomer with natural trepidation. Burnham, a thin, wiry and nervous man with dark, unruly hair, weather-beaten face, and the eyes of an eagle, was most friendly. He agreed to look at the telescope. After a moment he shook his head and said he doubted it would ever be any good. Then his eyes brightened. He happened to know, he said, of an excellent 4-inch telescope, one of the best ever made by Alvan Clark, that could be had secondhand. Immediately George decided he must have it. He rushed home and pleaded with his father to buy the instrument. But William Hale, a practical man, was not easily swayed. His son had inherited his father's high forehead and dark eyes deep set under heavy eyebrows; the father shared his son's tenacity and stubborn spirit. After asking what George planned to do with such a telescope, he listened to the flood of enthusiastic answers but said nothing. That was in November. On December 6, 1882, there was to be a rare event—a Transit of Venus when the cloud-covered planet crosses the sun's face. George explained why he must have the telescope by then—if possible, sooner. William Hale promised nothing. But one evening, when George had almost given up hope, his father drove up in his carriage, carrying the coveted telescope in a long wooden case. On the sixth of December, on one of the coldest days of winter, George mounted the telescope in his mother's room, opened the window, and carefully timed the beginning and end of the transit, while Mattie and Will stood by shivering, "getting colder and colder." [24]

After this, whenever it was clear, the observations went on, and the three young astronomers, like all astronomers, became increasingly conscious of the weather. In his red notebooks, in a large round hand, George kept a daily record of temperature and barometric pressure. In these notebooks, incidentally, he also kept his "cash accounts." These ranged from expenditures for his infusorial studies to the generous extravagance of five dollars for "Mama's birthday" and seventy-five cents for the circus; from the frivolity of the roller rink to a final item, simply "Gone—$.50." [25]

For some time he continued to make observations in the back yard. But already he had become interested in photography; one day he attached a camera to his telescope. But the telescope wobbled, distorting the images, and the results were disappointing. Early on a morning in 1883, he set to work to mount the telescope on the roof. With him were his "slaves." The day was intensely hot, and the work seemed unending. Will recalls, "In order to get a good foundation up there it was necessary to buy a heavy beam and to fasten it by huge expansion bolts into the brick of the inside wall close to the attic scuttle." All this had to be done from a swaying ladder. It was the heaviest job they had ever tackled.[26] Still, when it was finally done, George felt that at last, like Galileo, he was on the threshold of discovery. His first successful photograph of the craters on the moon was followed by observations of a partial eclipse of the sun, sunspots, and planets. He hung the photographs on the walls of his shop. "Thus I became an amateur astronomer." [27]

Soon afterward his interest in the stars was further enflamed by a visit with Burnham to the Dearborn Observatory, a large building with a castellated dome in Douglas Park. George Washington Hough, the director, with his long beard and venerable air, resembled one of the prophets. "To me," George wrote, "Professor Hough used to be the very embodiment of the idea of astronomy, and as he always appeared to be old, he hardly seemed to grow older." Hough, who was known for his observations of Jupiter and double stars, led them up into the tower to see the great 18½-inch telescope, one of the largest in the country. Looking up at it, George marveled, and wondered if the day would ever come when he could work with such an instrument. He could hardly believe it when a few minutes later Hough asked if he would like to help with the routine observations for determining time for the Chicago Board of Trade, and in running the feeble engine that turned the dome. So it was that he entered a new and wonderful world. In time, however, he came to realize that he could never be satisfied with the kind of work Hough and Burnham were doing. "The reason lay in the fact that I was born an experimentalist, and I was bound to find the way for combining physics and chemistry with astronomy." [28] Yet he was delighted by their interest, and grateful for the chance to help them in their work. After this, on cloudy evenings Burnham would often turn up at his shop. His feet propped up on the table, puffing at his stogie, he would describe his own astronomical beginnings and his observations at the Lick Observatory in California, contrasting the clarity of the California skies with the murky Chicago heavens. George, entranced, began to dream of the day when he too might work in California.

Still, astronomy was not his only interest. Excited over everything he was doing, he continued to rush from one scheme to the next, sweeping Will and Mattie along in his whirlwind path. A young cousin, Robert Hale, on the edge of the charmed circle, said later that he always regarded it as some learned society. Once a package arrived from London. Inside were bees' wings, frogs' legs, and a "remarkable assortment of other things." It was addressed to Dr. George Ellery Hale. Impressed, Robert concluded that George had gained an international reputation.[29]

In time the children noticed that George often became lost in a world of his own. When they spoke he did not answer. If they made a loud noise, he failed to notice. This was true even, as Will recalls, "when we were eating bread and butter and sugar in the kitchen pantry and he fell to reading an advertisement of St. Jacob's oil in the newspaper spread across the table." [30] This power of concentration would remain one of his outstanding traits. Once, on a trip to the West Coast, when he was thirteen, the family was staying at the Palace Hotel in San Francisco. The three children set out to investigate the Chinese quarter. Here, on a sidewalk table, George found a copy of the magazine *Golden Days*. He began to read. Will and Martha, thinking he would soon catch up, ambled on. But George failed to follow. Finally they decided to return to the hotel. There they waited anxiously. Two hours later he rushed in, looked at his little brother and sister in astonishment, and burst into tears. Certain they had been swallowed in some opium den, he had reported their disappearance to the police.[31]

In these years, William Hale's fortunes were booming. His elevator business had expanded—to London and to Paris, where he was erecting the elevators in the Eiffel Tower. He was busier than ever, yet George saw more of him, and the understanding between them grew. Often they would go together to watch the progress on the great edifices William Hale was building in downtown Chicago—especially the "Rookery," a pioneer example of skeleton construction, and a forerunner of the New York skyscraper.[32]

More and more William Hale hoped his son might follow in his footsteps. In line with this hope, George was taking the afternoon shop course at the Chicago Manual Training School. The course included molding and casting, forging and tempering, and machine-shop work, and here he built a steam engine and a 9-inch reflecting telescope. It was then that George often wished he could please his father by becoming an engineer. But his heart was not in the building of elevators, skyscrapers, or suspension bridges. Pure science was his goal.

Still, for some time he was puzzled over the way to that goal. He enjoyed looking through his microscope at the antenna of a gnat, the lung of a frog, the scale of a perch. Meticulously he recorded everything he observed. Yet he knew all this had been done before, and he was dissatisfied. "While the observation of natural objects, the collection of specimens, or the performance of simple experiments may content many beginners, there are those who yearn to advance into the work of original research." [33]

He attached his camera to his microscope and took microphotographs. He gazed through his telescope, and "the astonishing views it afforded of Saturn, the Moon, Jupiter and other objects excited an intense desire to carry on actual research." But how could he go about it?

"What I wanted," he wrote long after, when he had found what he was seeking, "was a description that I could follow of a *connected series of experiments*, leading step by step to the development of some branch of science and giving a clue to the nature of research." [34]

Suddenly, when he had almost given up hope, he found the answer in *Cassell's Book of Sports and Pastimes*, "a mine of delight" that his mother had given him on Christmas, 1881, when he was thirteen.[35] In these enticing pages he found a description of the spectrum, that beautiful rainbow of light first seen through a prism by Isaac Newton, which provides a key to the understanding of the physical universe.

The book described the nature of the spectrum, but even better, it explained how a spectroscope could be built. Eagerly he set to work to make the instrument, or rather instruments—for the book described two forms. One used a luster prism, the other a hollow one filled with carbon disulphide. He decided to try the luster prism first.

Behind an old trunk in the attic he discovered some lusters from a candelabrum. He washed them, split one of them in the way the book described, and mounted it on a tin stand. From a cardboard tube and a pair of old spectacle lenses, he made a collimating tube, and painted it black inside. Into this tube he fitted another tube "like a sword into its sheath"; into this second tube he fitted a slit made from a brass plate. Following the directions, he housed the whole instrument on a board. At one end he bored a hole "for the eye to spy into." The finished instrument looked just like the picture in the book!

That night he looked through his eyepiece at a candle flame. As the book had prophesied, the continuous spectrum shone out, with the red, green, blue, and violet "very vivid indeed." He shouted delightedly for Mattie and Will to "come and see." He tried other observations, but soon concluded he must have a larger image that would spread out (or disperse) the light of the spectrum more. He decided to build the large carbon disulphide prism. The experiment was odoriferously disastrous. But scientifically it was rewarding. In the light from the salted wick of a spirit lamp he could see the bright yellow sodium line. When he mounted a mirror in the window to reflect the sun's light into his instrument, he saw the dark lines in the sun's spectrum, named for the Austrian Joseph Fraunhofer, who first mapped and described them. From them, he soon learned, it is possible to identify the different elements in the sun. (Fraunhofer had marked the more prominent of these lines A, B, C, and so on, up to H and K with A and B at the red, and H and K at the violet end of the spectrum.)

Excitedly George showed his family this, his latest triumph. They held their noses. He read the description of the "dark lines in the solar spectrum" to Mattie and Will. "Each of these dark lines," said the book, "is found to correspond exactly with the bright lines obtained in the examination of colored flames. For example, a particular dark line which you will see with your spectroscope in the yellow corresponds completely with the yellow line you obtained in examining the salted wick of the spirit lamp." [36]

Now, the book continued, and George read, "This is only one of a host of facts which scientific men have studied, all tending to show that the sun is surrounded by an exceedingly hot absorbing atmosphere and by studying these dark lines in the spectrum, and carefully comparing them with bright lines you may obtain with substances here on earth, we have gained some insight into the composition of the sun."

It was clear, therefore, that if a prism could be arranged so that the sun's spectrum appeared in the upper half, while that of various earthly substances—iron, nickel, copper, zinc—appeared on the lower half, the two could be compared to find what elements the sun contains.

Soon, too, George learned that the cause of these dark lines in the solar spectrum had been discovered by Kirchhoff and Bunsen in 1859, the year of publication of the *Origin of Species*. From their experiments they had proved that the lines appear dark because the light coming from the sun's deep, intensely hot interior is absorbed in the sun's outer or "reversing" layer. This layer, while still intensely hot, is cooler than the sun's interior (which emits a continuous spectrum without lines). This discovery was to prove as revolutionary as Darwin's great discovery in its influence on the development of physics, astronomy and chemistry. It was a discovery that laid the foundations of Hale's own lifework.

As he began to see the possibilities of this extraordinary tool—the spectroscope—George was, as he said, "completely carried off my feet. From that moment my fate was sealed." [37] He saw that, like a vast suspension bridge, the spectroscope could help him travel the millions of miles from earth to sun, and so out to distant stars and nebulae. With it he could bring them into the laboratory to be analyzed as surely as any earthly substance. With it he might learn about the sun's behavior, possibly even discover what caused it to shine. At last he knew he had found what he had been seeking—in the world of light. He would never willingly leave that world as long as he lived.

"Even now," he wrote years later, "I cannot think without excitement of my first faint perception of the possibilities of the spectroscope and my first glimpse of the pathway then suggested for me. No other research can surpass in interest and importance that of interpreting the mysteries concealed in these lines. Their positions and intensities, and the extraordinary changes they may undergo as the result of variations in the physical and chemical state of the vapors that produce them, afford the chief clue to the nature and evolution of the sun and stars, and to the constitution of matter itself." [38]

Still, the gap between what the book described and what he could actually see with his homemade instrument was immense. He had to have a better instrument that would reveal the lines more clearly so that he could measure them with greater accuracy. Only in this way could he try to find for himself the elements in the sun. At first his father, to whom he appealed, did not give in. But in time, convinced, he helped him buy a small spectrometer. It was a vast improvement over the carbon disulphide horror. Soon, however, George decided that what he really needed was a grating, a plate of speculum metal marked with thousands of lines ruled closely together, which, like a prism, breaks light up into a rainbow. A grating, as he had learned from his reading, is a tool far more useful than a prism; it disperses or spreads out the spectrum more fully, separating the colors and the lines more clearly. The wavelengths of the lines could then be measured more accurately. For example, if he observed the sodium line, which he knew was double, his single prism barely showed the separation between the two lines. If he increased the number of prisms, the two lines would be separated more widely. If he could obtain a fine grating, the two lines would be separated so much that numerous other lines would appear between the two sodium lines.

He had read that these gratings were made by the noted physicist Henry A. Rowland of Johns Hopkins University in Baltimore, and sold by John A. Brashear, a telescope maker in Allegheny, a suburb of Pittsburgh. He decided to write to Brashear for a grating of "the smallest size." Soon, to his delight, a one-inch grating arrived, and with it a letter "overflowing with friendly feeling and encouragement." [39] With his new grating and his spectrometer, he plunged into the measuring of the dark Fraunhofer lines in the solar spectrum and recording them in his notebook. One such list is dated January 23, another, February 7, 1886. He was seventeen. At last he felt he was on his way to the fascinating, once hidden realm of research.

One day, some time after the arrival of that first small grating, he decided impulsively to board the next Pittsburgh-bound train. At Allegheny he asked his way to Brashear's shop. A wooden shack on the top of a steep hill was pointed out to him. Inside, he found a slim man with twinkling eyes deep set in a sparkling face. He introduced himself as "Mr. Hale of Chicago." Brashear looked at his visitor in astonishment. "If some fellow had taken a baseball club and hit me very, very hard, I should not have been any more surprised." "From the method of his writing," he recalled later, he had "judged his correspondent to be a man of about forty-five and to be certainly up in astronomical physics." Recovering from the initial shock, he welcomed "Mr. Hale" warmly. "I took a liking to the fellow at once." And George was equally taken with the genial man who received him as if he had been a well-known investigator. [40]

After this, they were to keep up a constant correspondence. Through the years, Brashear would be the first to counter any sign of discouragement, the first to respond enthusiastically to promising results. Through Brashear, too, George met Samuel P. Langley, a large man with a florid countenance, who was director of the Allegheny Observatory and a noted pioneer in solar physics. Brashear described Langley as a man of magnificent intellect, one never satisfied with a half-proved hypothesis, but always reaching out for final proof before he announced any of his great discoveries. Here young Hale had the chance to examine Langley's apparatus for studying the infrared solar spectrum and the marvelously delicate bolometer he had invented for measuring the heat from sun and stars, as well as his "aerodynamic" apparatus—a wonderful whirling machine on which he had suspended wheels, birds, and frictionless planes in his attempts to discover the laws of flight. But above all he was delighted to see Langley's magnificent drawings of sunspots. Some of these drawings, Langley said, were reproduced in his book *The New Astronomy*. Back in Chicago, George bought a copy. For days he was lost in its pages. As he read, a new vision of the universe emerged. "The prime object of astronomy until very lately indeed," Langley wrote, "has still been to say *where* any heavenly body is, and not *what* it is. But within a comparatively few years a new branch of astronomy has arisen, which studies sun, moon and stars for what they are in themselves, and in relation to ourselves . . ." [41] This was the "New Astronomy," later called astrophysics.

"The sun," Langley wrote, "is a star, and not a particularly large star." With the help of the drawings in the book, George studied the three chief layers in that star—first the photosphere, the "sphere of light" we see every day; then the chromosphere, the thin envelope "which rises here and there into irregular prominences, some orange scarlet, some orange pink." This thin shell, Langley noted, is composed "mainly of crimson and scarlet tints, invisible even to the telescope except at the time of a total eclipse, when alone its true colors are discernible, but seen as to its form at all times by the spectroscope. It is always there, not hidden in any way, and yet not seen, only because it is overpowered by the intenser brilliancy of the photosphere, as a glow worm's shine would be if it were put beside an electric light." [42] Above these layers lies the "unsubstantial" corona, a strange, mysterious shape then visible only during a solar eclipse.

From this time, George read everything he could find on our little star, and the more he read, the more fascinated he became. In *The Story of the Heavens*, Robert Ball drew a graphic picture of the sun's vast size and distance from the earth. "If a railway were laid round the sun, and if we were to start in an express train moving sixty miles an hour, we should have to travel night and day for five years without intermission before we had accomplished our journey." Compared to the earth, Ball noted and George read, the sun's bulk is extraordinary. "Suppose his globe were cut up into one million parts; each of these parts would appreciably exceed the bulk of our earth. Were the sun placed in one pan of a mighty weighing balance, and 300,000 bodies as heavy as our earth placed in the other, the sun would still turn the scale." [43] Yet, compared to other stars, the sun is but "a stone's throw from the earth." Above all, as Hale would realize increasingly, it is the only star near enough to us for close study of its surface. It is, as he would often say, a typical star, yet a very important one, even if it is only an insignificant speck in the vastness of the universe.

As time passed, Hale became increasingly anxious to investigate it for himself. His eagerness grew when he acquired a copy of Lockyer's *Solar Physics* and then his *Studies in Spectrum Analysis,* which more than anything helped him "turn from aimless observation and experiment toward actual research." If his enthusiasm needed any further kindling, it came in the summer of 1886 when the Hale family sailed for Europe with Burton Holmes and his family. The worlds opened to him by his telescope and spectroscope had "fired" George Hale with the desire to see as much as possible of the land "from which flowed the books and papers of Huggins and Schuster, Lockyer and Dewar." "I was still only a boy and could not venture to seek these semigods in person. But I could go about London, and see in the Strand the shop of Browning, where the spectroscope with which Lockyer had first observed solar prominences without an eclipse had been made, and . . . the source of the powerful induction coils used by the leaders of Spectroscopy. I could also read above the windows of a simple optician, where you can still have your spectacles repaired, the name of Dollond, famous as the inventor of the achromatic lens." [44]

But if he could not seek the British "semigods" in person, he was amazed by his reception by one such "god" at the Meudon Observatory outside Paris. This was the astronomer Jules Janssen, pioneer in solar spectroscopy, who had been a chief player in one of those dramatic cases of coincidence that so often occur in the history of science. It happened in 1868, the year of George Hale's birth.

Janssen had gone to Guntur in eastern India to observe a solar eclipse. During the eclipse, which occurred in August, he noticed that the spectral lines in the prominences (those flaming objects that Langley had described and that rise from the chromosphere at the limb or edge of the sun's apparent disk) were bright, and that the hydrogen lines were exceptionally so. As the sun reappeared and the prominence lines faded, Janssen exclaimed, "I shall see those lines outside an eclipse!" That day clouds prevented further observation. But the next morning he was up by daybreak. The sun had hardly risen above the horizon before he turned his instrument on the region of the sun's limb where the most brilliant prominence had appeared the day before during the eclipse. The same lines now shone out, clear and bright, and he succeeded in seeing the prominences quite distinctly through the slit of his spectroscope. They appeared, however, entirely different from those of the preceding day. Of one great prominence "scarcely a trace remained." For seventeen days Janssen continued to observe and make drawings of the prominences, and his drawings showed that these huge flames change their form and shape with extraordinary rapidity. By moving his telescope so that the sun's limb occupied different positions with reference to his spectroscope slit, he discovered that he could trace the form and measure the dimensions of a prominence. These observations were so absorbing that not until September 19th—a month after his first observation—did he send off a paper describing his results to the French Academy of Sciences.

Just five minutes before the arrival of that paper at the Academy, a letter was read from the English astronomer Norman Lockyer. In it, Lockyer described his successful observation of the solar prominences with his spectroscope on October 16, 1868, and his subsequent observation of prominences around the entire solar circumference. While, therefore, Janssen was actually the first to see the prominences outside an eclipse, he is given joint credit with Lockyer for the discovery. In honor of that discovery, France struck a medal. Janssen's portrait appears on one side, Lockyer's on the other.

Whether Janssen retold this story on that day when Hale visited him at Meudon we do not know. We do know, however, that the older astronomer invited him in, showed him over the observatory, gave him copies of his astronomical papers, and presented him with a precious glass positive of one of his photographs that showed the granulation on the sun's surface.

Only one other event on this memorable trip was more outstanding. It was one that Burton Holmes was destined to remember always. As soon as the two boys arrived in London, they hurried off to the shop from which they had ordered their magical paraphernalia. Burton, after lengthy consideration, bought "five pounds worth of new tricks." George looked on with interest, but bought nothing. From the magic store they hurried on to a shop where scientific instruments were sold. Here, while Burton looked on in amazement, George "squandered forty pounds on funny looking things which meant nothing to me. One thing he called a 'spectroscope.'" That spectroscope, bought at the famous firm of Browning, was for George Hale the climax of the trip. For many years it would remain his most prized possession.

Sixty years later, Holmes recalled the occasion: "I marvelled at the generosity of George's Dad, letting his boy spend money like that. But father Hale knew what he was doing—or did he? At any rate he was staking his son to a magnificent career. George Hale knew as a boy what he wanted. He had found a new magic in astronomy." [45]

CHAPTER TWO

College Years

MADISON, CONNECTICUT. Through all George Hale's early years, this old New England town continued to be the place of all places to spend the summer. Sometimes he even resented the European trips that deprived him of Madison. Then, when he was about thirteen, its charms were enhanced by a new and overwhelming attraction in the form of a young girl with brown hair cropped short, and blue eyes. Her name was Evelina Conklin. George had been invited to a party given by her cousin, and to the surprise of everyone he had accepted. To his own surprise he had enjoyed himself "hugely." Thereafter, on the flimsiest excuse, he would dash off to Clinton where "Lina," as she was called, was spending the summer.

In time, the Conklins, who lived in Brooklyn, rented a cottage on the Madison beach. After this, whenever they got the chance, George and Lina played tennis and went on surrey rides, and sometimes in the evenings he would point out the constellations and the planets with a captain's spyglass. Lina was fascinated by her handsome companion who looked so young yet knew so much. For his part, George was equally entranced.

When his steamer docked in New York on a sunny morning late in August, 1886, his first thought was of Madison where, he knew, Evelina was waiting. Over the years their attachment had grown, and George was now convinced that this was the girl he wanted to marry. Almost as soon as he reached Madison, he proposed and she accepted. Yet they decided to wait the four years until he was graduated from college. Sure of the wisdom of his choice, he went on to Boston for the opening of the Massachusetts Institute of Technology. It had been recommended to William Hale by D. H. Burnham, the architect. On the way he scribbled a note to his family to tell them the "great news."

To his surprise, his father replied by return mail, expressing alarm at the thought of a marriage that might interfere with his son's college career, and fear that he was taking on too much responsibility too early. While William and Mary Hale liked Evelina, they felt they were both young and immature. George had never really known any other girls. While he liked a good time, his work was and would remain his passion. Would Evelina ever understand such devotion? With little interest in science, would she comprehend why George would always prefer to explore the mysteries of a sunspot than to go to a dance? Even if, after four years, they should decide to marry, their life together might be far from easy.

Knowing his son's stubborn spirit and hypersensitive nature, William Hale concluded a long letter by cautiously proposing "an understanding, instead of an acknowledged cast iron engagement." [1] George was upset but unconvinced. He loved Evelina and she loved him. He felt that was enough.

On the same day he received this letter, he entered the Massachusetts Institute of Technology. Old Boston seemed to him quiet and staid. Its higgledy-piggledy streets, lined with red brick houses, and dotted with old churches and crumbling graveyards, reflected a long tradition unknown in Chicago. At the corner of Boylston and Clarendon streets on Copley Square stood the Institute. Next to it was the imposing Boston Society of Natural History. Across the street was the fashionable Hotel Brunswick; at the far end of the square rose the spire of Trinity Church, where the powerful Phillips Brooks was the preacher.

In the institute, a large gray building with Corinthian columns, George began a new life. Twice a week he wrote long letters home, which his mother preserved in a special box. Unfortunately, these were apparently destroyed when the house in Chicago was vacated, and only fragments, repeated in letters from his mother to her stepmother, now exist. Nor do any of his letters to Evelina remain. One source only, his letters to Harry Goodwin, whom he met the first day and who became his lifelong friend, is available. [2] The evidence indicates that throughout his college career he was extremely busy, but also, particularly at the beginning, extremely lonely. He missed Evelina, his family, and his Chicago laboratory. He wondered if the institute was really the place for him. As the years passed, these doubts grew.

In later years he spoke little of his "Tech" career. It was perhaps the part of his life that contributed least to his creative development. Yet it gave him insight into the methods of teaching in technical schools, an insight that resulted in his realization of the need for a broader outlook in scientific education. Many years later this realization would play a significant role in his building of the California Institute of Technology.

George boarded with a Dorchester family, the Bumpuses, at 3 Wheatland Avenue in that semirural suburb. Each morning he made the long trip into the city aboard the horsecar. At Roxbury, Harry Goodwin joined him for the rest of the ride on that five-mile, five-cent route. One morning, George confided the "astonishing fact" (as Harry called it) that he was engaged to a girl named Evelina Conklin.[3] The previous day he had received a most disturbing letter from William Conklin, Evelina's father, who was a teller in a Brooklyn bank. He had to explode to someone. Harry was the target.

As the horsecar jogged along, George produced the letter.[4] "My dear Mr. Hale," it began. Then, in reply to George's request for permission to correspond with Evelina, it went on: "I have no objection whatever to your *continuing* a correspondence with her. I have learned, however, (to my great surprise) that you had made a declaration of love to her. I had no suspicion that you intended any warmer feelings than that of friendship for her. I cannot help a feeling of regret and that you were overhasty in this matter, not on account of any personal objection to you but you are both so young, many years must intervene, and changes may take place which would make the consummation of your wishes undesirable, so that while I place no obstacle in your way of cultivating her acquaintance, I must request that you will not attempt to bind her by any promise, but patiently wait, at least two years; at the expiration of that time should you desire it, I will cordially grant you an interview."

As George read the letter to Harry he became increasingly perturbed. What could he do? There was obviously no good answer, except the odious "Nothing at all.

In this way, through the exchange of confidences, George Hale and Harry Goodwin became warm friends. They learned the value of a friendship that meant shared emotions and experiences, and ideas shared not only in science but also in poetry, literature, and music. From the first, Harry was impressed by George's plans for work on the sun and his dreams for the future of astrophysics.

"George knew at the beginning of his college days—indeed long before—exactly what he wanted to do in life, namely to follow a career of scientific research in astrophysics. He was a born investigator; to experiment was for him an exciting adventure." [5] Harry himself was brilliant, but had no such driving ambition.

At the institute their classmates composed a ditty that followed the "twins" through all their college days:

> If you know Georgie Hale
> You can see without fail
> If Goodwin's the comet
> That he is its tail. [6]

Which shows, as a classmate comments, how far they were from realizing Hale's genius. In retrospect other classmates would speak of his analytical and inquiring mind—his far-reaching imagination "in a relatively frail body." They would describe his generous nature, his charm and friendliness, his sense of humor, his modesty, above all his restless energy and intensity. On the rare occasions when he escaped from the laboratory to play tennis, he played, as he worked, at fever pitch, and he played well. When he walked, he seemed to run, and one of their most vivid memories was of a small, slight figure tearing through the halls of "Tech," across Copley Square, his dark cape flapping in the wind.

This absorption in a future goal meant he had less time for trifling than some of his more gregarious classmates. "He never appeared at Charlie or Jake's or even in the old Elm." [7] Perhaps he had indeed inherited some strain from his teetotaling grandfather, Benjamin Hale, once the editor of the temperance newspaper *The Fountain*. Years later his more convivial brother Will commented, "The trouble with George is that he never had any bar-room experience." [8] Yet he loved a good time. Often, when something struck his funny bone, he would laugh till the tears rolled down his cheeks.

Before entering the institute, George had read in the catalogue this statement of its aims, "It is sought to equip the Pupil with such an amount of practical and technical knowledge . . . as to qualify him immediately upon graduation to take a place in the industrial order." This, he soon realized, was what most of the students wanted and this was what they got. But he had already tasted the wonder of pure research, and nothing could ever persuade him to abandon it for the "industrial order." "I was born a free lance," he had said. As time passed, he found himself rebelling increasingly against the study of dull facts and routine techniques that characterized the majority of his courses. He found little here to fire his imagination. A born experimentalist, he was happiest in his laboratory, working with his hands, designing his own apparatus, trying to solve unsolved problems. This overwhelming interest must have been recognized even by his English professor who, in addition to essays on Scott and Milton, received others titled "Astronomical Photography," "Science," even "Diffraction Gratings."

A classmate, Frank Greenlaw, recalls that Hale arrived at the institute with "a suitcase in one hand, his precious grating in the other." The grating, Greenlaw notes, was a most valuable object. No one else at "Tech" owned such a thing; hardly anyone would have known what to do with it. To the majority, who had little idea what they wanted in life, it must have seemed extraordinary that George Hale could be devoted to such an "impractical" subject as astronomy. A few, like Willis R. Whitney, later to become head of research of the General Electric Laboratory, envied him not only his interest but also the ease with which he seemed to get "good marks and honors without trying." Neither Hale nor Goodwin was forced, as Whitney was, to earn his way through the college course. "I was in earnest in being jealous of both of them," he said.[9]

Hale himself, unconscious of his classmates' opinion, had other concerns. Throughout his college days, he was torn between his love for research and his knowledge that somehow he must complete four dreary years at the institute. He was faced, too, by another conflict that added to his inner turmoil. This was the problem of reconciling the fundamentalist faith his mother had taught with his growing belief in Darwinian evolution, a theory she considered contrary to biblical teaching.[10]

On his arrival in Boston, he had gone to stay with the "Bumpi" in Dorchester, where his father and mother thought he would be safe from the temptations of a big city. Laurin Bumpus was a city missionary who considered tobacco a sin and alcohol the product of the devil. He was devoted to a cause—the spiritual salvation of his fellowmen. Every morning there were family prayers. Every Sunday, morning and evening, George was expected to attend services at the Second Congregational Church where Laurin was deacon.[11]

Often then, pondering the meaning of religion, Hale discussed his background with Harry Goodwin. His paternal grandfather, Benjamin Hale, like his maternal grandfather, Gardner Browne, and Erastus Scranton, had been a minister in Connecticut. But, unlike the brimstone-haunted Erastus, he had been a man of high good humor and a jokester, able to make friends even of the saloonkeepers he was trying to put out of business. From his father, William Hale had inherited a strong religious sense. He was active in the Congregational Church and on the American Board of Foreign Missions, and was a deacon in the Church.[12] With such a background it was inevitable that George should have had profound feelings of guilt when he first began to doubt the literal truth of fundamentalism.

Harry, a liberal Unitarian, found it inconceivable that a boy who had picked up fossil trilobites along the Chicago breakwater should ever doubt the truth of the doctrine of evolution. As they argued, George began to see the weakness of his position.

George Hale was born just nine years after the publication of the *Origin of Species*. When he was three, the *Descent of Man*, even more devastating in its effect on organized religion, appeared. The turmoil created is hard for later generations to comprehend. It is equally hard to understand how a boy like George Hale, whose entire philosophy would become based on evolutionary theory, could ever have doubted its truth. He himself later found it hard to recall these doubts. But Harry never forgot the intensity of their discussions or the depression into which his friend was thrown in his search for the "truth."

In the fall of their freshman year an event occurred that must have helped to sway George in his thinking. Alfred Russel Wallace, who had discovered the theory of natural selection simultaneously with Darwin, arrived to lecture at the Lowell Institute on "Darwinism and Some of Its Applications." The bewhiskered, long-bearded biologist explained that, just as Copernicus and Galileo had dislodged the earth from its central place in the universe, so Darwin had eliminated man as an object of special creation and made him merely a part of the vast evolutionary scheme of the universe.

After this, Hale read everything he could find on evolution, especially the luminous essays of T. H. Huxley. Huxley had declared that a deep sense of religion is quite compatible with an entire absence of theology. George was impressed. He could not help realizing that the faith his mother had taught was built, not on a searching spirit, but on dogmatic authority.

When the first year at "Tech" ended, he hurried back to Chicago. The previous Christmas his family had moved to a magnificent new house on fashionable Drexel Boulevard. It was designed by the noted architect D. H. Burnham. When the carriage drove up under the porte cochère, George must have felt he was entering a medieval castle. A heavy carved door led into a vestibule decorated with paneled woodwork, furnished by William Hale's friend, the railroad magnate George Pullman. The parlor, with its green marble fireplace, was decorated with mirrors and a Venus de Milo statue; the drawing room was adorned with blue and gold wallpaper; the library was filled with glassed-in bookcases, and a Raphael Madonna hung on the wall. In every room there were chairs upholstered in velour, and heavy brocade portieres adorned with dangling balls and tassels.[13]

An elevator led to the second floor, then to the attic, which, on his arrival home, was George's goal. Here, in the laboratory he had planned for his spectroscopic and astronomical work, he immediately began setting up his equipment. On a long bench in the middle of the room he mounted the grating spectrometer, now equipped with a camera. On the wall he hung Lewis Rutherfurd's long photographic map of the solar spectrum, "my chief incentive toward high dispersion (i.e. increased spreading out of the light in that spectrum) until I obtained Rowland's classic map." On a shelf in one of the south windows he placed a small heliostat (or mirror) to reflect the sun's rays into the spectrometer. On a pier built on the south side of the house, at the edge of an unroofed porch, he mounted his telescope. Soon he was ready to photograph the sun and its spectrum. His observations of the sunspots, whose cause was still unknown, and of the faculae that lie like brilliant clouds above the sun's surface, are recorded in notes like these:

1887

July 21—Arranged shutter and camera for photographing Sun with telescope.

July 22—Took photo of sun. Poor seeing negative—spoiled by telescope moving. No spots on sun. Small faculae near limb.

July 25–29—Observed on two nights. Jupiter, Moon etc. Made photos of Moon—poor.

The record was interrupted by a family trip to the Yellowstone. On his return it continued:

> *August 18*—Observed Sun. 2 large and 1 small spot on disk. Observed solar spectrum with spectrometer mounted on tripod. Easily saw Ni (nickel) line between D's in 3rd spectrum. Photographed solar spectrum as below (using Rowland flat grating.) [14]

Here he reproduced his first photograph with the new apparatus, the first step on a long path that would lead him deep into the exploration of that spectrum. Instead of the single spectrum that he had obtained with his prisms, several spectra now appeared. Nearest the slit lay the first-order spectrum; the second order, twice as long as the first, which it partially overlapped, lay farther from the slit; still farther out lay the third order, in which the colors, and therefore the spectral lines, were spread out or dispersed even more. The photograph was poor, but with such a grating Hale saw that as he was able to separate the lines in the spectrum more, he would be able to measure their positions more accurately, and so determine their origin more exactly. The possibilities, he felt, were limitless.

As September approached, he became increasingly disturbed by Harry Goodwin's apparent decision not to major in physics—the course he himself had elected. He bombarded the weakening Harry with a stream of letters. "You *must* take physics for your own good, for my good, for the Tech—in fact—for everything," said one. "Cap! Please take physics!" was the heartbroken plea of another.[15] Like Martha and Will before him, like so many of George's friends afterward, Harry finally gave in. That fall, they returned for their sophomore year, ready to suffer through physics together. Long afterward Harry would head the Physics and Electrochemistry Departments and become dean of M.I.T.'s Graduate School.

Charles Cross, the Physics Department head, was a dour, bald-headed individual dressed always in a black cutaway coat. He had made notable contributions to the infant science of telephony and had founded what was said to be the first electrical engineering course in the country. He was a rigid disciplinarian with a sharp tongue, a terror to the dull-witted. But gradually, as he realized the ability of the two physics majors, "Charlie," as they called him, "modified the austerity of his manner," and gave them "access to the more valuable pieces of apparatus which were kept in locked cases."

The physics course, however, held little allure. In their sophomore year there was, in fact, only one science course that roused Hale's imagination at all. This was a course in qualitative chemistry, taught by a young assistant, an Institute graduate, who had taken his M.S. in Germany. His name was Arthur Noyes. He was quiet, dark, gaunt, intense. Noyes, Hale soon discovered, came closer to sharing his views on research than anyone else at the institute. He shared his passion for science in a way that Harry never quite approached. Hale was delighted by his quiet wit, his love of experiment, his belief in research, his vivid imagination.

It was the beginning of a long friendship, and George entered into it, as he entered into everything, with total intensity. In college and in afteryears, there would be periods when he would withdraw into himself. Yet all his life he craved companionship. His friendships with Harry and "Arturo," as he called Noyes, were the first in a small chain that would compensate for lacks in his social life. Whenever he had an idea or made a discovery, whenever a crisis occurred, he would write to these few close friends for advice and understanding. Years later, when he looked back on his college life, it was his memory of these two friends that stood out against a bleak background. "As you know," he wrote to Harry, "I am not the kind that can pour out their souls to many friends, and if it had not been for you, I would have been a solitary recluse at Tech." [16]

In his sophomore year one other event of far-reaching importance occurred. One day in January, 1888, he was walking near the Harvard College Observatory in Cambridge. Why, he wondered, couldn't he help there as he had done at Dearborn? The director was Edward C. Pickering, who at M.I.T. had founded the country's first physical laboratory for students. Late in February, George wrote to Pickering to ask if he could, by chance, use a volunteer assistant. On the 28th of February the answer came:

DEAR SIR,

Your letter of yesterday is at hand, and I shall be very happy to have you come here on Saturday from one to eleven, as you propose.

Yours respectfully,
EDWARD C. PICKERING [17]

On the following Saturday, Hale boarded the horsecar that ran from Bowdoin Square to Cambridge. He stuck his feet in the straw in an effort to keep them warm. At Harvard Square he got off, then, "passing under the Washington Elm," hurried up Garden Street toward the two frame buildings that housed the observatory. In the larger building he found Dr. Pickering, a huge and powerful man who welcomed him warmly. They toured the observatory—the big dome that housed the 15-inch telescope, the smaller 8-inch dome, the library, and the computing rooms. It was a fascinating place. In contrast to Hough's and Burnham's observatories in Chicago, it was a place where spectra were actually studied. This in itself was a great thing. Pickering, one of the first to realize the immense possibilities of photography in astronomy, was a leader in spectral classification and photometry. He was also, as Hale would soon learn, a man of tremendous energy.

In the next two and a half years, Hale would come to know Pickering well. Occasionally he would be invited to his house to dine. The dinners were enhanced by the majestic presence of Mrs. Pickering, a veritable "grande dame," a true Bostonian from Beacon Hill, the daughter of Jared Sparks, former President of Harvard. She was accustomed to entertaining the elite of Boston, as well as the luminaries of the scientific world. She even used her talents as a pianist to entertain her guests.

But for Hale the best part of this introduction to Harvard was the chance to work in an observatory and eventually to carry on his solar experiments there. Pickering had said he could work from one to eleven on Saturdays. Actually, it was often much later when they closed the dome and he started home through the deserted gas-lit streets. Often he reached Dorchester in the early dawn when the stars were dimming, or even as the first rays of the sun were appearing in the east.

In that first year he spent most of his time on routine photography. But he was gaining valuable experience and he was happy. He lived for those hours at the observatory and for the time he could snatch from his work at "Tech" to spend at the Boston Public Library. There he devoured everything he could find on the sun, on spectroscopy, on anything remotely connected with astronomy, and especially with its physical side. He read such classic papers as those of Faraday and Maxwell, the current *Transactions* and *Proceedings of the Royal Society*, and the *Comptes Rendus* of the Paris Academy of Sciences. Always he made careful notes.[18]

Years later, looking back on the importance of these hours, he was to write: "To the would-be investigator, dimly perceiving for the first time the boundless possibilities of original research, there is no such stimulus as that to be derived from direct contact with the masters of his subject. He should not be content with some second-hand description of their work, but go straight to the original sources, and read what was written in the very heat of progress. Thus he may feel with them the dawning consciousness of new lands to be explored and experience some of the delight which they felt when they first passed from the known into the unknown. More-over, he may best realize in this way how simple are the tools so often employed to attain the most important ends, and recognize the advantage of cultivating an alert and optimistic mind, sensitive to new impressions and aware of the opportunities which are ever open to the ready inquirer." [19]

At the library, as he read, he forgot everything else. At other times, however, Evelina absorbed his thought. Whenever he could he journeyed to Brooklyn. After one such visit he reported to Harry: "I got here ok (like some of my calculus examples) and have *not* had a *bad* time since. Tuesday evening we went over to New York to see 'The Wife' which is playing at the Lyceum, 23rd Street and 4th Avenue." Then he added, "Yesterday morning, I enjoyed (?) some dry goods stores with Evelina." [20]

This, however, in that memorable year of 1888, was not his most memorable visit. He had decided to surprise her on her birthday, March 18th. The weather had been balmy, and everyone was look-ing forward to an early spring. Then, on the night of March 11th, two storms swept into the north and collided, and everything changed. By four in the morning the snow was falling so fast in New York that in five minutes a man's footsteps disappeared en-tirely.

George, in Boston, had decided to start a few days early. He had no idea what was happening in New York—all telegraph lines were down, all railroads paralyzed. Therefore he took the Fall River Line boat. By the time he started, the snow had stopped falling, but the wind was still blowing. The boat reached the New York pier in a gale. From there he plowed his way through downtown Broadway to the Brooklyn Bridge. At the bridge a narrow footpath had been cleared. Finally he reached the Brooklyn shore. The streets, piled high with drifts, were almost impassable. He finally reached Lina's house, close to collapse. Drifts were piled high to the second story of the Conklin house. He groped his way through a tunnel to the lower story, and knocked. The door was opened by one of Lina's brothers, who looked at the apparition on the doorstep as if he were seeing a ghost. Lina was equally astonished.[21]

That summer Lina visited the Hale family during the summer vacation—first in the magnificent United States Hotel at Saratoga, then in the Adirondacks. They were idyllic weeks. Yet George was restless. Even when boating with Evelina he talked enthusiastically of his work. In April, 1888, on a lot adjoining the stable of the Drexel Boulevard house, he was building a small red-brick spectroscopic laboratory to which he had given the grand name "The Kenwood Physical Laboratory." It contained a darkroom, a laboratory and shop, and a room large enough to house a Rowland concave grating of ten feet focal length. Now, even though he was enjoying himself in the Adirondacks, he could hardly wait to get back to Chicago.

He reached home early in August. At first nothing went well. Exasperated, he wrote to Harry, "Dear Prof. In reply to your various questions I must 'freely admit' that I know no more about the spectrum of the sun than I did when you were here." After endless difficulty he got a fair-looking plate of the solar spectrum. Then came an exasperating bout of cloudy weather. Just before he was supposed to leave for Madison it cleared. Frantically he set to work. To his joy his efforts were rewarded. For the first time he was able to identify carbon in the sun's spectrum—the first nonmetallic element discovered in the sun. This question of the presence of carbon in the sun had been under debate for some time. Lockyer in England claimed it was present. Trowbridge at Harvard agreed. But other astronomers opposed the claim. "Hence," George wrote to Harry, "I have not actually made a discovery of anything of the kind, but simply corroborated the results of Lockyer and Trowbridge." [22]

While he disclaimed any special discovery, this confirmation of the existence of carbon in the sun and identification of carbon flutings was, in fact, his first scientific triumph, the beginning of his real astrophysical research. It must have given him a heady feeling to know that, shortly after his twentieth birthday, he was joining the ranks of those who had contributed to scientific knowledge.

Back in Boston he found himself burdened with a course that left little time for sungazing. It included electricity, integral calculus, literature, heat and thermodynamics. There was also something called Physical Laboratory, and a course in applied mechanics that was so painful he was sure he would never pull through. At the year's end, when he found he had passed, he was surprised and relieved. Still, he knew that more lay ahead the following year. "To think," he exclaimed, "of Physics Lab. calculations in store for *me!* That is the bitter pill I dread. . . . How shall I ever finish the miserable work!" [23]

On his way home at the end of his junior year, he stopped first in Brooklyn, then in Baltimore to call on the redoubtable physicist Henry Augustus Rowland. Three years earlier he had written to ask the cost of such apparatus as Rowland had used in photographing the solar spectrum with a 6-inch grating. Rowland's reply had been brief and pointed. It probably would have been even more curt if he had known that his correspondent was only seventeen. "In reply to your letter of inquiry, I may state that large gratings suitable for photographing the spectrum in the manner I have done are very rare indeed. There are only three in the world. One I have; another is at Harvard College. The third is going to Lockyer." [24] That, he doubtless concluded, was that. But he had not reckoned with his young customer's persistence. Hale had written back, explaining his reasons for wanting such a grating. To this second plea, no answer had come. Brashear had said that Rowland was a notoriously bad correspondent, but this was little consolation.

This time, instead of writing, Hale had decided to drop in on Rowland in his laboratory. Through the open door of his office he could see him working at his desk. He knocked. Rowland did not move. After a while he knocked again. Still there was no sign of life. While he waited he thought of all the things Brashear had told him about Rowland, especially the famous tale of the time he was called into court as an expert witness. "Who," the judge demanded, "is the greatest authority in the field of physics in the United States?" "I am," was Rowland's simple, direct reply under oath.[25]

Now Hale knocked again. This time, with startling suddenness, Rowland wheeled. "What can I do for you, young man?" he demanded.

Hale told him his name, gave him Brashear's letter of introduction, and said meekly that he would like to discuss a certain imperfection in his 4-inch grating. Rowland rose. Like Pickering, he was a big man. He looked down at his youthful visitor. "That grating is good enough for any infant," the great man growled. Hale, intimidated, nevertheless plunged ahead with an explanation of what he was trying to do and why he needed a better grating. This time he saw that Rowland was listening intently. Instead of the short interview he had expected, he was there for over three hours. By the end they were on "the best of terms." Afterward Hale wrote to Goodwin, "I talked as freely and easily to him of my plans for future work, etc. as I would to Sile." [26] Rowland even praised his photographs and admitted that they were probably as good as could be expected with his 4-inch grating.

Rowland led him all over his laboratory and showed him everything he was doing. Hale was entranced. In the underground constant-temperature vault he saw the ruling engines with which Rowland had made his famous gratings that had helped to revolutionize spectroscopy.

In his laboratory, Rowland also demonstrated the concave grating, which he had invented in 1882 and which had inspired Hale to build his own 10-inch concave grating at Kenwood. With this invention Rowland had shown that if the grating were given a concave spherical surface, no lenses would be needed. With such a grating it was possible to produce a spectrum many feet in length; as a result of the great dispersion, he had been able to separate and analyze countless lines in the solar spectrum. Many of these corresponded to the iron lines produced by the electric arc in his laboratory; others corresponded to other substances. Evidently the sun's chemical composition must closely resemble that of the earth.

That day in Rowland's laboratory remained a red-letter day in Hale's life. Henceforth his desire to photograph the solar spectrum on the scale of Rowland's great map grew. He also began to dream of photographing other stars on the same magnificent scale. Only in this way could he learn more about the nature of sun and stars. As this dream came to dominate his life, it would lead to his building of greater and greater telescopes with larger and finer gratings.[27]

The meeting with Rowland took place on a Monday. The next day Hale returned to Brooklyn. But that afternoon, with Evelina, he was off again to Princeton to see the great solar pioneer, Charles Young. Young's *The Sun* had long been one of his favorite books. He had read it over and over until its edges were dog-eared and its pages worn. In it, Young had given a vivid description of his method of observing the solar prominences visually (see illustration following page 112) with a slit. He showed how, as the slit is widened, the prominences could be observed, and how, if not too large, a whole prominence could be seen at once: "With the widening of the slit, however, the brightness of the background increases, so that the finer details of the object are less clearly seen, and a limit is soon reached beyond which further widening is disadvantageous. The higher the dispersive power of the spectroscope the wider the slit that can be used, and the larger the protuberance that can be examined as a whole."

Now, at Princeton, Hale was able to see this instrument at work. From the first, Young was as cordial as Rowland had been gruff. First he showed his visitors his 9½-inch telescope, focused it on the sun, and attached the spectroscope. Unfortunately, it was too cloudy to observe any prominences. Still, they could easily see the "arrow-headed C-line" of hydrogen at the edge of the solar disc where the dark Fraunhofer spectrum merged into the bright chromospheric spectrum. Meanwhile Young described prominences he had observed through the spectroscope slit. Sometimes, he noted, they appeared so much like terrestrial clouds in form and texture that the resemblance was startling. It was like looking out through a partly opened door on a sunset sky, except that the solar "cloudlets" were of a single pure scarlet hue. Once, he said, he had observed an ordinary "horn" which in half an hour became intensely brilliant and doubled its height; in the next hour it reached the enormous altitude of 350,000 miles, then broke into filaments and faded away.

He also recalled Angelo Secchi's vivid description of the beauty of prominences. In *Le Soleil* Secchi had noted that it is impossible to reproduce completely "the vivacity of color of these enormous masses, or to depict their rapid motions when they are shot by eruptions from the interior above the surface of the sun. The best drawings are inert and lifeless when compared with the actual phenomena. These incandescent masses are vivified by internal forces which seem to endow them with life, and their colors are so characteristic that they enable us to determine spectroscopically the chemical nature of their constituent gases."

Later, when Hale had a chance to see these prominences for himself, he, too, recalled Secchi's account. "It is difficult," he would write, "to convey any conception of the beauty of the prominences as seen with the red hydrogen line."

At this point, however, as he told Young, his efforts to observe these prominences with his spectroscope had failed miserably because of the smallness of his grating and the feebleness of the light it provided.

As they talked, Young led his visitors to the 23-inch telescope, one of the largest in the country. Before they left, he gave Hale "innumerable points" on his solar work, and offered to help in any possible way. "I can tell you he was *fine*," [28] Hale wrote to Goodwin with an enthusiasm he was never to lose for his new-found friend whom he was to consider his "patron saint." If there had been any doubt about the branch of astronomy he would follow, that doubt now vanished forever.

Early in June he ordered from Brashear a large "telespectroscope" that could be attached to the 15-inch at Harvard and used for photographic work. Afterward Brashear wrote to Holden, director of the Lick Observatory in California, about "the young man Hale. He is a sharp young fellow and you'll *hear from him* some of these days!" [29]

In Chicago once more, Hale hardly took time to greet his family before he was back in his lab. "Before I had been home an hour," he told Goodwin, "things began to look natural. . . . I leveled up the tracks, set up the slit, and mounted the grating. In a short time, thanks to Rowland's direction, I had it in pretty good shape." [30]

Just then, unfortunately, the weather intervened. He waited disconsolately for the sun to show its "formerly jovial face." He wrote to Harry, "Die Himmels sind mit wolke gecovered, and it thunders, lightens, and rains continuously. . . . It looks now as if this summer's work would amount to nothing." [31]

On June 29th he celebrated his twenty-first birthday. In honor of the occasion his father gave him a block of stock in the new Reliance building he was going to "put up" in Chicago. He was also elected a director. Soon afterward he decided to tackle his father again on the question of his marriage, which he hoped to celebrate right after his graduation the following June. When the ordeal was over, he reported joyfully to Harry, "I can hardly wait to tell you the results of a little interview I had with my père. . . . He agreed to June, and as my mother acquiesced the next morning, the thing is settled! It took a *fearful* (count 1000!) brace to ask him."

The strain of the interview had been great. Afterward he was prostrated by a splitting headache and nervous indigestion. These were complaints that were to plague him the rest of his life. From childhood, his mother had worried about his "constitution." As a result, he also worried about himself. He was tense; he found it impossible to relax. This tenseness would undoubtedly prove an advantage in his research—it would keep him constantly dissatisfied, constantly racing from one problem to the next. But when bound up with personal stress, it would often lead to temporary collapse. Eventually it would, on occasion, result in complete physical breakdown.

Yet, on the whole, the summer of 1889 went well. In the interludes, when the skies were cloudy and the sun was invisible, he spent his time writing an article for the *Beacon*, a small Chicago journal. Excited by all he had learned in his reading in the Boston Public Library, and by "the almost unexplored domain of spectroscopic astronomy," he described the instrument he had first discovered in the magical pages of *Cassell's Book of Sports and Pastimes* —the spectroscope:

"Attached to the telescope, it brings to view gigantic gas streams upon the solar surface, never seen before its invention, except at the rare occurrence of a total eclipse; with it their velocity is measured and their composition determined; it tells us that iron, magnesium, calcium, hydrogen and a score of other elements exist in a state of vapor in the solar atmosphere; it measures the velocity of rotation of the sun about its axis, and reveals the nature of the spots and faculae. Applied to the study of the stars, it discovers their composition and velocity of motion toward or away from the earth; it explains the sudden appearance of new stars in the heavens; it tells us that the nebulae are vast masses of glowing gas, and helps to unravel the mystery of comets."

As he traced the history of spectroscopy, he went back to the "memorable experiments of Kirchhoff and Bunsen in 1859, when the prism first disclosed to their astonished gaze the secrets of the sun." He showed how, as a result of these experiments, a "new science" had sprung up "which offers problems not to be solved by the astronomer alone, but by the combined skill of the astronomer, the physicist and the chemist."

"The New Astronomy," he concluded, "seems assured of a most brilliant future. Questions of all degrees of complexity remain to be answered, and every day sees their number increased. The variations observed in spectra with every change of temperature or pressure; the mysterious 1474 line in the corona spectrum; and ultimately the true nature of matter itself—these problems and many like them, are yet to be solved by a careful comparison of terrestrial and celestial phenomena." [32]

This article, aside from an earlier one on stellar photography, was George Hale's first publication. It appeared in the *Beacon* in July, 1889, just after the celebration of his twenty-first birthday. At that time, too, another event occurred that would make that summer one of the most memorable of his life. On a hot day at the end of July he boarded a cable car that ran along Cottage Grove Avenue, and sat down on a front seat. As he rode along he was thinking of his recent visit to Princeton and of Young's futile attempts to photograph a prominence in full sunlight. By setting the spectroscope on the C line and attaching a small camera to the eyepiece, Young had been able to photograph a bright prominence. But the light was so feeble, the image so small, the time of exposure so long, the requisite accuracy of motion in the clockwork so hard to obtain, that he had been unable to produce a picture of any real value. He could see the bare outlines of the prominence but no details.

As he rode along, Hale pondered these questions: How could he photograph solar prominences without waiting for the short and uncertain seconds of an eclipse in some remote corner of the earth? How could he cut down the sun's tremendous glare and the resultant fogging in order to bring out the necessary details in a photograph? How could he do this over the sun's entire circumference at one time?

As the cable car rattled along he looked out the window. By the side of the road he saw a white picket fence. As he watched it pass, his thoughts flashed to the lines in the solar spectrum. In that instant an idea came to him "out of the blue." Out of that idea his invention of the "spectroheliograph," as he was to call it, was born.

A few days later, on August 5th, he wrote to Harry: "Of scientific work I have accomplished but one thing this summer, and even that did not involve much labor. It is a scheme for photographing the prominences, and after a good deal of thought I can see no reason why it will not work. The idea occurred to me when I was coming home from uptown the other day, and it amounts to this. Stop the clock of the equatorial [telescope] * and let the sun transit across the slit, which is placed radial to the limb. Bring H (in the blue) into

* I.e., stop the mechanism by which the telescope was driven to follow the apparent motion of the stars.

the field of the observing telescope, and replace the eye-piece by a plate-holder held in a suitable frame, and drawn by clockwork across the field at the same rate as the sun crosses the slit. As the H line lengthens and shortens—as it will do with the variable height of the prominence, the plate will photograph its varying lengths side by side and thus produce an image of the prominence. That is the idea in the rough, but I have studied it out in detail, and designed a travelling plate holder, which I will have Brashear make. I have also got an arrangement by which all fog is avoided and I have great hopes that the thing will be a success. If it is, new chances for work on the prominences will be opened, as in this way the changes during short intervals of time can be noted with much greater accuracy than in drawings.

"Let me know what you think of the plan," he asked, then noted cautiously, "I am not going to say anything to Charles or even Prof. Pickering about it until I find out whether it will work, but I am going to write to Prof. Young today, so as to have a record of it in case it should amount to anything." [33] With his letter he enclosed a blueprint of his invention.

This was George Hale's first account of the invention that would bring him worldwide fame, and revolutionize not only the study of the prominences but also that of the sun's entire surface. With it an immense leap forward would be taken. As Robert Ball, the Astronomer Royal of Ireland, commented: "It is impossible to overestimate the stimulus that has thus been given to solar research. In a few minutes photographic processes will produce a far more accurate picture of the chromosphere than eye observations could have produced in hours." [34]

"The principle of this instrument is very simple," Hale was to explain. "Its object is to build up on a photographic plate a picture of the solar flames, by recording side by side images of the bright spectral lines which characterize the luminous gases. In the first place, an image of the sun is formed by a telescope on the slit of the spectroscope. The light of the sun, after transmission through the spectroscope, is spread out into a long band of color, crossed by lines representing the various elements. At points where the slit of the spectroscope happens to intersect a gaseous prominence, the bright lines of hydrogen may be seen extending from the base of the prominence to its outer boundary. If a series of such lines, corresponding to different positions of the slit on the image of the prominence, were registered side by side on a photographic plate, it is obvious that they would give a representation of the form of the prominence itself.

"To accomplish this result, it is necessary to cause the solar image to move at a uniform rate across the first slit of the spectroscope, and, with the aid of a second slit (which occupies the place of the ordinary eyepiece of the spectroscope), to isolate one of the lines, permitting the light from this line, and from no other portion of the spectrum to pass through the second slit to a photographic plate. The principle of the instrument thus lies in photographing the prominence through a narrow slit, from which all light is excluded except that which is characteristic of the prominence itself. It is evidently immaterial whether the solar image and photographic plate are moved with respect to the spectroheliograph slits, or the slits with respect to fixed solar image and plate." [35]

Only long afterward did Hale learn of the unsuccessful attempts of Braun and Lohse to make a similar instrument in 1872 and 1880 respectively. Later, he also discovered that the *invention* of the principle had been published first by Janssen as early as 1869, then by William Huggins. But principle is one thing, practice another. If the method was simple, the process of applying it, he knew, would be difficult. This practical application of the principle would be his first great contribution to solar research.*

Young, with characteristic generosity, was delighted by this new and promising step in a field in which he had pioneered. "Mechanically it is all right," he wrote back, "and will work perfectly." [36]

Hale quickly realized that it would be impossible to try out his new idea with his present equipment. He would have to wait until he got back to Harvard. He decided, therefore, to forget the sun entirely, and rushed off to Madison, Connecticut. There, on arrival, after commandeering a buggy at the livery, he urged his "beast" over the four miles of muddy road to Clinton as fast as he could go.[37] The next afternoon Evelina arrived to spend several days in Madison.

Summer was almost over. Yet, as September approached, he remembered sadly that all the books he was supposed to have read remained unread. "I must take a brace, but *when*—I have no idea,"

* For an account of Hale's priority in the invention of the spectroheliograph, see Chapter Three.

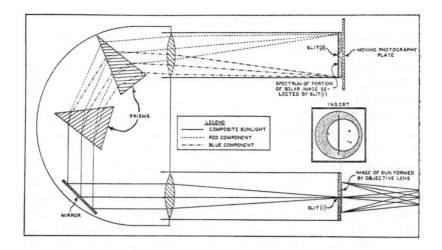

The principle of the spectroheliograph. Slit (1) selects some particular segment of the solar image (see insert); slit (2) isolates a particular wavelength in the spectrum of that segment and allows it to impress its image on the photographic plate. As the sun's image is made to move across slit (1), the photographic plate moves in synchronism past slit (2). Thus a photographic image of the sun, in a particular wavelength of its radiation, is composed segment by segment. (Courtesy of Sky and Telescope)

he told Harry. He had tried to force himself through the dry pages of Whewell's *Optics* and the equally dry Grant's *History of Astronomy,* but admitted he had read 250 pages of the latter "with absolute ignorance of their contents!" [38] Long afterward he was to become deeply interested in the history of science, but this introduction nearly discouraged him for life.

September marked the beginning of his last year at "Tech." It included a full academic schedule; the excitement (and distraction) of experimenting with his new "spectroheliograph" at the Harvard Observatory; and an extracurricular program that included the presidency of the Electric Club, membership on the Executive Committee of the Photography Society, and an interest in the Literary and Science Society.[39] Still, the time he could spend in such diversion was slight. He had persuaded Professor Cross to allow him to choose "The Photography of Solar Prominences" as the subject of his thesis. As a result he spent every available moment at the Harvard

Observatory—to the detriment of his work at "Tech." At the year's end his report indicated he had passed all his courses, but had barely scraped through organic chemistry. He received a "50" in the course.[40]

At Harvard, Pickering was intensely interested in Hale's invention. Eager to do anything he could to help, he offered the use of the 15-inch refractor—the "grand old instrument" that had once been the largest telescope in the country, and had been used by the Bonds to make the first daguerreotype of a star. It soon turned out, however, that the large spectroscope built by Brashear was too heavy to be attached to the end of the old wooden telescope tube. Therefore, they decided to adapt it to the 12-inch horizontal telescope. But a series of discouraging events, including a spell of "abominable" weather, hampered progress. When the sun did shine, it heated the mirror so much that the prominences were hopelessly distorted. Sometimes, because of the diffuse light that was reflected by the mirror into the tube, they were entirely invisible.

Finally, on April 14, 1890, after weeks of failure, Hale was successful—to a degree. In his thesis, published in part in the *Technology Quarterly*, he wrote: "On April 14 a cool breeze was blowing, making the seeing fair in spite of a little whiteness in the sky. A hasty examination of the limb discovered a prominence in a good position for the work, and a photograph was made through F, the slit being about 0.0005 inch wide. On developing the plate, the outlines of two prominences could be seen rising above the limb. As only one prominence had been noticed in observing the point in question, I returned to the telescope, and found that there were in fact two prominences in the exact position shown in the photograph." [41]

The images were still not much to brag about, but they indicated that the method had worked. Therefore he concluded his thesis on a hopeful note: "Given a good refracting equatorial and a plate very sensitive to the longer waves of light, I am confident that the spectroscope and attachments described in this paper will be sufficient to produce prominence photographs of real value for study and measurement."

In this way the spectroheliograph, conceived in a cable car on Cottage Grove Avenue, was born in the Harvard Observatory. Its invention, as Harry Goodwin was to comment forty-six years later at a symposium in Hale's honor at the observatory, "unquestionably contributed more to advance our knowledge of solar phenomena than any other invention since the time of Galileo." [42]

On May 28, 1890, George Hale was recommended for the degree in physics, known as Course VIII. His graduation portrait shows a handsome young man with sideburns and a small moustache, grown perhaps in an attempt to add years to his still boyish appearance.[43] His face appears longer, his forehead higher than in earlier photographs. But in the eyes the same quizzical, searching look remains.

On the first of June the baccalaureate sermon was given by Phillips Brooks in Trinity Church, and the Hale family was there to hear it. At first his mother had thought herself too unwell to come. She had, in large measure, become a recluse. "I go out of the house so rarely that dresses do not have much consideration." At the last moment, however, she came.

Two days after their graduation, George and Harry Goodwin found themselves in Brooklyn, preparing for another ceremony in the Conklin house on Adelphi Street. The thermometer registered 93 degrees, and thunder resounded overhead. Evelina was described as a lovely bride, and George a dashing groom.

After the wedding and two nights at the fashionable Windsor Hotel in New York, the young couple boarded the steamer for the trip up the Hudson to Albany. Then, in true honeymoon fashion, they took the night train from Albany to Niagara Falls, to begin a tour that would take them all the way to the West Coast.[44]

Yet, even during these supposedly carefree days, George could not forget his work entirely. When they arrived in Chicago for a brief stay, Mary Hale reported to her mother: "George is just as much absorbed in his studies as ever, and went to Beloit Monday to look over the Observatory there." [45] The head of the Astronomy Department was to be away for six months, and President Eaton had offered him the use of the 9½-inch telescope. George thought he might accept the offer. Evelina went along. Mary Hale's letter expressed a wish the bride must also have shared. "I wish he cared a *little* more for Society, but now he cannot be induced to make calls or do anything in that line that is not *absolutely* necessary. He is as absorbed in his studies as his Father in business—otherwise a model son . . ."

On their way west, George and Evelina stopped at Manitou, the

Garden of the Gods, and the Yosemite. They continued on to San Francisco, where they took a guided tour through Chinatown, then started for the spot that for George Hale was the goal of the entire trip, the Lick Observatory on Mount Hamilton. This observatory, with its great 36-inch refractor, given by the eccentric James Lick, and only recently completed, was the largest in the world. After a night at the Hotel Vendome in San Jose, they boarded the stage for Mount Hamilton. After a precarious climb over a road "full of twists and turns" through "rugged, wild" scenery, they reached the top. Waiting to receive them was Hale's old friend from Chicago, S. W. Burnham. With him was James Keeler, who had been with Brashear and Langley at Allegheny and was now a member of the Lick staff. More than anyone else in the country, Keeler shared Hale's interest and belief in the embryonic science of astrophysics. There was no one Hale was more anxious to meet.

That night he joined Keeler at the 36-inch telescope, while Burnham did his best to entertain Evelina. But his bull's-eye lantern spattered oil, and the "fine objects" he showed her meant little. At the time she said nothing, but she never forgot the strangeness of that night on Mount Hamilton.

In contrast, Hale wrote enthusiastically to Goodwin, "The Lick Observatory took the cake." [46] Later he described the long tube "reaching up toward the heavens in the great dome," and his examination of the spectrum with which Keeler was making his classic observation of the motions in the line of sight of planetary nebulae. Here, at last, was a modern stellar spectroscope, attached to the most powerful telescope in the world. Enchanted by the possibilities, George was deeply impressed. [47]

The next morning, the director, Edward S. Holden, offered him the use of the 36-inch for work with his spectroheliograph. It was a great opportunity. Only one thing deterred him. That morning he tested the solar definition with the 12-inch. The results were disappointing. Therefore, on Keeler's advice, he decided to "reserve judgment." Young had offered him the use of the 23-inch refractor, and there was also the problem of his Ph.D. He had considered going to Johns Hopkins to work with Rowland and also to Germany. Back in Chicago he debated the problem with William Hale. When his father offered him a 12-inch telescope, to which he could attach his spectroheliograph, the inducement was too strong to resist. Early in August the order for a 12-inch glass went off to Brashear, while the

order for a mounting stiff enough to carry his heavy spectroheliograph went to Warner and Swasey. He decided, too, to add on to
the south end of the Kenwood Physical Laboratory in the back
yard. Regretfully, he refused all other offers of jobs. Later, when his
telescope was delayed, he changed his mind about Holden's offer,
but by then there was no room for him at Lick. Nevertheless his
visit to Lick had shown him the possibilities of a great telescope like
the 36-inch, especially for stellar and nebular work and for increased
spectroscopic dispersion. His dreams for greater and greater telescopes would continue to grow.

CHAPTER THREE

Life and Work at Kenwood

✳ *1890-1892*

FOR GEORGE HALE the return to Chicago was exciting. In the back yard the Kenwood Observatory was slowly rising. In Pittsburgh, Brashear was grinding the lenses for the 12-inch. In Cleveland, the Warner and Swasey Company was working on the mounting. Hale himself had plenty to do. One day he wrote in haste to Goodwin: "I never saw time pass so rapidly, and the worst of it is, tho' I slave steadily, I don't accomplish a thing." [1] Four days later he scrawled again, "This week has been like the others—gone almost before it commenced." [2]

For Evelina Hale, in sharp contrast, the days in the cavernous gray house on Drexel Boulevard seemed endless. She yearned for the gaiety of her Brooklyn home. How different that house from this where Mary Hale, describing the daily routine, told of the "children" rising at six in the morning and "counting all the moments" until ten at night. [3] George worked enthusiastically in his laboratory the day long. After school Mattie and Will would hurry out to help him as they had always done. At times George tried to explain to Evelina what they were doing in the clear and vivid way he had, so that even the most difficult concepts became comprehensible. At times, too, she would sit near him while he worked. Still she found it hard to understand his predilection for spending every available moment in his "lab"—even on holidays.

Nor could she find in the companionship of his mother any real solace. Day after day Mary Hale stayed in her room, suffering from migraine. The shades were drawn; the house was still. Any exertion seemed too great, any foray into the world too strenuous; only at times did she feel able to go for the afternoon carriage drives in which Evelina would join her. [4] "I should be so glad if I would be strong enough and not give out when most needed," she lamented. In self-defense, Evelina began to read the many books available in the house that might increase her knowledge and make her more a part of this household where books ruled and ideas dominated. She also joined a "Guild" of young women working for the Hahnemann Hospital.

In time, too, despite her innate reserve and fear of outward emotional expression, she proposed to George that they set up "housekeeping on their own." Immediately, however, Mother Hale claimed that people would think she had turned them out, and would not hear of it. So Lina continued "to endure existence in that forlorn house." While George must inevitably have been conscious of her feelings, he could do little about it. Yet, with his sensitive nature, he must have felt the conflict keenly.

That first winter he lectured at Beloit, at Northwestern, and at Illinois College in Jacksonville, Illinois. Evelina went with him to Jacksonville, and listened proudly to his talk on the "Principles of Spectrum Analysis." The local reporter praised the "rising young astronomer" as a "talented speaker." At the end of the series, when a resolution of thanks was passed and published, Evelina cut it out and saved it. "In fact I was a *great* hero all around," [5] he told Harry Goodwin. Yet, with his inborn shyness, he never really enjoyed public speaking, and felt always that he was not born to teach.

Despite these doubts of his ability, however, he was interested in a letter that arrived one day in which President William Rainey Harper of the new University of Chicago proposed a meeting. In February, 1890, while he was still at M.I.T., Hale had read in *The Tech* this announcement: "Mr. J. D. Rockefeller of New York City has given $600,000 for the establishment of a University of Chicago." As he had read the notice, he had realized what this new venture might mean to Chicago. The first attempt to found a university in the city had been a fiasco. With Rockefeller money, the institution might now become a permanent reality. Accounts of President Harper's plans, which would "revolutionize College and University work in this country," continued to appear. Said Harper, "It is 'bran splinter new' and yet as solid as the ancient hills." That summer Hale had explored the university site in fields "dotted by scrub oaks" where, as a boy, he had picked wild strawberries.[6]

He had also learned from Pickering that Harper had written to Harvard, seeking information on "George E. Hale and his work." To this request Pickering had replied: "Mr. Hale's work at this observatory was an experimental investigation, and gave me a very favorable opinion of his talents for such researches. I had no opportunity to form an opinion with regard to his mathematical knowledge. I have no hesitation, however, in saying that I consider him well qualified to enter upon the profession of astronomy." [7]

"Everything," Harper wrote to Hale's photographer friend, Gayton A. Douglass, president of the Chicago branch of the Astronomical Society of the Pacific, "shows that he is a young man of very great promise. I am quite sure that the University of Chicago would like to have him take hold in some way with its work." [8] Douglass forwarded this note to Hale with the comment, "The enclosed looks like business." [9]

As soon as they met, Hale discovered why William Harper had been chosen president of the new university. He was a large, stocky man with a round face and brilliant eyes that gleamed from behind thick, gold-rimmed spectacles. He had earned a reputation as a "true prodigy" by graduating from college when he was fourteen and receiving his Ph.D. when he was eighteen. But, unlike some prodigies who vanish into obscurity, he continued at forty-five to be a dynamo of energy. His vision of the university's future was so far-reaching and his drive so impelling that anything he believed in seemed possible. "It seems a great pity to wait when we might be born full-fledged," he said. A spark of understanding quickly ignited between the two men. Afterward Harper wrote, "I feel that between us we ought to be able to accomplish a good deal." [10] Hale, unexpectedly impressed, agreed.

At that first meeting nothing definite was decided. Soon, however, Hale discovered that Harper's object was the annexation not only of him but also of his observatory. On May 19th the president wrote, listing his hopes: for the transfer of the Kenwood Observatory to the university grounds; for its enlargement to include all astronomical work; and for its endowment "with a liberal sum sufficient to pay salaries of two professors etc." He proposed to Hale the "putting of the whole plan under your charge and advancing you as rapidly as your work and age might seem to you and to us to call for advancement." He recognized that this was a high ideal, but he did not think it "too much to hope for, nor do I suppose this to be more than you and your father would care to undertake." He added, "I should like to call the Observatory, of course, the 'Hale Observatory.' This is only proper and would be appreciated by everyone concerned." [11]

In this way, he noted, "You would be placed in a position where for life you could carry on your investigations and at the same time have as large a number of pupils as you might care to instruct." The result would be a great thing for the university: "A magnificent work would be accomplished for the city of Chicago. And, greatest of all, the cause of science would be materially advanced."

When this letter arrived, Hale read Harper's proposals, couched in patronizing terms, with growing anger. He finally replied: "The possibility of securing a position for me by gift or lease of this observatory and additions cannot be considered favorably by us. If I am not competent to obtain a place on my own merits at present, it will probably be best for me to wait until I shall have gained experience by future study. Possibly I shall not be able to fit myself for an important position; if not, I certainly should not desire to obtain one." [12]

Harper was in New Haven when he received Hale's letter. Immediately he sent off a conciliatory reply. Hale was mollified but not convinced. And there, for some time, the question rested.

Meanwhile his own work went on. The observatory, finally finished at the end of April, was an impressive sight. Passersby stopped to stare at the strange domed brick building. At night its electric lights, reflected on the stained-glass windows, gave it the look of a Chinese pagoda. One neighbor, in particular, watched its completion with interest. Lincoln Ellsworth, later famous as a polar explorer, wrote fifty years afterward: "I used to stare across at that dome in awe, wondering what mystery it concealed. We children were told that George Hale looked at the stars there, through a telescope. It didn't seem to me that Chicago's stars were worth looking at. Even the gas lights dimmed them." [13]

The 12-inch lenses had arrived from Brashear in March. "I do think you have a very beautiful glass," he said. "We have done all that mortal could do to expedite the work and I don't believe Clark or any person living ever got out a 12″ quicker. . . ."

March passed into April. With a sunless sky Hale had no chance to test the new lenses. It was not until May 7, 1891, that he succeeded in taking his first photograph that showed the actual form of a prominence. (He used the strong blue hydrogen line H_γ.) The image was faint, and he was sure he could do better. He decided to look for other lines in the violet and ultraviolet regions of the spectrum. In the extreme violet he discovered that two brilliant lines (H and K), owing to calcium vapor, were present in every prominence he examined. They also proved to be particularly valuable for photographing those prominences. They were exceptionally bright; they lay at the center of two broad, dark bands that helped to cut down the brightness of the sky spectrum; and the photographic plates then available appeared to be particularly sensitive to light at this wavelength in the violet.

A week after taking his first photograph, Hale tried again, using the bright K line. This time the prominence appeared vividly on his photographic plate. "Yesterday," he wrote exuberantly to Harry Goodwin, "I got a prominence photo good enough to prove the success of the method, and the result is that I am just now feeling pretty *neat*." [14] Long afterward he recalled, "The principle of the spectroheliograph had thus been applied in practice." Immediately he sent a note to the *Sidereal Messenger*. It appeared in June, 1891. [15]

As soon as they heard the great news, Young congratulated him, and Brashear wrote warmly: "I congratulate you on your success which I knew would be yours if personal interest, bull dog tenacity, and knowledge of what you were after had its just reward. Of course," he asked, "you won't forget *us* when you make a print of that photo, even if it is not all you expect, for it will become historical, and I should prize it very much." [16]

While he was exploring the sun's depths, Hale was absorbed in numerous other projects. One of these was the publication of papers in scientific journals both in the United States and abroad. "I am getting to be quite a littérateur," he informed Goodwin in November, 1890. The idea amused him. One paper, "The Principal Line in the Spectra of the Nebulae," included an account of his work on magnesium; another, on prominence photography, written for the *Sidereal Messenger*, was read at a meeting of the Royal Astronomical Society in London. Even before its dedication, he was at work on Volume I, Part 1, of the *Publications of the Kenwood Physical Observatory*. "Isn't that fancy," he exclaimed in a hurried note to Goodwin. "It will be new business for me." [17]

In June, 1891, at the age of twenty-two, he was elected a "Fellow" of the Royal Astronomical Society of Great Britain. Just a year had passed since his graduation. It had been a rewarding year, as even the impatient Hale must have had to admit. On June 3, 1891, the Kenwood Physical Observatory was incorporated as a "legally organized Corporation" under the laws of the State of Illinois. Its object, according to the certificate, was to "carry on scientific investigation, especially in astronomy and physics."

On June 15th came the day of dedication. For days beforehand Hale had rushed around getting everything in order. On the appointed day Charles Young arrived from Princeton, John Brashear from Pittsburgh, Charles Hastings, professor of physics and optical authority, from Yale; and President Edward Eaton from Beloit. Gayton Douglass of Chicago opened the proceedings. He welcomed the large audience of one hundred gathered around the telescope and introduced Professor Young, who spoke glowingly of the "new astronomy," of Hale's work and its possibilities. "If Mr. Hale thinks it well to try an experiment, with one chance in ten that it will prove a success, he is likely to try it, and he has nobody to answer to, and that is a great advantage that a private institution like this has over a public one."

Next came Brashear. He recalled their first meeting, then described how Hale had made his first spectroscope with candelabra prisms. "It is in these little beginnings that we may look always for something in the future." "I believe," he continued, his eyes twinkling, "that the name of this young man will be known when we are forgotten. He has started in a new field; he has branched out in a new direction."

Brashear was followed by old "Doc" Williams, who looked back to the early days at 3989, to the first "school of science" in the back yard—"a little house of boards crowded just as full of simple apparatus as it could be where some wonderful work was done." In such beginnings he saw future triumph. "When you can see that done day after day and year after year without any lagging interest, you may be sure that there will be something developed by and by. . . ."

Afterward Hale was surprised to find accounts of the celebration in such leading journals as *Science*. "The new Kenwood Observatory," it declared, "bids fair to be a worthy successor to the earlier observatories of Rutherfurd in this city, and of Dr. Henry Draper at Hastings, as a place for experiment and the pursuit of untried paths as roads to discovery." [18]

Soon after the dedication the time arrived for the trip abroad on which Evelina had counted so long. George read his passport over. His age was given as twenty-two years; his "stature" five feet eight inches; his "forehead" medium, his "nose" straight, his mouth medium, his hair brown, his complexion fair, his face oval. In all this, as in the accompanying portrait, he found it hard to recognize himself.

A little over a month later, armed with letters of introduction to famous European scientists, he sailed with Evelina on the *Heavel* for Southampton. In Chicago he left his "Esteemed Assistants" Will and Martha to carry on. He urged them to continue a search for large prominences and for hydrogen lines in the ultraviolet.[19] They set to work to please him. One day a huge prominence appeared, the biggest they had ever seen. They photographed it, as he had directed, and, intensely excited, sent off the result. Some months later, their photograph was published. Underneath they read their names. Over fifty years later Martha would speak of this observation as one of the most thrilling experiences of her life.

In London, Hale made his headquarters at the American Elevator Company at 4 Queen Victoria Street in the heart of the City. Soon after their arrival a letter arrived from the noted pioneer in astrophysics, William Huggins, inviting Hale to call. He was amazed. On his arrival at Huggins's house on Upper Tulse Hill he was met by a man with a large head and nose and high forehead crowned by a shock of white hair, and by his wife, her husband's assistant in all his astronomical work. In the observatory in the back yard they showed him the 15-inch telescope and the 8-inch made by Alvan Clark, the American telescope maker, with which Huggins had laid the "cornerstone of astrophysics."

Here Huggins had analyzed the spectra of stars and nebulae. Here, for the first time, an observatory, as he said, "took on the appearance of a laboratory." It became, as Huggins had written and Hale had read in his hours in the Boston Public Library, "a meeting place where terrestrial chemistry was brought into direct touch with celestial chemistry. The characteristic light rays from earthly hydrogen shone side by side with the corresponding radiations from starry hydrogen, the dark lines due to the absorption of hydrogen in Sirius or in Vega. Iron from our mines was line matched, light for dark with stellar iron from opposite parts of the celestial sphere. Sodium which upon the earth is always present with us was found to be widely diffused through the celestial spaces.

"The time was indeed one of strained expectation and of scientific exaltation for the astronomer almost without parallel; for nearly every observation revealed a new fact, and almost every night's work was red-lettered by some discovery." [20]

As Huggins now talked of these discoveries, and Hale listened, he must have felt he was reliving those nights with the great English astronomer. In 1863 (five years before Hale was born), Huggins had taken the first spectrogram of a star. The image was foggy and blurred and quite useless; but it foreshadowed a time when, with invention of the faster, more sensitive dry or gelatin plate, thousands of stars and nebulae would record their spectra for astronomers to read and interpret. The following year Huggins had turned to those "mysterious bodies," the nebulae, whose nature was still an "unread riddle." As telescopes had increased in size, more and more of the nebulae had been resolved into stars. Astronomers believed that, in time, all of them would be so resolved and that all evidence of nebulosity would disappear. To find the answer, Huggins directed his telescope to the planetary nebula in Draco. With a "feeling of excited suspense mingled with awe," he put his eye to the spectroscope. "Was I not about to look into a secret place of creation? I looked into the spectroscope. No spectrum such as I expected! A single bright line only . . . The riddle of the nebula was solved. The answer which had come to us in the light itself read: Not an aggregation of stars, but a luminous gas." [21]

Sir William Herschel had regarded the nebulae as regions of fiery mist, out of which the stars had been formed. Now Huggins with his spectroscope had apparently proved it. He had discovered the first and strongest of the nebular lines produced by a gas which he called "nebulium." But while he could give it a name, he had no idea what the gas really was.

Yet, if Huggins was unable to solve this riddle, he had made another far-reaching discovery in 1868, the year of Hale's birth. This was his observation of "motion in the line of sight" of a star (or its radial velocity, as it would come to be called). It was based on a principle that would become a fundamental tool of astronomers concerned with motions in the universe. It was one, therefore, in which Hale was intensely interested.

Everyone is familiar with the variation in pitch of a train whistle as it goes away, stands still, or approaches us. This principle was discovered by Christian Doppler, who soon found that the same

principle applies to light. He saw that, as positions of observer and
light source change, the color of the light will shift. Huggins had
taken Doppler's observation and had applied it to the spectrum of a
star. If the wavelength or position of any line in the spectrum of a
star is known, he reasoned, then any change in the star's distance
relative to the earth should show up in a shift of that line in the
star's spectrum. This was exactly what he found when he compared
the spectra of certain stars with spectra observed in the laboratory.
The shifts were small and difficult to measure, but they were quite
real; and, in accord with the Doppler principle, they gave the exact
velocity of approach or recession of a star (or in time of any other
celestial object).

This discovery would become the basis for thousands of observations of "radial velocities." It would lead to important studies of
double stars and their motions. It would help in the solution of the
difficult problem of solar rotation. It would even point to observation of the now famous "red shift" in the spectra of spiral nebulae.

Toward the end of his life Huggins would comment, "It would be
scarcely possible, without the appearance of great exaggeration, to
attempt to sketch out even in broad outline the many glorious
achievements which doubtless lie before this method of research in
the immediate future." [22] On that day in 1891 when Hale and Huggins talked in the observatory on Upper Tulse Hill, they could only
guess at those achievements.

Yet, as they talked, Huggins must have realized how intensely
Hale shared his own excitement over such discoveries and his faith
in the future of astrophysics. As it turned out, this was the beginning
of a long and meaningful relationship between the older astronomer
who had broken ground in this new field and the younger one who
was to follow eagerly in his footsteps to discover even greater frontiers in the universe.

Nineteen years later, when Sir William died at the venerable age
of eighty-six, Hale recalled the way he had "thrown himself into his
work with all the ardor of the ancient alchemist." "Science," he commented, "would advance at a faster pace if more investigators
shared Sir William's enthusiasm for research. The pessimism of some
men and the half-heartedness of others were alike foreign to him.
When past eighty his boyish delight in science was in no wise
abated. . . . To the best of my knowledge, the fire burned steadily
to the end." [23]

In 1891, this visit to William Huggins was only the first in an exciting round, not only in England but also on the Continent. "To say that I have been busy is to put it very mildly," Hale wrote to Goodwin, "for I have done nothing but hustle around at the top of my speed from morning till night." [24] He visited the Royal Observatory at Greenwich; he went to Common's Observatory, to James Dewar's laboratory, to the Radcliffe and the University Observatory in Oxford. He was welcomed by Norman Lockyer, a thickset man of medium height, with a commanding appearance and striking features. Lockyer's observation of solar prominences in full daylight (with Janssen), his discovery of the element "helium" in the sun, and his pioneer ideas on the relation of stellar development to temperature were major astronomical contributions. In these experiments he showed that by changing the temperature of a single element, its spectrum can be changed—an epoch-making idea in a time when the spectrum of an element was still believed to be unchangeable and the atom indestructible.

Hale, prejudiced by Lockyer's autocratic reputation, nevertheless confessed that he liked his restless energy and tremendous enthusiasm for solar research. "Though I took the opposite side in several protracted discussions with him, he certainly treated me very well, and put the whole lab at my disposal." [25] In that "lab" Hale measured his latest photographs, and "ground out the material" for the paper Dewar and Huggins had persuaded him to give at the British Association meeting at Cardiff, Wales, in August.

That meeting proved to be another high point in this magnificent tour. From Cardiff he described their extraordinary reception to Goodwin: "I am treated like a *Grand-Duke!* Made a member of the General Committee, sit on the platform etc. etc.! Lina sits with Mrs. Huggins and *Lord* Bute's party at the evening lectures etc., and all together you would think we were celebrated people. Of course it is all Dr. Huggins' doings, for he is all-powerful. He took a fancy to us and to our work and hence the result." [26]

Hale's paper on "The Ultra-Violet Spectrum in Prominences" was listed third on the program on the third day. His audience was composed largely of bearded, white-haired, distinguished-looking scientists. As he bounded up on the platform, many of these ancients must have gasped. At twenty-three he appeared "like a mere infant." At the end they applauded when, in speaking of the prominences, he said, "There will evidently be no lack of opportunity in the new and interesting field thus opened to investigation." [27]

Robert Ball, the Astronomer Royal of Ireland, "a most jovial Irishman," and Dr. Copeland, the Astronomer Royal of Scotland, carried him off to lunch. In the evening Huggins introduced him to Sir William Crookes, the noted English physicist, who invited him to visit his laboratory. In conservative England, where strangers were not easily accepted, this reception was extraordinary.

At Cardiff, too, Hale had some "great discussions" with Agnes Mary Clerke, the brilliant historian of science. He discussed with her his dream of founding an international journal of astrophysics that would bring together the best results in astronomy and physics from every part of the world. She was enthusiastic, and he became more eager than ever to get to the Continent to discuss his plan with astronomers and physicists there. But, just at this point, a disturbing event occurred that forced him to cut short his European visit. Soon after the British Association meeting, he was in the rooms of the Royal Astronomical Society in London, waiting for Arthur Ranyard, a barrister with a profound interest in astronomy, who was editor of *Knowledge*. While he waited he picked up the *Comptes Rendus* for August 17th. Leafing through it idly, he came upon a paper that "knocked me cold." "Read it and you will understand," [28] he told Goodwin. It was by Henri Alexandre Deslandres, director of the Observatory at Meudon, just outside Paris. In it he described his photographing of the prominence spectrum "which he had been at since May last."

Hale immediately decided he must "hustle over to Paris" to see Deslandres's apparatus for himself. When Ranyard arrived a few minutes later, he urged him to go along. Ranyard, carried away by Hale's enthusiasm, agreed. The following day they caught the train for Dover. In Paris, they hurried out to Meudon. Deslandres, who at first sight seemed "a very nice fellow indeed," greeted them cordially. It appeared that his interest in prominences had been roused accidentally. "It seems that he was testing the correction of a photographic 12-inch object glass and happened to find the H and K lines bright. Of course he at once took up the work, and you will see that his results are practically the same as mine." However, Deslandres's results were confined entirely to the spectrum. He had made no use of the spectroheliograph principle. Still, Hale "at once decided that unless I wanted to lose the whole field of work I had better come home and sail in again. . . ."

He booked passage on the *Fürst Bismarck* for September 26th. Then, in the interests of his proposed international journal, he made a whirlwind tour, visiting Cologne, Berlin, Leipzig, Frankfurt, Basel, Milan, Genoa, and Nice before racing back to Paris. Everywhere he had "fine luck." Astronomers and physicists everywhere seemed eager to help this impetuous young man who appeared suddenly at their doors, and as suddenly departed. By the time he was ready to start for home, he had the endorsement of most of the leaders in the field, both on the Continent and in England.

In Paris, Deslandres invited the Hales to dinner and the opera. Their relations were amicable. While Hale was in Germany, Deslandres had fixed up a grating and "using it as I had told him" he got seven hydrogen lines. "He had 2 before, and we have 4, so he is a little ahead."

"But I am inclined to think he will have to hustle to keep ahead, if I know myself," Hale wrote to Goodwin, then added a significant comment he would have cause to remember often in later years: "Deslandres is first rate about the thing, and admits that I made the first photo (I was about three weeks ahead of him), that I published first, and that our photos are better than his." [29]

It was not long, however, before Deslandres forgot or chose to forget Hale's priority in the invention of the spectroheliograph and in the first successful results. In time, he even insisted that his spectroheliograph had been the first. The result has been that in most textbooks Hale and Deslandres are given joint credit for the invention. As years passed, Deslandres would become increasingly difficult, and Hale, with his hatred of controversy, would do everything in his power to avoid an open quarrel. Yet, two or three times, in the interests of historic accuracy, he felt forced to explain the actual circumstances of the invention. Thus he wrote to H. F. Newall, the English astronomer, who was writing a biographical article: "All the books also say that he [Deslandres] and I invented or developed the first spectroheliograph simultaneously, but of course this is not true." He looked back to that memorable day in 1889 when the idea suddenly flashed into his head as he rode along on the front seat of the Chicago cable car. "It seemed to come straight from the blue, but perhaps it didn't. . . ." [30]

If it was a question of priority in the actual invention, it did not, he noted, "belong either to Deslandres or to me." But if it was a question of making the idea work, this was quite different. Janssen, as well as Braun and Lohse, had all been unsuccessful. It was therefore not until the winter of 1889–1890, when Hale "tried the principle crudely" at the Harvard Observatory, that the invention became a functioning reality. It was not until the spring of 1891 at Kenwood that he proved its success.

After this, as Hale told Newall, "So far as I have ever known, Deslandres' first reference to the principle was in the Comptes Rendus of August 17, 1891, in which he mentions my papers. But he did not build and use a spectroheliograph until the summer of 1893. Meanwhile, Evershed had built one and confirmed my results."

Yet, anxious as he was to set the record straight, he cautioned, "For heaven's sake, don't go into the thing in your Nature article, as I detest polemics, especially with a man like Deslandres. What he did first describe and use was his 'spectroenregisteur des vitesses,' though I employed the same principle in 1891 (not with an automatic spacing device) to obtain photographs of the K line in various parts of prominences and on the disc."

Still he stressed that he wanted no quarrel with Deslandres or anyone else: "How often I have refrained from answering his caustic and unfair criticisms because of my dislike of quarrels."

In the summer of 1891, however, all these difficulties lay in the future. Hale was eager only to get back to his newly designed spectroheliograph. "Every day counts," he wrote to Harry Goodwin. He could have no idea that Deslandres would turn from a "very nice fellow" into a vindictive rival.

While he was abroad, Brashear had made a new pair of moving slits with a hydraulic device to drive them. "I had no less than six men working so I could get your things off, because I do want you to get 'first blood' in your new work," he wrote. "Every boy in my shop likes to work for 'Hale.'" If any modifications to the new apparatus were needed, he urged, "Do not fail to let us know in good time for I do want you to get ahead of Deslandres—as *you* were the one who *worked* up the problem, he stumbled into it." [31]

Shortly after this, Hale, running like a racehorse to reach the finish line first, ordered a small spectroscope from Brashear. He wanted it "at once" in order to mount it on his 4-inch telescope for eye observation of prominences. He could then use the 12-inch for photography alone. "When we have such eruptions as the two I observed and photographed yesterday there must be several instruments at work as at an eclipse," [32] he noted expansively.

The first eruption that morning lasted an hour. After lunch he was surprised to find a second eruption that looked exactly like a great geyser "with the cloud of spray blown to a great distance by the wind." He opened the slit as wide as it would go—and photographed it. The result was beautiful. "As the upper part of the prominence descended toward the limb, the chromosphere became full of bright lines." Where the dark Fraunhofer lines had previously appeared, bright lines "like a bursting rocket shooting out stars" now shone out in their place. In other words the lines had been "reversed." Moreover, these reversals appeared not only at the sun's edge and near sunspots but also "almost anywhere on its disc."

These results offered a fertile field of observation. If now he could photograph the *forms* of these bright calcium regions on the sun's surface, he might fathom the still unknown relationship between spots, prominences, and faculae (those bright cloudlike objects that rise above the sun's surface and shine out when near its edge, but are nearly invisible when they lie near the center of the disk. They seemed to be connected with sun spot development.) Perhaps he could then answer questions such as these: Do these regions change in form, and if so, what relation do the changes show to simultaneous changes in neighboring spots? Are the reversed regions like prominences in the sense of rising above the surface, or are they shallow layers on or below the general level? If they are elevations, they ought, as he wrote to C. A. Young, "to show it by their form as they approach the limb, and they ought to project above the limb, when passing around. —But perhaps I am building castles in Spain, for it will certainly be no easy matter to secure successful photos in the face of the brilliant spectrum." [33]

Sooner than he expected, however, his "castles in Spain" came down to earth. The new apparatus with moving slits arrived from Brashear two weeks later. After an absorbing period of observation, he reported that, moving the slits by hand, he had been able to photograph the *forms* of the bright calcium regions around a spot for the first time in history. Up to this time these bright clouds or faculae ("little torches"), which may cover vast areas larger than any of our continents, had offered too little contrast with the photosphere to be visible except at the sun's limbs; as they moved across the sun they had disappeared against its brilliant background. Ordinary photographic methods had been no more successful than the eye in following them across the disk. Now, on these latest photographs, Hale could actually watch the changes in these bright calcium regions and could see them as beautifully at the sun's center as at the limb.

His "great luck" continued. A month later he succeeded in photographing the prominences all the way around the sun with a single exposure. "This is what I always wanted to do, but hardly dared hope that I could." [34] "The work," he added, "is becoming less experimental and more practical." Before the development of the spectroheliograph, it had been possible to make single drawings of prominences. But those drawings could not compare in accuracy with this new and magnificent *photographic* record made in two minutes.

Looking back on the far-reaching importance of this period, the British astronomer H. H. Turner was to comment: "Hale . . . was the first to realise that the faculae could be photographed all over the Sun's disc. . . . His special achievement was not only to 'exhibit to us the rose-coloured prominences . . . as plainly as the faculae,' but to photograph the faculae as plainly as the prominences, and in this achievement of 1892 the key note is struck of the theme which he later developed in his observation of the calcium 'flocculi' above the sun's surface." [35]

These results, like so many others George Hale would make in succeeding years, were due directly to the invention of new methods and the improvement of instruments as well to his experimental skill and bold ingenuity. At that time little was known of atomic forces in the sun. The atom was still considered a hard, indivisible particle. Yet he hoped that by observation of the sun's changing phenomena day in and day out, he might learn something of its fundamental nature and discover the causes for the extraordinary changes he was recording. In the next three years he was to take thousands of photographs in an attempt to carry out this aim.

Among the most extraordinary phenomena seen on the sun are the sudden, short-lived eruptions that are now called solar flares. On July 15, 1892, soon after Hale had mounted his spectroheliograph on the new 12-inch photographic telescope, he was observing as usual. Suddenly a flaming object appeared above a double sunspot. At eleven o'clock he photographed it. He looked at the spots but saw nothing remarkable. Thirteen minutes later he photographed the object again. When he developed the plate, he found an intensely bright, hook-shaped form extending across the bridge that joined the double spots. Twenty-seven minutes later he took another photograph. The dark areas at the spots' center were almost completely covered by the brilliant outbursts. By 1:34 P.M. they had disappeared entirely. Such extraordinary activity, he believed, should affect terrestrial magnetism. He sent out an S.O.S. for reports of magnetic disturbances as well as for auroral observations that might be related to this solar disturbance that had covered the fantastic area of four billion miles! [36]

Meanwhile he had made one other curious observation. All during this extraordinary exhibition the faculae had not changed materially in form. The phenomena, he concluded, must have occurred in a region high above them. "In other words we seem to be dealing with an extremely brilliant eruptive prominence." Yet the first outbreak had occurred exactly on the line of separation of the two umbrae (or inner, dark regions) at the center of the double spot. Further observation would be needed before he could reach any conclusion. No one could tell when another such flare would occur. He could only keep on observing in the hope of catching it when it came. Still, he could think of no better occupation. Watching the sun as one brilliant event followed another was like watching a magnificent display of fireworks—but far more wonderful.

Observing the sun was, however, only one of many projects that kept him "hustling" in these months. Another absorbing job was the organization of his international journal, which everyone, or at least nearly everyone, considered a "great scheme."

There was, however, one unexpected snag that would delay for some years publication of a journal actually called "Astrophysical." In the fall of 1891 Hale discovered that W. W. Payne, editor of the *Sidereal Messenger*, published in Northfield, Minnesota, was greatly agitated by the idea of another astronomical journal, even if it would not compete and could not be considered in any sense "popular." Finally, after a meeting in Chicago, Hale agreed to join Payne in the editing of the *Sidereal Messenger*.[37] The name was changed to *Astronomy and Astrophysics*.

On January 1, 1892, the first number of the new journal appeared. Hale looked at it in disgust. It was anything but what he had

dreamed. Yet, if he found the first issue disappointing, he was grati-
fied by the friendly response. From Lick, Holden wrote: "It is full of
interesting things, and you have promises which ought to bring
you many more. Then you are a host in yourself!" [38] From England,
R. A. Gregory, the illustrious editor of the British scientific journal
Nature, wrote: "It is by means of such a journal as yours that the
cause of the New Astronomy will be advanced." [39]

All this was encouraging. But as the demands of the job grew,
Hale became increasingly concerned. Letters and manuscripts
flooded his desk. "Let me advise you on one point," he wrote in des-
peration to Goodwin. "If you ever want any time at all to yourself,
do *not* take up editorial work of any kind. No sooner is one number
out of the way at the end of the month than it is time to take up the
work again." [40] He listed some of the problems—the difficulties of
printing in the "one horse" office in Northfield; the translations he
had to make himself; the drudgery of proofreading, and an
enormous correspondence that ranged from offers of papers from
eminent scientists to requests from crackpots who wanted telescopes
to set up on streetcorners.

"George," Mary Hale wrote to her mother, "is hurried in his work
and more successful than he dared to hope—but as his work is on
the sun, it is not of much interest outside the Astronomical world.
He is Associate Editor of an Astronomical Magazine and this gives
him plenty to do when the sun does not shine. . . ." [41]

About this time, however, help appeared unexpectedly in the
form of a young man who worked for James S. Kirk and Company,
"Soap Makers, Perfumers and Chemists," on Chicago's North Side.
One day this jovial, bright-eyed man with a goatee appeared at the
observatory door. His name was Ferdinand Ellerman. He was, he
said, interested in astronomy, and wondered if there was anything
he could do in his spare time. He had no telescope, only an opera
glass. He was, he added, an amateur photographer. Hale, attracted
by the young man, invited him in and gave him a job. [42] So began a
lasting association. Ellerman's remarkable ability as a photographer
and his skill with instruments were to prove indispensable.

Meanwhile, in April, 1892, ten months after their original disastrous correspondence, President Harper, who evidently had never
given up hope of making Hale and his observatory part of the new
university, wrote again to say that arrangements had been made

with Professor Albert Michelson of Clark University to take charge of the Physics Department.[43] Hale heard this news with "surprise and delight." He had met the brilliant physicist four years earlier at a meeting of the Association for the Advancement of Science in Cleveland. In his vice-presidential address the black-haired, hazel-eyed Michelson had discussed the possibilities of the new instrument called the interferometer in the making of very fine measurements. "The beauty of the instrument," Hale noted, "could not fail to impress everyone who appreciated in any degree the value to science of such new and powerful methods of research." He considered Michelson one of the most remarkable men he had ever met.

Later he learned something of Michelson's extraordinary early life. He had been born in Strelno, a small Prussian town near the Polish frontier. When he was only two his family had migrated to San Francisco. His earliest memories, however, were not of San Francisco but of a mining camp in Calaveras County. Like countless others, his father had joined the gold rush. From Murphy's Camp they moved on to Virginia City, Nevada, near the booming silver mines. Here Michelson had begun his education in a one-room schoolhouse. Later his family returned to San Francisco, then little more than a frontier town. In San Francisco he went to high school and prepared to enter the United States Naval Academy.

As Hale came to know him better, he must often have felt that Michelson had been strongly influenced by his early surroundings. He retained always the simplicity and independence that character-ized the western pioneer. An extreme individualist, he was often misunderstood and disliked. His singleness of purpose and total absorp-tion in his work resulted in an indifference to others less dedicated. An extreme perfectionist, he was violently intolerant of slipshod work, and many of his assistants rebelled at the discipline he exacted, despite the benefit of contact with his genius.

As a student at Annapolis, Michelson had made his first measure-ment of the velocity of light. Henceforth his interest in the study of light grew, until in his brilliant Lowell lectures "Light Waves and Their Uses," he would exhibit his vivid, poetic understanding of this phenomenon. In time Michelson would show his interest in light and color, not only in science and poetry but also in painting. In both fields, he felt, as did Hale, that similar qualities of imagination are needed. "Like Einstein and almost every other deeply original thinker, he belied the strange and widespread public belief in an antagonism between art and science." [44]

In 1892, when the handsome, elegant Michelson arrived at the university, Hale knew little of all these things. He did, however, appreciate the importance of his acceptance of Harper's offer. It was a tribute to the president's persuasive powers and to the vivid portrait he must have painted of the possibilities of this new university in a part of the world that was still a scientific wilderness. In the end, it proved a decisive factor in Hale's own decision to join the faculty. Later he learned that Michelson had written to Harper, "I sincerely hope that every *reasonable* effort will be made to secure Mr. Hale's services. I should say that even if he insists on the condition that the experimental and observational part of the work be chiefly devoted to astro-physics, such a condition would not impair the usefulness of the Observatory—but—on the contrary—by concentrating work in this field—in which there are few competitors—results of great importance are likely to be obtained." [45] If Harper needed any further convincing, this letter, written in July, 1892, must have removed any lingering doubt.

Meanwhile Hale himself, in answer to a request from Harper, suggested a meeting at Kenwood. The meeting took place, and after another two-month silence Harper wrote once more: "I wonder whether the time has not come for you to take a place in the Faculty of the University. What do you think?" [46] This time Hale agreed. Since their first meeting he had watched the university's progress with interest and admiration. He had noted its "broad and liberal policy" and its evident plan to "make much of original investigation."

William Hale, also impressed, wrote to Harper: "If the first year's services of my son, George E. Hale, in connection with your University prove satisfactory to you and agreeable and pleasant to him, I will, at that time, agree to give to the University, the movable apparatus and instruments, at the Kenwood Observatory, estimated to be worth $25,000, on condition that within one year from that time, the University shall secure good subscriptions to an Astronomical Fund, to be expended for an Observatory of not less than two hundred and fifty thousand dollars, including the twenty five thousand to be given from the Kenwood Observatory." [47]

This letter was an expression of George Hale's hopes for a big telescope and of his father's faith that with such an instrument the frontiers of the universe would be pushed back. For visual observation of sunspots, Hale was sure no instrument could be better than the 12-inch, when used with the grating spectrograph. But his observation of the character of the spot spectrum had convinced him of the need for a photographic study of that spectrum. The 2-inch image he could get with his 12-inch telescope was too small for such a purpose. Therefore he had long been dreaming of a larger telescope. This, now, was the primary reason for his willingness to join the university faculty. "I would not consider the thing *for a moment* were it not for the prospect of some day getting the use of a big telescope to carry out some of my pet schemes." [48]

It was this dream that led him now to consider Harper's offer. His appointment came on July 26, 1892. "At a meeting of the Board of Trustees of the University of Chicago held yesterday," it read, "you were elected Associate Professor of *Astral* Physics and Director of the Observatory. The agreement made between yourself and President Harper was approved and the appointment was made on that basis." [49]

It was a notable group that Harper gathered for his original faculty. Among its members Hale would count many lifelong friends—notably Henry H. Donaldson, the neurologist; Eliakim Moore, the mathematician; James Breasted, the archaeologist, who was head of the Oriental Department, and, of course, Albert Michelson. "I doubt," Hale was to write, "whether this country has ever seen in any of its cities a more stimulating educational and research development in an equal period of time, unless it was during the rise of the Johns Hopkins University."

Moreover, members of this group would form a nucleus of friends, not only for George but also for Evelina Hale. "Once a week," in the mornings, the wives of these men would meet. Thus a welcome social element entered into the life of the lonely bride. Occasionally, too, she would entertain at dinner, and the guests included other faculty members, who became good friends—the Harry Judsons, the George Goodspeeds, the Frederick Ives Carpenters. Sometimes, too, astronomical guests were invited. Mary Hale wrote to her stepmother, "We had Prof. and Mrs. Crew from the Northwestern University (lately of Lick Observatory) and Mr. and Mrs. Warner from Cleveland to take dinner with us last Tuesday." At another time Charles Young and John Brashear of Princeton arrived to spend a week (at the time of the Kenwood dedication). All this activity brought welcome diversion into Evelina's life.

Meanwhile, as the new University of Chicago began its long career, there were those, Hale knew, who criticized Harper's methods. They said he was extravagant, that he overexpanded, that he jeopardized the university's financial state. Although, at times, Hale disagreed with him, he never shared such feelings. "While the impatience of Harper's boundless imagination carried him on faster than his financial means would warrant, he did not fail in the most exacting test of the University President—the ability to judge men. A scholar himself, he brought together a band of investigators which any university, at home or abroad, would surely covet. To me, at work in my small observatory, the coming of these men was a delight unparalleled." [50]

In the summer of 1892 George Hale started with Evelina for a vacation at the Cascade Lake House not far from Saranac. Even here, however, he found it hard to relax. He cast his fly for trout from an "Adirondack canoe." Yet he could think of nothing but the great telescope and the "charms of solar research."

After a few days he ended his "vacation" with a trip to Rochester where he was to address the American Association for the Advancement of Science on "The Spectroheliograph of the Kenwood Astrophysical Observatory, Chicago, and the Results Obtained in the Study of the Sun." [51]

In the audience were Albert Michelson and Edward Morley, as well as those firm adherents of the "old astronomy" Simon Newcomb and Lewis Boss, who, Hale knew, had little use for his "new-fangled notions." In general, however, the audience was enthusiastic and the press was glowing. "Not the least remarkable feature of the address," one reporter commented, "was the sense assured that a new era has come to the science of astronomy, and a distinctive and efficient new method of research has been introduced."

One evening, after the meetings, Hale was sitting on the Powers Hotel veranda with Eliakim Moore, the mathematician, and Edwin Frost, the Dartmouth astronomer. Near by sat Alvan G. Clark, the well-known optical expert, who was telling a group of astronomers the history of the 40-inch blanks for the lenses which had been ordered from his firm by the University of Southern California. The blanks were now in his shop in Cambridgeport. E. F. Spence had authorized the university to order these lenses in 1889, hoping to surpass the recently finished Lick Observatory on Mount Hamilton by building the largest observatory in the world on Mount Wilson, a peak above Pasadena. After many failures the 40-inch blanks had been successfully made by Mantois in Paris. But just then the land bubble in Southern California had burst; Mr. Spence's gift of land which was to pay for the lenses was worthless. Mantois was "vainly seeking payment of the $16,000 at which the lenses were valued." Here, ready made, was the opportunity Hale had been seeking. It seemed miraculous. Apparently all he needed was $300,000 for the lenses, the mounting, and a suitable observatory building. In those enchanted moments such a sum seemed not only possible but easy to find.[52]

He hurried to his room, packed his suitcase, and boarded the next train for the Adirondacks, where he had left Evelina. There, in equal haste, he packed his fishing tackle, and before she knew what was happening they were on their way back to Chicago. There George told the fabulous story to his father, who promised to do what he could to help.

Soon, however, he discovered that the money he craved was not so easily come by. "Begging money is a new thing for me," he wrote to Goodwin, "but I am willing to do any amount of such unpleasant work if I can only succeed." [53] Day after day, "working like a slave," he tramped the streets, calling on the wealthy men of Chicago, feeling much more like a beggar than like an astronomer. He met only rebuffs. Then, one Saturday, he called on Charles Hutchinson at the Corn Exchange Bank. He was a university trustee, a public-spirited man, and "the enthusiastic friend and supporter of every such effort." Hale asked for suggestions. "Why don't you try Mr. Yerkes?" Hutchinson asked. "He has talked of the possibility of making some gift to the University and might be attracted by this scheme."

For months past, the trustees, and President Harper had been wooing the traction magnate Charles Tyson Yerkes, in the hope that he would give a biology building to the university. Everything had appeared promising. Then, at the last moment, Yerkes, apparently influenced by his wife, a former actress (whom T. W. Goodspeed, secretary of the trustees, called the "most gorgeously beautiful woman I have seen for years"), refused point-blank to have anything to do with the project. Just possibly, said Hutchinson, the idea of an observatory might appeal to Yerkes's imagination. Hale hurried off to find Harper. The president, obviously discouraged by previous experience, nevertheless agreed to try once more. He suggested that Hale write out a statement they could send to Yerkes.[54]

Hale wrote immediately. "I am sure you will be pleased to learn of an opportunity which now exists of securing the largest telescope and the largest and best equipped observatory in the world."

He gave the history of the 40-inch disks, then commented, "You can readily see what an exceptional opportunity lies open to any one with the necessary funds." As he compared the 40-inch with the 36-inch at Lick, he pointed out, "The difference between a diameter of 40 inches and one of 36 inches does not seem great, but when it is remembered that the efficiency of a telescope depends upon its light-collecting power, and hence upon the *area* of its objective, the advantage of the 40-inch at once becomes apparent. Its area is one quarter greater than that of the Lick glass, and consequently it will gather 25% more light. A higher magnifying power can be used with it, and in every way its advantage over the world's largest instrument will be very marked."

When compared with the Kenwood telescope, the size of the sun's image in a 40-inch would be startling. In the 12-inch it was only two inches. In a 40-inch it would be a magnificent six and a half or seven. Theoretically, the new telescope would have a magnifying power of 4000; in actuality, because of the plaguing effects of our atmosphere, powers of more than 1000 would rarely be used. Yet, even more staggering, even more important would be the increase in light-gathering power; it would be more than 35,000 times greater than that of the human eye. Moreover its resolving power—its ability to distinguish and separate very closely adjacent stars and its power to discern the most minute structure in sun, moon, and planets—promised to be superb. With a micrometer the angle between two stars separated by a tenth of a second of arc could be measured. Or, more graphically, it should be possible to see a quarter dollar at a distance of 300 miles.

The opportunity was indeed an extraordinary one. "The installation of such an instrument in the Observatory of the University of Chicago would immediately give it the first rank among the observatories of the world, and allow it to make most substantial contributions to the progress of astronomical science. It would become the mecca of thousands of science-loving pilgrims, as the Lick observatory, even in its isolated position, is today. And the donor could have no more enduring monument. It is certain that Mr. Lick's name would not have been nearly so widely known today were it not for the famous observatory established as a result of his munificence." [55]

As Hale had hoped, this approach appealed to Yerkes's vanity. A reply came back inviting Harper and Hale to call at his office.

CHAPTER FOUR

The Founding of the Yerkes Observatory

✳ *1892-1894*

TUESDAY MORNING, October 2, 1892: 10:30 A.M.

Charles Tyson Yerkes, ruler of the vast Chicago traction empire, sat at his desk at 444 North Clark Street. As Harper and Hale entered, he rose to greet them. His manner was suave, his charm undeniable. His large moustache gave him the look of an amiable walrus. Hale, looking into his clear blue eyes, found it hard to believe that this was "Yerkes the Boodler." Yet he knew that in Philadelphia, Yerkes had been jailed on a technical charge of embezzlement and that in Chicago he was notorious for watering stock, defying city ordinances, and operating under dubious franchises. His reputation was in fact sinister; but none of this could be discerned in his warm and friendly greeting.*

As Hale amplified the arguments in his letter, Yerkes listened quietly. He looked up at the question, "What could be more magnificent than an exhibit of the great mounting for the telescope at the Columbian Exposition next year? What could bring more lasting fame to the donor?" He nodded thoughtfully when Hale exclaimed, "Chicago must have this telescope, and the name of the great Observatory could be The Yerkes Observatory. The donor could have no more enduring monument." Hale knew that many wealthy people find the idea of having their names perpetuated on a great building irresistible. He was delighted to note by the gleam in Yerkes's eye that he was no exception.

Suddenly Yerkes rose and walked to the window. He stood looking out over the city with its streets crisscrossed by his spreading trolley lines. Then slowly he began to speak, half aloud, and his lis-

---

* For more detailed accounts of Yerkes's history see the books on Chicago listed in the bibliography.

teners had to strain to hear. He spoke of his school days in Philadelphia. "I boasted then," he said, turning to face them, "that some day
I would be the owner of the largest telescope in the world." Smiling,
he murmured, "The Yerkes Telescope." His voice reflected his pleasure in the name. Then he exclaimed: "Yes, I'll do it. Build the observatory, gentlemen. Let it be the largest and the best in the world
and send the bill to me." But it must, he insisted, be the largest. He
would not give a thousand dollars for the telescope if it could not
surpass all others in every respect—buildings, instruments, and
equipment.

In retelling the story Hale said that the news seemed too wonderful to be true. Out on the street, he looked at Harper and Harper
looked at him. Then, like two demented beings, they laughed and
danced a jig, and Harper cried, "I'd like to go on the top of a hill
and yell." [1]

After this, events moved rapidly. The following Saturday, Alvan
G. Clark arrived from Cambridge to discuss details of figuring the
lens. On October 10th Harper reported to Frederick T. Gates, Rockefeller's influential associate: "The enterprise will cost Mr. Yerkes
certainly half a million dollars. He is red hot and does not hesitate
on any particular. It is a great pleasure to do business with such a
man." [2] Everything appeared so wonderful that neither Harper nor
Hale foresaw any difficulty. They failed to realize that Yerkes's eel-
like reputation was well deserved.

Afterward there were many who claimed that Yerkes had an ulterior motive in his gift. His business was expanding and he needed
credit. When he found he would not have to advance funds for the
building for some time, he was said to have told a crony that his
fortune would then be made or he would be completely "busted." He
decided to gamble.

As rumors of the gift spread, the newspapers were agog. S. W.
Burnham, Hale's old astronomer friend, in whom he had confided,
in answer to a reporter's question, blurted out the truth. Like everyone else, he was jubilant. "Hurrah for we'uns," he exclaimed. [3] President Harper, forestalled in his announcement, was chagrined. "I feel
just a little wronged. We intended to publish it all in a few days.
Since it is all out, I can only confirm it." [4] Yerkes himself was quoted
by the *Tribune* as having said, "Here's a million dollars; if you want
more say so. You shall have all you need if you'll only lick the Lick."
On December 5th he wrote a letter for publication, giving the terms

of his gift, which included the lens, the mounting, and a "home" for the telescope. It must, he repeated, be the "largest in the world."

At this point, however, no one knew what was happening in California. For days no answer came to Harper's telegram in which he had offered to buy the 40-inch lenses. Hale, worried, warned the newspapers without avail that the scheme was not yet assured. They had a field day. A cartoon in the *Daily News* showed "Colonel Herschel K. Yerkes" asking Hale, "Say, Professor, do you think the new telescope will be powerful enough to discover a right of way on Mars? There are none left down here."

The Chicago *Post* was more sympathetic. "Many good people have thought bitter things and said harsh words . . . concerning Mr. Yerkes' cable lines. The pleasant things now said of him will soften, it is hoped, any reproachful temper on the part of Mr. Yerkes." The New York *Herald* was more concerned with the scientific importance of the telescope and the "genius of the young director." "The best equipment that instrument makers and intelligent benefaction can bestow is not too good for such a man."

The morning after the announcements appeared, Hale sat down with a heavy heart to write to Brashear, the telescope maker, who had long been dreaming of building a great telescope. This wound, he knew, would cut deep. "I am sure that you will believe me when I say that under ordinary circumstances I would *much* rather give you an order for a telescope than Clark." [5]

Before this letter could reach Pittsburgh, the telegram he had feared arrived from Brashear; "Dear Hale is There Anything in the Big Telescope Scheme as Reported in the Newspapers?" [6] When Hale's letter arrived, Brashear read it with bitter disappointment. It was hard to see his dream vanishing. Still, he sent Hale a warmhearted note of congratulation. "You have a big responsibility on your hands, and you will have enough to bear and fight through, and the only thing I beg you to look out for, *don't overwork yourself*. You have a bright future before you, and no one can wish you success more than I do. All will be well that ends well. So delegate all the work you can. Save yourself for that—which you can *do better than anyone can do for you*." [7]

Days passed; still there was no word from California. "The Los Angeles people," Burnham commented, "seem to be playing some sort of a game." [8] Clark received a telegram, from President Widney of the University of Southern California "Mail copy of contract.

Ours implied. Probably arrange to begin work." [9] What this meant no one was sure. Perhaps once again they had hopes of mounting the telescope in California. Hoping to force a showdown, Hale advised Harper to telegraph withdrawing the offer.[10] In reply, a telegram came extending the option by two weeks. Finally, a week later, President Widney telegraphed, "Will sell you lenses. Terms by mail. Answer." [11]

There was, of course, no doubt about the answer. Yerkes paid the necessary deposit, and a week later the deal was closed. The 40-inch lenses belonged to the University of Chicago, and George Hale, at twenty-four, was director of the Yerkes Observatory, in prospect the largest in the world.

Letters of congratulation poured in, and with them others that echoed Brashear's caution against overwork. Indeed, the strain of these past months was telling. With his high-keyed nature, his inclination to worry, Hale had suffered from the long uncertainty over the "great observatory." He had also been driving himself beyond his strength.

At this point, too, he obtained his first real insight into the quixotic character of Yerkes, who could be the most charming of men and yet the most difficult. The poet Harriet Monroe aptly described Yerkes as a "man of might then at the beginning of his high-handed reign over city politics and his vain efforts to reign also, with his wife, in 'society'; but always a strange combination of guile and glamor." [12]

Never dreaming of trouble, Hale submitted a list of specifications for the observatory on which Yerkes had promised to spare no expense. The list included parts needed for the great telescope, as well as for smaller instruments essential to regular observatory work. The total came to $285,375.[13] Yerkes's response was immediate and violent. Sure Hale was trying to exploit him, he was furious.

Brashear, who had been commissioned to build much of the auxiliary equipment, came on from Pittsburgh for a hurried conference with Yerkes. Afterward he noted, "He seems to have gotten the idea that you fellows did not know what you wanted and that you would put on enough so that it would be useful sometime (like the second hand pulpit was in the old story)." [14] At such a charge Hale, who had spent endless hours figuring ways to keep down costs, exploded. Yerkes, with his limited astronomical knowledge, unable to understand the need for such essential auxiliaries as spectroscopes, could not see why he should pay for them. He had originally promised to furnish the best equipment obtainable. Apparently he had now forgotten that promise. He had given the money for the 40-inch in order to attach his name to the telescope and so to the observatory. Evidently he now considered this sufficient. The rest should be paid for by someone else.

William Harper, who had been to see him, reported back, "He was, to use his own words, 'entirely out of patience.' . . . He has the feeling that you have made the whole matter cost him more every time; that you act upon the theory that no economy is to be practiced, and to use his own words that you are 'ready to ride a free horse to death.' . . . It took me nearly two hours to quiet him down." [15]

Hale, who at their first interview had considered the money problem solved, was astounded. Yerkes's present stand, he knew, was as ridiculous as it was irritating. He was stung, too, by Harper's apparent acquiescence to it.

Meanwhile the gentle Brashear, accused of bad faith and of overcharging for his spectroscopes, commented to Hale with unaccustomed bitterness, "The fact is, if *you* or some good spectroscopist like you does not have something to do with this work, I confess I have no heart to make it for such a man as Yerkes, because he has absolutely no interest in its scientific value." [16]

Months passed, and the depression of 1893 deepened. On May 6th the headlines said: "Wild Panic: Stock Exchange Trembles." The market was in ruins; everywhere men went bankrupt. Yerkes was no exception. He suffered heavy losses and became increasingly loath to contribute to *his* observatory. Yet he continued to insist that his name alone be connected with it. Once, when he feared that a smaller telescope might be named after someone else, he wrote to Harper, objecting strenuously.

There was another serious obstacle to the completion of the observatory. In Harper's plans for the university, it was only one of dozens of projects. When he could consider this difficult period more objectively, Hale realized that "nothing short of his overwhelming optimism could have induced Harper to favor an undertaking which could play but little part in his educational scheme. With limitless funds for a great university, but without the lecture halls, dormitories, libraries, laboratories, museums needed for the thousands of students that peopled his dreams; without the large faculty required for instruction and research; and without the many millions essential to meet current bills, it is no wonder he could not see, except in Mr. Yerkes, any source of funds for a large and expensive observatory, situated far from the University campus and devoted almost solely to research." [17]

On his side, Yerkes refused to appreciate the general problem. He felt that "Rockefeller's millions should immediately build, equip and maintain the Observatory he had initiated." [18] Again and again Hale tried to remind him of his original promise and to reconcile these conflicting viewpoints. For months there was a complete stalemate.

Still, Hale was absorbing experience that would prove invaluable in future dealings with those other multimillionaires Andrew Carnegie and Henry Huntington. They were of an entirely different cast from Yerkes, yet they shared some of his difficult characteristics.

Fortunately, through all this period, Hale had one great consolation—his first and greatest love, the sun. At every opportunity he hurried back to his Kenwood telescope—impatient to get on with his own research. In these glorious hours he was excited by everything he did. One day he wrote to his college friend Harry Goodwin: "Tell me what you will of the charms of physical chemistry. I would not exchange subjects with you for anything in the world. Some think that solar work is pretty well played out—in reality it is only beginning, and if I am not mistaken the next ten years will see some remarkable advances in this direction." [19] In those advances he hoped to play a leading role.

In the summer of 1892 he decided to tackle one unsolved problem that had long attracted him—that of the sun's corona. This flaming halo is familiar to everyone who has ever seen an eclipse. Sometimes it is symmetrical in form. Again, like a four-rayed star, its brilliant streamers extend far beyond the more brilliant prominences.

Whenever the problems of the Yerkes Observatory allowed, Hale thought of the corona. In March, 1892, he had written to Huggins of his hopes of detecting it with experiments "based on your method" but using his spectroheliograph. Yet where failure had been so common he had little hope of success. Still, he had photographed the prominences. Perhaps he would again be lucky. "I have as yet tried no experiments," he told Goodwin, "but when I do I shall not give up without a struggle. It took a long time to get the first photographs of prominences." [20] His work at Harvard had made him used to failure, "so if I don't get anything I shall not be very much disappointed."

In May, 1892, he made his first attempt, using his spectroheliograph, but making a much longer exposure than on the prominences. When he developed the plate he saw something that looked like the corona, but he suspected the images might be instrumental in origin or caused by haze or passing clouds. He decided, therefore, to build a special instrument which he called a coronagraph, in which he hoped to eliminate the sun's dazzling light so that the corona alone would be visible. A clearer atmosphere than that of smoky Chicago was essential for his experiments. He decided to head west.

Finally, after weighing the pros and cons of different sites at altitudes of over 11,000 feet, he decided on Pikes Peak, which he had first seen years before on a trip to Colorado with his family. With his enthusiasm for any new venture and his desire to share it with others—especially one that would take him away from Chicago's gloom—he urged various friends, John Brashear, Ambrose Swasey, and James Keeler, to go along. If Brashear would only consider it, he argued, he would actually save time, since he would "return a new man—capable of twice as much work." Despite this plea Brashear refused. Swasey, too, turned down the enticing offer, but offered a 6-inch telescope for the "anti-eclipse expedition." Keeler, however, jumped at the chance.

On a hot day in 1893, the expedition set out for Manitou. Its members included George and Evelina Hale and James Keeler. But Evelina, unable to stand the altitude, was forced to return to Manitou. Hale must have been glad she had done so, because the following day, as they were adjusting the spectroheliograph, a snowstorm came up with hail and heavy winds, thunder and lightning.

There were also other difficulties. On his first night on the mountain, Hale fell wearily into bed, only to find he was sharing it with creeping "varmints." Unable to sleep, he found that Keeler was squirming too. Finally they got up. It was frigid. The wind was blowing a gale. Snow blanketed the landscape. They piled wood into the potbellied stove and soon had a roaring fire. Beside that fire they huddled, talking till morning.[21]

As Keeler described his boyhood, which was in many ways like Hale's own, he told of his first shop in Mayport, Florida, where he too had owned a scroll saw and made his first telescope. Like Hale, he had given his shop a name—the Mayport Astronomical Observatory.[22] As the night passed, Hale listened entranced while his charming and witty companion told of his boyhood in the South, of his days at Johns Hopkins University, and of the expedition to Mount Whitney with Samuel Langley. Hale forgot entirely their bleak surroundings at 14,000 feet.

The night finally passed. Eager to get to work, they hurried outside. The snow had stopped falling. The wind had died. But as they looked out over the distant ranges they saw smoke from forest fires that had broken out around the mountain base. "A sky not much better than that of Chicago was the result." Moreover, the peak with its steep sides acted like a chimney. Heat waves rose from the plain, and the "seeing" was miserable. Nevertheless Hale decided to take some plates. They were badly fogged. Yet, under clearer skies, he still hoped to trap the elusive corona.

None of those who went on that expedition ever forgot it. Years later Hale recalled Pikes Peak as "the strangest place I ever lived in." In addition to the "varmints," they were plagued by mosquitoes. The winds reached seventy-four miles an hour. The altitude was devastating. "The grub was poor, the water poorer, and nearly everyone on the expedition was very sick."

Yet, if the expedition was a failure scientifically, it proved the beginning of a close friendship between the two men who had shared these hardships.

After this rugged interlude the Hales hurried back to Chicago. Plans for the Columbian Exposition in 1893 had been burgeoning, and Hale had been pondering the possibility of holding an astronomical congress in connection with the exposition. Chicago was determined to make the fair the most magnificent event in American history. For a while, therefore, as he pushed his plans, he forgot the corona and even Mr. Yerkes.

Two months earlier he had written to Edward Holden at Lick: "If we could only arrange to get together the best astronomers (in fact, *all* astronomers in this country and abroad), what a congress it would make! I do not know how practicable such a scheme is, but I would give a good deal to see it carried out." [23]

Now Hale continued his campaign. He wrote to all European astronomers who had any interest in the "new astronomy." His aim was "to cover the existing field of astro-physics." Most of the answers were enthusiastic. The prospects appeared bright, and he had high hopes for this Congress of Astrophysics—the first, as it turned out, of many international meetings for which he would be responsible.

One day in 1892 he had invited the physicist Henry Crew, who had recently resigned from Lick to join the Northwestern faculty, to go down to the fair site. They went in a barouche. The mud was knee-deep; the horses could hardly plow their way through the morass. The ground was covered with wreckage. Said Crew, "It was impossible to think that it would be done the following year, but it was!" [24]

For months Chicago had talked of little else. For months William Hale had spent most of his time there, supervising exhibits as well as the installation of his elevators. Charles Yerkes, in his capacity of art collector, was appointed to the committee on ceremonies. George Hale had also spent endless hours at meetings with all sorts of people who, in later years, would again cross his path, among them James Ellsworth, member of the board of directors, and D. H. Burnham, the architect. When nearly everyone had argued that it was impossible to have a fair in two years on the "sandy waste" of Jackson Park, Burnham had insisted on and proved its feasibility. With its lagoons and canals, its courts and walks, its basins and islands, its great palaces, it was indeed a fairyland. People came from everywhere to admire the Ceylon building, the Irish castle, the Buddhist temple. They wandered through the streets of Cairo to a village where unhappy Eskimos, wearing fur parkas, sweltered in the blistering Chicago summer. In the great Manufacturers' Building they gawked at the gigantic mounting for the 40-inch telescope, built by Warner and Swasey. Despite Hale's impatience over the delays, it had been completed in what now seems a remarkably short time. It was called "a symbol of the future."

The newspapers, too, were impressed. The Chicago *Tribune* described the mounting in detail. It told of the rigidity and strength needed in the 62-foot tube that was to carry the 40-inch lens, which, with its cell, would weigh half a ton. It described the moving parts that would weigh over twenty tons, yet could be perfectly controlled by the "new" method of electricity (never before used in operating large telescopes). "By touching the buttons on a little keyboard the great instrument may be made to move slowly or rapidly in either direction to any part of the sky." It told of the remarkable floor of the "observing room," which would be seventy-five feet in diameter and would be so arranged that it could rise and fall through a distance of 23 feet by means of electrically controlled hydraulic rams.[25]

The fair opened in May, 1893. In August the Astronomical Congress convened. It was an enthusiastic gathering, although, as a result of distance and expense, the attendance was smaller than Hale had hoped. From Heidelberg came Max Wolf; from Rome, the noted solar physicist Pietro Tacchini (with whom Hale had formed a friendship through two years of correspondence); from Berlin, the great Hermann von Helmholtz; and from Paris, Eleuthère Mascart. All agreed that such international meetings were invaluable and the exchange of ideas essential. They looked forward to further meetings.

Since leaving college, Hale had become increasingly aware of the gaps in his scientific knowledge. He felt he was not keeping abreast of developments in optics, thermodynamics, and other aspects of physics which had a bearing on astrophysics. The fountainhead of scientific education was in Germany, and he knew this was the place for him to go. Moreover, Harry Goodwin was already in Leipzig. With his usual feverish energy, therefore, Hale prepared for a year abroad.

As the sailing date in the fall of 1893 approached, Evelina felt she was going on a spree. On landing in Liverpool, however, George became involved in his usual astronomical round. He visited Huggins outside London, gave a paper before the Royal Astronomical Society, then went with Evelina for a brief stay with the astronomer Hugh Newall, at his estate Madingley Rise, near the Cambridge Observatory. Soon afterward they crossed to Paris with Hale's good friend Arthur Ranyard. The two men spent an afternoon with Deslandres "without bloodshed." "As I expected," Hale wrote significantly to Goodwin, "the 'rotating spectroscope' has never been constructed but he has adopted one of my spectroheliographs. The question of the corona was not discussed." [26]

In Berlin they were welcomed exuberantly by Harry Goodwin. He led them to Fraulein Strom's pension at Schönberger Strasse 12. At first they were favorably impressed by the large, clean room with its huge bed and vast porcelain stove. But soon the dark fall days turned into the even darker, shorter winter days. The porcelain stove might "in the course of time achieve warmth sufficient to be detected by prolonged application of one's palm." The "barbarous diet" proved highly upsetting to Hale's digestion. "Choice morsels brought before us were 'Gänse Klein,' consisting of the skull, beak and feet of geese: 'Blut Wurst,' raw fish, and other German dainties." [27]

Germany was, Hale concluded, a bloody, "warlike nation." "Lectures were attended by hordes of students, many of them emitting strong odors of disinfectants from the bandages on their heads, which concealed the wounds received the evening before in Corps duels." In his walks along Unter den Linden he watched the "Garde du Corps officers, in resplendent corselets, crowding ordinary mortals off the sidewalk when their fancy willed," while the Kaiser, "changing his uniform as often as his whims," rode arrogantly up and down. One memorable day the troop display increased, roaring crowds lined the boulevard, and the air resounded with cries of "Bismarck!" It was the day of the Kaiser's so-called reconciliation with Bismarck. These and other impressions sank deep. Before the outbreak of World War I, Hale would recall them vividly.

Quite different was the atmosphere in the laboratories. At the Physikalisches Institut and the university, he found an intellectual excitement he had never known at "Tech." At the Institute he listened to lectures on experimental physics and optics. At the university he attended the course in thermodynamics given by Max Planck. At the Potsdam Observatory he discussed the latest advances in spectroscopy with H. C. Vogel, who had followed up the work of Huggins on radial velocities and had been the first to photograph these shifts in the line of sight that provide the key to stellar motions. In Hanover he visited the noted spectroscopists H. Kayser and Karl Runge, who had helped to lay the foundations of the science. Wherever he went he discussed his plans and hopes for the Yerkes Observatory.

For a while he had hoped to carry on his solar research in Berlin. But the German sun was elusive, and soon he turned to a more promising subject. The pioneer work of Julius Elster and Hans Friedrich Geitel had shown that surfaces of metals such as sodium and potassium, sealed within glass cells, would emit electrified particles when exposed to light. "To a student of astrophysics the possibilities of this curious emission seemed numerous, and we attempted to explore the effect in different parts of the spectrum in a laboratory over the great lecture hall." Both August Kundt and the

great Hermann von Helmholtz, with his towering brow and penetrating eyes, were helpful; but various problems, including the lack
of a proper electrometer, nullified the attempt.

As the first semester passed, Hale became increasingly restless.
The dark winter days were depressing; the boardinghouse was
dreary. His research was at a standstill. Moreover, a stream of letters
from Chicago kept reminding him of the need for his presence
there. As Yerkes still refused to provide funds for the observatory,
Henry Ives Cobb, the architect, suggested that an attractive set of
plans might revive Yerkes's lagging interest. (At this point the traction magnate was buying the first piece of sculpture by Rodin ever
to go to America. For the moment his interest in art seemed greater
than his devotion to astronomy.) When Cobb's blueprints arrived,
Hale set to work to improve them. For days he immersed himself in
the job, neglecting his classes entirely. The building was to be in the
shape of a Roman cross, with three domes and a meridian room.
The long axis of the building was to measure 326 feet; the large
tower to contain the 40-inch was to be 92 feet in diameter. The
edifice was to be built of gray Roman brick, with gray terra cotta
and stone trimmings (in conformity with the desires of Yerkes and
the architect rather than in line with scientific needs.) In the design
Hale had to cope with an enormous mass of practical detail. He
wanted an observatory ready to meet not only immediate problems
but also the needs of the astrophysical future he envisioned. His
aims were expressed in a vivid statement:

"One of the principal aims of the Observatory is to bring together
the physical and astronomical sides of the work. Therefore the
building will be provided with laboratories for optical, spectroscopic
and chemical work, concave grating rooms, large dark rooms, developing, emulsion and enlarging rooms, a galvanometer room and the
large heliostat room.

"The Observatory will be in reality a large physical laboratory as
well as an astronomical establishment. All kinds of spectroscopic,
bolometric, photographic and other optical work will be done in
these laboratories under much better conditions than those that prevail in cities. . . ."[28]

This concept of an entire observatory as a physical laboratory was
revolutionary. To the majority of astronomers an observatory was
still simply a place for a telescope and an observer. When, some
years later, Hale compared the old and the new view, he would

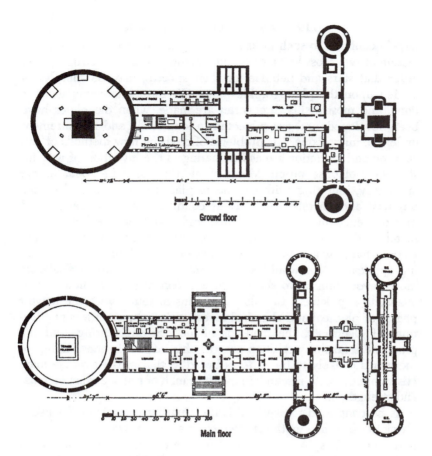

Plans of the Yerkes Observatory, incorporating Hale's revolutionary ideas for an observatory that would be a physical laboratory. In contrast, the original plan for the Lick Observatory, completed in 1889, had no provision for a darkroom or for a spectroscopic laboratory.

write: "According to the old view, the astronomer, soon after the setting of the sun, retires to a lofty tower, from whose summit he gazes at the heavens through the long watches of the night. His eye, fixed to the end of a telescope tube, perceives wonders untold, while his mind sweeps with his vision through the very confines of the universe. . . ."

In contrast the "present day student of astrophysics" is an observer of an entirely different sort. "His work at the telescope is largely confined to such tasks as keeping a star at the precise inter-section of two cross-hairs, or on the narrow slit of a spectrograph, in order that stars and nebulae, or their spectra, may be sharply re-corded upon the photographic plate. His most interesting work is done, and most of his discoveries are made, when the plates have been developed, and are subjected to long study and measurement under the microscope. His problems of devising new methods of cal-culation and reduction are as fascinating as the invention of new in-struments of observation. Much of his time may be spent in the laboratory, imitating, with the means placed at his disposal by the physicist and chemist, the various conditions of temperature and pressure encountered in the stars, and watching the behavior of metals and gases in these uncommon environments. If, in the con-viction that new and promising means of research are always await-ing application, he would advance into still unoccupied fields, he must devote himself to the design and construction of new instru-ments, to supplement the old. Kept thus in touch with the newest phases of physical and chemical investigation, the countless applica-tions of electricity, the methods of modern engineering, and the practical details of workshop practice, his interest in these things of the world is likely to be quite as broad as that of the average man. His sympathy with research in every branch of science must increase and strengthen."

As he continued to sketch this illuminating portrait of the astro-physicist and his methods (which was in a way a vivid self-portrait), Hale spoke of the need for the controlled imagination in all his research: "His dreams run far ahead of his accomplishments and his work of today is part of the development of a plan projected years ago. He perceives that only a few generations hence many of the instruments and methods of his time are to be replaced by better ones, and he strains his vision to obtain some glimpse, imperfect though it be, into the obscurities of the future. As he sits in his laboratory, surrounded by lenses and prisms, gratings and mirrors, and the other elementary apparatus of a science that subsists on light, he cannot fail to entertain the alluring thought that the intelli-gent recognition of some well-known principal of optics might suffice to construct, from these very elements, new instruments of enormous power. He learns of some advance in engineering or the art of the glass-maker, and dreams of new possibilities in its applica-tion to the construction of his telescopes or the equipment of his laboratory. He reads of discoveries in physics or chemistry, and at once his mind is busy in its endeavor to apply the new knowledge to the solution of long-standing cosmical problems.

"But here again, we see the need of control; for with such a multiplicity of interests, and such constant stimulus to the imagination, the danger of mere dilettantism is obvious. With scores of problems suggesting themselves for solution, and with attractions at every hand, each rivalling the other in its apparent possibilities of development, the chief difficulty is to choose wisely. It is not a question of searching for something to do, but of picking out those things which are most worthy of pursuit. Here the importance of having a definite and logical plan of research becomes apparent. Such a plan may involve a single investigation, continued along systematic lines over a long period of years, or it may comprise several investigations, carried on simultaneously. . . ." [29]

With this far-ranging philosophy, Hale's plan for the Yerkes Observatory was so precise in detail, so broad in scope, that it would revolutionize the old view of the astronomer. Fifty years later, Yerkes astronomers would say: "Only now are we carrying out many of the ideas Hale planned for. His vision was extraordinary."

Meanwhile, however, the actual work of planning, with all its inherent difficulties, went on. From his far-off post in Germany, Hale followed the developments anxiously. On December 9, 1893, Burnham wrote that the exact site for the observatory, on Lake Geneva near the town of Williams Bay in Wisconsin, had been chosen. Countless other sites had been turned down—Washington Park, Hinsdale, Lake Forest, even Mount Wilson in California, which was unacceptable because Yerkes wanted a site nearer Chicago. T. C. Chamberlin, the noted geologist on the university faculty, concluded that Lake Geneva had more advantages than any other site in the vicinity.[30] It was eighty miles north of Chicago; its latitude was 42° 34′ N.[31]

But just as this vital question was settled, an ominous letter arrived from President Harper. In it he described a near disaster on the Exposition grounds. A fire had broken out and completely destroyed the Casino and the Peristyle. The fire spread to the Manufacturers' Building, endangering the giant telescope tube and mounting of the 40-inch. With the help of horse teams and twenty-five men, they had succeeded in getting out most of the parts of the telescope, except the five heaviest pieces. Harper concluded, "I left the building still burning at 11:30, but I think we have saved the telescope." After this there was no further news for some time. Hale could only hope fervently that Harper's last remark was right—as it turned out to be. Still, disturbed by these and other problems, he decided he must end his stay in Berlin. In doing so, he made a decision he was later to regret. It had been his plan to work for a doctor's degree in Germany. In abandoning this aim, he was able to speed his research at the expense of an advanced scientific education. All his life, therefore, he preferred the plain title "Mister"— although he was to receive many honorary doctor's degrees, including one from Berlin.

Before returning to Chicago, however, he decided on a southern tour, with Mount Etna as his objective. There he hoped to try again to trap the corona. In the spring of 1894, "joyful at the prospect of the Italian sun," he headed south with Evelina. Recalling the journey, he wrote: "If you have spent a winter in an odoriferous German pension, you may have shared our relief as we hastened southward in the early spring. Vienna cheered us by its gaiety. Prague puzzled us by its mysteries, Italy enchanted us by its welcome. We can never forget our delight in Venice, after threading its dark canals at midnight and awakening to look out upon a blossoming peach tree against a deep blue sky." [32]

The days in Venice were followed by days equally enchanted in Florence with the astronomer Antonio Abetti, who was building his new observatory at Arcetri. There a large roll of blueprints arrived from the architect Henry Ives Cobb. Now, for the first time, Hale could see his ideas worked out as they would appear in brick and mortar. "My delight was so great that I could hardly leave the blueprints behind me even when visiting picture galleries!"

After a month in Rome they journeyed to Monte Cavo, Paestum, Ravello, and Amalfi. At Amalfi, Hale spent a morning working on an astrophysical paper, drinking in the scent of orange blossoms, gazing out over the blue sea. How wonderful it would be, he reflected, to live forever in a place like this. "Although of New England descent, I am after all a true son of the south."

Spring and summer were late that year. Delay in the arrival of the apparatus to be used in the observations on Mount Etna forced Hale to postpone their ascent until July 7th. With a wonderful sense of freedom, he wandered on with Evelina to Taormina, Syracuse, and Palermo. In Palermo the highlight was a visit to a puppet theater

William Ellery Hale.

Mary Browne Hale and
George Ellery Hale, 1869.

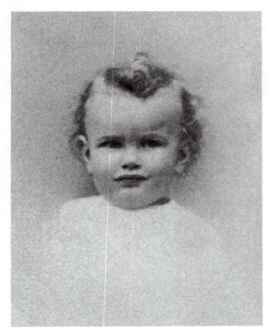

George Ellery Hale,
Christmas, 1870.

George E. Hale,
photograph taken in Boston,
October 22, 1887.

Family photograph taken by George E. Hale, June, 1890. *Front row:* William
E. Hale, Mary B. Hale and "Grandma" Scranton. *Back row:* George W. Hale
(Uncle George), William B. Hale, Evelina C. Hale and Martha B. Hale.

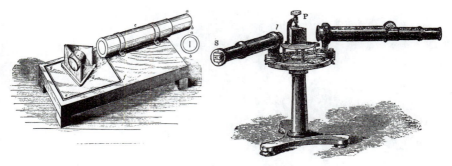

Homemade spectroscope such as young George Hale made according to directions in *Cassell's Book of Sports and Pastimes,* from which this drawing is reproduced.

Spectroscope. (*From* Cassell's Book of Sports and Pastimes)

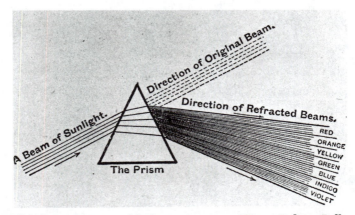

The decomposition of light in a prism. (*From Robert Ball, The Story of the Heavens*)

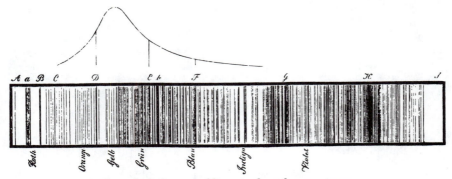

Fraunhofer's map of lines in the solar spectrum.

The Kenwood Observatory, Chicago, Illinois. *(Courtesy of Yerkes Observatory)*

George E. Hale in the Kenwood laboratory.

Spectroheliograph attached to the 12-inch Kenwood Telescope.

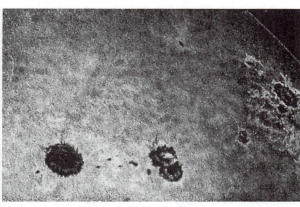

Spots and faculae on the sun, from a photograph taken by Warren de la Rue, September 20, 1861.

Visual observation of prominences with spectroscope slit. (*From Charles A. Young,* The Sun)

Solar prominences, photographed with the spectroheliograph in full sunlight. (*Courtesy of Mount Wilson and Palomar Observatories*)

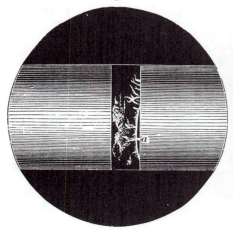

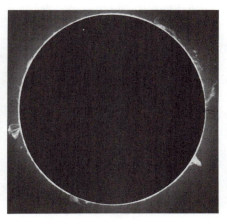

The Yerkes Observatory, Williams Bay, Wisconsin. *(Courtesy of Yerkes Observatory)*

The 40-inch refracting telescope at the Yerkes Observatory.

Rumford spectroheliograph attached to 40-inch telescope. *(Photos courtesy of Yerkes Observatory)*

The Hale house at Williams Bay, with Lake Geneva in the background.

Evelina C. Hale with William Hale and Margaret Hale.

George E. Hale with Margaret Hale and William Hale at the Yerkes Observatory.

where Tasso's *Rinaldo and the Giant* was being presented before a fiery Sicilian audience.

On July 7th the long-planned climb by pack train began. With the Hales was Antonio Ricco, director of the observatories of Catania and Etna. At 4:00 A.M. they mounted their donkeys on saddles of wood and carpet. The trail meandered through an enchanted classic landscape, through vineyards, and into the thickly wooded groves where Theocritus had written his pastoral poems, then into the arid region above. It was an extraordinary change. "We might have been on one of the mountains of the moon, passing here and there such craterlets as one detects with a powerful telescope on the similar slopes of Copernicus. . . ."

Suddenly a cloud cap descended, and as suddenly they were transported from the 100-degree temperature of the lower slopes to a frigid zone with snow patches on the ground. The mules balked, but Hale, inspired by Rinaldo in the puppet show, shouted in his best Italian, "*Coraggio, Rinaldo, coraggio!*" The words had a magical effect. Before long, men and animals were standing on the top.

The Bellini Observatory, built of blocks of lava, stood at an altitude of 10,000 feet at the base of the central crater. The astronomers decided to climb the rim. Ricco and the others chose the easier spiral ascent. The more adventurous and fearless Hale climbed straight up the eastern slope. "The ashes formed an easy footing." Soon a cloud of sulfurous steam from the crater poured down the slope, swallowing him in its vapor. He got down on his hands and knees and tried to crawl through the obscurity. Suddenly, to his horror, he found himself on the crater's edge, with his head projecting over the precipitous brink. "Here I was in actual danger, as great masses break off from time to time and fall into the hot abyss." Finally he managed to maneuver to safety. As night came on and he looked deep into the cone of the crater, where bright patches of fire glowed, he was reminded of Dante's Inferno. "I could imagine the spirits of the damned circling below me."

And what of the elusive corona? He set up his instruments, attached the spectroheliograph to the 12-inch telescope, and day after day tried to catch it. But again, as on Pikes Peak, he failed completely. The sulfurous steam that had almost smothered him destroyed the seeing and ruined the polish on his speculum mirror. The wind, blowing always in the wrong direction, made photography impossible. "I may say," he concluded, "that the investigation has been a fascinating one, in spite of its succession of failures." He left the task of pursuing the corona on Etna in Ricco's able hands. But unfortunately the Italian astronomer had no better success.

Leaving Etna, the Hales started the long descent. On the way, with his keen musical ear, Hale heard a familiar air. A string of donkeys, faggot-laden, was moving homeward under the setting sun. The peasant drivers were singing an air from *Cavalleria Rusticana*. As he stopped to listen, he wondered whether the peasants had heard the opera, or whether Mascagni had found his inspiration in their folk songs.

CHAPTER FIVE

A Period of Waiting

* 1894-1897

BACK IN CHICAGO in August, 1894, Hale buried himself in all the jobs that had been awaiting his return—his research at Kenwood, the Yerkes Observatory, his journal, and his seemingly endless task of trying to find money for all these projects. One of the most pressing of his jobs was that of editor. Three years had passed since he had compromised with William Payne on the publication of *Astronomy and Astrophysics*. They had proved to be hard, dissatisfying years. The only thing to do, he concluded, was to "strike out" on his own.

Soon after his return, he wrote to Payne, inviting him to Chicago. On a hot night they met at the Rome, one of the astronomer S. W. Burnham's favorite bohemian restaurants. At the dinner's end they turned to a discussion of the journal. On one side was Payne, supported by Burnham, upholding the cause of the "old" astronomy. On the other were Hale and the astronomer at Northwestern, Henry Crew. At first, when Hale suggested forming separate journals, Payne "seemed to feel a little injured." He said he would turn over the present journal to Hale, and not try to continue publication of his department alone. But Burnham leaped to his defense, and, "after a number of remarks not altogether complimentary to astrophysics, said he didn't think the present journal would lose anything if the astrophysical department were withdrawn." Gradually however, Payne, flattered by Hale, succumbed and agreed to Hale's proposal for a separate astrophysical journal. Crew, who had been listening quietly, while supporting Hale, was amazed by his power to turn things in the direction he wanted. "Hale had the ability," he commented, "of putting a knife in your side, and of making you think he was doing you a favor." [1]

So *Astronomy and Astrophysics* died and *Popular Astronomy* came into being, with Payne as editor. Under Hale, the *Astrophysical Journal* was born to a distinguished if turbulent career. Especially in its early years, Hale had to go out again and again, hat in hand, to find money to keep it going. He begged from friends in Chicago; from his father and Uncle George Hale, from that "noble woman" the philanthropist Catherine Bruce, who had done so much for astronomy in America. In the beginning its board included the leading physicists in America. As a result, many important physical papers by such pioneers as Rowland and Michelson, were published. As the field of astrophysics grew, the journal's influence on the then suspect and embryonic science was to be a powerful one. Yet it was some time before the astronomical world in general could accept Hale's belief in astrophysics, the "most fascinating" department of astronomical research, or would support it financially.[2]

On November 2, 1894, the first meeting of the editorial board was held at the fashionable Fifth Avenue Hotel on Madison Square in New York. It was a notable group. The redoubtable Henry Rowland of Johns Hopkins appeared together with Albert Michelson, Charles Young, Charles Hastings, E. C. Pickering, and James Keeler. After some discussion, the title *The Astrophysical Journal, an International Review of Spectroscopy and Astronomical Physics* was adopted. The *Journal's* scope, embodying a definition of astrophysics, was discussed. The question of adopting standards of measurement, such as Rowland's scale of wavelengths, was considered. At the end, Michelson moved that they adjourn to meet again in a year.[3]

So Hale settled down to edit a purely astrophysical journal. The first issue appeared in 1895. Soon, however, he discovered the difficulties inherent in the task. It was unpleasant to have to reject papers of dubious value, like some submitted by Percival Lowell on the canals of Mars. Yet the most difficult part of the job was the lack of subscribers and therefore of money. By February, Hale was in despair. The contents were considered too technical by many subscribers, who soon dropped out. He foresaw that by the year's end they would be $2,000 short. Yet he would not compromise by printing sensational material. In a letter to Keeler he concluded dejectedly, "I suppose I must begin my much dreaded task of looking for funds in the city." [4] Keeler, knowing the potential audience was small, was more philosophical. "Above all," he urged, "don't worry about the matter. You can't afford to lose your health for 50 *Astrophysical Journals.*" [5]

In the midst of these difficulties Keeler himself made a discovery that aroused widespread public interest and for a time placed him on a popular pedestal with Lowell. A modest man, he was perturbed by the publicity given his discovery—that Saturn's rings are meteoric in nature.[6] Hale was delighted. "It is by all odds the prettiest application of Doppler's principle I have seen, and is certainly the most important discovery made in a long time. What a lot of fine fish there are in the sea—for the right kind of fishermen!" [7]

At this time, too, the world was electrified by an even more far-reaching discovery that was to lead to a revolution in physics and to change the entire concept of the nature of matter. In 1895 Konrad Roentgen published his paper "A New Kind of Rays." These amazing rays could penetrate wood and metal, flesh and cloth. Immediately scientists throughout the world set to work to duplicate Roentgen's discovery. In America the "Immigrant Inventor," Michael Pupin, photographed a gunshot in a human hand. "The practical applicability of this method of photographing to surgery seems certain," he said. In 1896 Hale wrote excitedly to his father in Europe to describe a photograph of the bones of a living hand taken by Elihu Thomson at the Western Electric Company. "I think," he concluded, "I will get one of the tubes, and try the thing for myself."

At this time little was known of the true properties of these remarkable rays. Some astronomers even wondered if they could be detected in sun and stars, and, if so, whether they might be used for penetrating the sun's depths.

Meanwhile the Frenchman Henri Becquerel made another spectacular discovery—by accident. He happened to place a carefully wrapped photographic plate in his desk drawer, and on it a piece of uranium. Some time later he opened the drawer and discovered that the plate was fogged (owing to the then unknown property of radioactivity). From this discovery the isolation of radium by the Curies was to stem, and from these three giant steps the Atomic Age was born. Soon scientists everywhere were forced to recognize that the atom was not indestructible after all. This was a shattering idea whose implications, even today, we are far from having fully explored.

In 1895, these discoveries lay in the future. Yet already Hale realized how much these developments might mean to astronomy. This realization, which would grow with the years, was to become a guiding force in his experimental planning.

At the moment, however, prospects for experiment were discouraging. The Yerkes Observatory was still far from completion. Chicago, plagued by strikes and haunted by an army of unemployed, remained under a cloud of depression. In such an atmosphere there was little hope of finding money for the observatory, which Yerkes had promised but was now unwilling to provide. Nevertheless Hale wrote to Harper, urging that E. E. Barnard and James Keeler be invited to join the staff immediately, and that Schumann of Leipzig be engaged in the near future. He asked for $15,000 for the astronomers' salaries. The president replied that, instead of being increased, expenses must be cut. "We must recognize that astronomy is *one* department of the university." [8] He proposed that Barnard be the only addition to the staff. He suggested that on certain nights T. J. J. See and Kurt Laves from the university's Astronomy Department be allowed to use the 40-inch.

At the time Hale could not know that the real opposition to his plans had come from the astronomer Thomas Jefferson Jackson See, who, against Hale's advice, had been invited by Harper to join the university faculty. See had come from the University of Missouri, whose president, George D. Purinton, had characterized him in a letter to Harper as "visionary, dangerous, a genius in presenting his own claims to preferment, and to sum up, largely devoid of moral principles." This extraordinary man was responsible for the huge volume grandiloquently entitled *The Unparalleled Discoveries of Thomas Jefferson Jackson See*. His seal and letterhead consisted of a circle and various geometrical figures. The border contained the legend in Greek: "God always geometrizes—T.J.J., the geometer."

As soon as Hale had sailed for Europe in 1893, See had approached Harper with "claims of preferment," even urged his qualifications as director of the Yerkes Observatory. He had gone to Kenwood and demanded an observatory key from Ellerman. Ellerman had refused, then had bombarded Hale with frantic letters.[9]

Now it appeared that See had convinced Harper that the expenditure of $15,000 for salaries at the Yerkes Observatory was a gross waste of money. The facts were revealed a year later when See wrote to the president, applying for advancement to Associate Professor of Astronomy "on the sole ground of scientific attainments and of conspicuous services rendered the University since 1893—not on the grounds of social intrigue practiced by others." The letter stated further that except for his intervention "the University would have suffered dreadful financial disaster and the entire department would have been involved in chaos and ruin." [10] The letter, which compared "the mediocre attainments of others in high places" with his own epoch-making discoveries, was so obviously the work of an egomaniac that it opened Harper's eyes. He showed the letter to Hale, who was astonished not only at the jealousy he had generated but also at the fact that Harper could have taken stock in claims to what See described as "extensive researches—unequalled by anyone else." Hale found See's animosity incomprehensible. He was also disturbed by Harper's actions.

Some time later Harper himself was made acutely aware of the condition of See's mind. When he finally insisted that See must resign from the university, See appeared menacingly at the president's door. Harper, he shouted, had no sense of value! [11] (If, over sixty years later, Harper and Hale could have lived to read See's obituary, they would have been astounded by the claim that See was the joint founder of the Yerkes Observatory.)

Still, the year closed on an encouraging note. Shortly before Christmas, 1894, Hale received the Janssen Medal, highest astronomical award of the Paris Academy of Sciences. Jules Janssen, the astronomer whom Hale had met at Meudon in 1886, forwarded the medal with his congratulations on the spectroheliograph method "of which I indicated only the principle and which in your hands gave such important results."

Meanwhile Hale had made another important observation, one that was linked with one of the most exciting scientific detective stories of the nineteenth century. Two of the most conspicuous lines in the solar spectrum are the yellow lines, D_1 and D_2, whose identity with sodium was first proved by Gustav Kirchhoff. At the Indian eclipse of 1868, Janssen had observed another bright line close to D_1 and D_2. At first it was thought that this line, which he called D_3, was due to hydrogen. But Frankland and Lockyer in England, after showing that it had nothing to do with hydrogen, attributed it to an element never before seen on earth. They named it after the Greek *helios*, sun—helium.

In 1895 the dramatic announcement had come from England that William Ramsay had found "helium" on earth. The previous year he had discovered argon. Now, while examining the mineral cleveite in the hope of discovering other companions of argon, he found a brilliant line in its spectrum which coincided exactly with the line which Frankland and Lockyer had called helium in the sun. As soon as Hale learned of the discovery, he wrote to Ramsay, asking for a small tube of cleveite. Meanwhile a letter arrived from Germany from Karl Runge, who had observed this line in his laboratory and detected a faint, close companion. He pointed out that the bright yellow line Ramsay had discovered could not be positively identified with helium unless the solar D_3 line were also proved to be double.[12]

On June 11, 1895, Hale turned his spectroheliograph on the D_3 line in a bright solar prominence and photographed it. As soon as he had developed the plate, he discovered that the line could indeed be resolved into a "delicate, unequal pair." As Ramsay had made an earthly confirmation of a solar observation, so Hale had now reversed the process. He reported his discovery to Runge. There was no doubt, he said, "about the coincidence of the vacuum tube line with that found in the sun—seen double." [13] When Hale published his discovery, he noted, "Our possession of helium as a truly indigenous element was rendered incontrovertible." Ramsay's discovery was dramatically proved.

This observation was one of the last Hale was to make at Kenwood. Soon afterward the 12-inch telescope was dismounted, and many months would pass before it was remounted at Williams Bay. For several years, too, there would be delay in obtaining a large spectroheliograph to attach to the 40-inch. As a result, and as Hale fortunately could not foresee in 1895, he would be forced to neglect the sun for a long time to come.

By the summer's end Alvan G. Clark finished grinding the 40-inch lenses. In Cambridge, Hale called them "the most beautiful glass I ever laid eyes on." The time for final testing came. On a clear, windless night in October a small knot of observers clustered around an improvised telescope set up in the yard behind the shop. With Hale were S. W. Burnham, James Keeler, and his old college friend Harry Goodwin. They turned the instrument on the Orion nebula. Hale was the first to look. He gasped. Never had he seen anything so magnificent. In smaller telescopes the nebula had appeared to be a whitish cloud. Through the 40-inch it was a brilliant fluorescent green.

Back in Chicago he wrote to Yerkes, "I beg to congratulate you that through your generosity a telescope has been completed which fully equals in definition, and excels in light-collecting power the most powerful refractor hitherto constructed" [14] Yerkes was elated. If he could have foreseen that his telescope would remain the largest refractor in the world, he would have been even more delighted. Now he asked for a letter to which he could send an answer—for publication.[15] Everyone was gay that day. The time seemed propitious to ask again for some of the minor but very necessary instruments.[16] This time Yerkes promised to consider the matter.

Weeks passed. No word came. Once again Hale wrote, "It will be a most serious loss if the great telescope must remain another year unused." He also wrote to Keeler, "You can imagine our predicament—no transit, no clocks, no comet-seeker, no chronograph, no anything. . . ." [17] Keeler replied, "The predicament in which you are left would be funny if it were not serious; in fact I think it is funny anyway." [18]

Meanwhile, too, there were delays on the house building at Williams Bay just as there were on the observatory. Many months would pass before Evelina Hale's wish for a home of her own could be fulfilled. In the summer of 1895, however, George rented "Love Cottage" near the lake, and there they spent a happy interlude.

A year later, in August, 1896—over six years after their marriage —great happiness came in the birth of their first child. Like most fathers, George Hale was, he confessed, "in a fearful state of mind" for two days before the event. But when the baby arrived, he was ecstatic. "Ever since the great event," he wrote to Harry Goodwin, "I have been in the seventh heaven. *Now* I know that a man must not only marry to realize the full measure of happiness—he must also be willing to go through the heavy ordeal which brings him a little helpless child. Beside this, the pleasures and rewards of scientific research, great though they be, are as nothing." [19]

As for Evelina Hale, she asked the doctor whether the newborn baby would be an astronomer like his father. Referring to the first woman astronomer in America, he answered, "Well, yes, but of the Maria Mitchell kind." [20]

For weeks it was uncertain whether the baby, Margaret, would live. She was underweight and sickly. During these anxious weeks her father felt that if she failed to live, he too would die. Nor could he ever overcome these original fears. All his life he would worry about Margaret as his mother had worried about him. Her slightest illness would plunge him into despair; her smallest accident would bring on a wave of depression.

From the time of her birth he tried to spend as much time as possible in Chicago. It was difficult. He had to be at Yerkes to supervise the building. He wanted to continue his coronal experiments. Early each morning he would board the train for Williams Bay. Each evening he would rush back to Chicago, usually catching the train on a flying run down the hill. Often he felt he was spending more time on the train than on his experiments. He fretted at his lack of progress. But finally his apparatus and bolometers were in first-class adjustment. "If I can only get a really good sky," he exclaimed, "I shall hope for some results. Surely it is time to get some." Once again, however, his coronal observations were doomed to failure.[21]

By the fall of 1896, the observatory was still surrounded by a sea of mud. Yet the offices, though chaotic, were usable. In December the Kenwood dome was moved to Williams Bay and the 12-inch telescope was installed. By early December the Hales could move into their new house.

For over six years Evelina Hale had waited for this event. Yet she soon found that life at Williams Bay was not as idyllic as she had hoped it would be. It was hard to adapt to conditions in this "outpost of civilization." She loathed the isolation. Once her husband had planned to be away for a few days. She arranged for her mother to come and stay. At the last moment Mrs. Conklin was unable to come. Hale canceled his trip, including a lecture in Pittsburgh. At another time, when he was due at a meeting of the university faculty in Chicago, he wrote, "I have to forego the pleasure, as the baby is far from well and I could hardly leave Mrs. Hale alone in this desolate place over Sunday." [22] It was an agonizing time. He was constantly torn between his responsibility to his work and his devotion to his family.

That first winter after Margaret was born, Evelina, fearful of the effect of the long, cold winter on the baby, decided to go home to Brooklyn. Hale moved into a little room in the observatory. That winter, a hard one, must also have been an extremely lonely one, busy as he was. In the following spring he went to Brooklyn to bring Evelina and Margaret home to Williams Bay. As luck would have it, they arrived there in the midst of one of the worst snowstorms of the year.

Spring deepened into summer. Around Lake Geneva the big houses were opened and the lights twinkled again. Life was gayer and Evelina was happier. Gregarious by nature, she liked (as her daughter-in-law was to comment many years afterward), "being with people who were fun for her and interested in the same kinds of things—music, lectures, philanthropy, parties." [23] Occasionally the Edward Ayers, at the opposite end of the lake, would invite the Hales to their magnificent house, crowded with books and Egyptian antiquities. More rarely they would visit the Charles Hutchinsons' great mansion or tour the lake aboard the Martin Ryersons' beautiful steam yacht.

Yet, as time passed, Hale was to realize increasingly that Yerkes was a wonderful place for astronomers who wanted to do their work without distraction. It was a place for the independent, the self-sufficient. But for those who had no stars to watch, no figures to conjure with, no spectra to measure, and no other absorbing occupation, it was, especially in winter, a desolate place where deep banks of snow often shut the inhabitants in for weeks on end. Even in summer he found that personality conflicts, especially among the women of the small community, seemed inevitable. Soon, therefore, whenever a candidate for the staff was proposed, he would ask first about his personality, "as it is of the greatest importance that everyone connected with the staff of this out of the way observatory should be able to get on well with everybody else." Unfortunately, he found that, while he could choose the astronomers, it was impossible to choose their wives. In time, therefore, he vowed that if it should ever fall on him to found another observatory, he would set up the living arrangements in a different way.

Gradually, during her first summer in the invigorating climate of the Bay, little Margaret gained strength. Soon she was "outgrowing everything." Doting Grandfather Hale came to visit. He watched proudly as she crept along on all fours and tried to climb the stairs. Her father was delighted by the improvement in her condition. He himself was charmed by everything at the Bay: the small village where butter and eggs were taken in barter; the view of Lake Geneva, along whose shores the Pottawotomie Indians had once lived in wigwams; the spectacle of the changing seasons, so much more vivid in the country than in the city; the migrating wild geese flying high overhead in V- formation. He thought of himself as a true countryman, and decided to buy some livestock. Soon four fine carriage horses and two cows joined Sirius, the black collie, and Pegram, the cat. Now, on those rare occasions when he did go to Chicago, George disliked the noise of the city. "It seemed to worry him," said William Hale. He seemed anxious to get back to the country. "He is sure it is the best place to live."

By this time, too, other astronomers had joined him there, and already they were making plans for the years ahead. First to come was the faithful Ellerman, whose ingenuity and uncanny observational and photographic skills had proved so invaluable at Kenwood.

George W. Ritchey, who had also worked at Kenwood, was of quite a different sort. Hale had first known him as head of the Woodworking Department at the Chicago Manual Training School. His father, James Ritchey, an ardent amateur astronomer, had bought an 8½-inch telescope from Brashear and had inspired his son with a love of astronomy and an interest in telescope making. For a while he had been an assistant in the Cincinnati Observatory. At Kenwood, working with the 12-inch refractor, Ritchey had taken some of the most beautiful moon photographs ever made up to that time. Yet, if his optical skill was notable, Hale soon discovered that Ritchey had, as a colleague put it, "the temperament of an artist and a thousand prima donnas." Nothing delighted him more than a beautiful astronomical photograph. After spending hours getting the telescope adjusted and the lens focused, he would wait patiently for the best possible seeing. Yet when he finally took the photograph he would become so fascinated by the artistic result that he would fail to record the date or time or condition of the sky. The result might be a work of art, but it was scientifically useless.[24] Still, despite his unorthodox methods and other more disturbing personal traits, Hale valued his skill as an optician, and for many years was to feel that his abilities outweighed his idiosyncrasies.

Before long, too, the noted astronomer Edward Emerson Barnard arrived. Months before, at Hale's urging, he had resigned from the Lick Observatory and come to Chicago. There he had impatiently awaited the completion of the Yerkes Observatory. Barnard, a completely self-taught astronomer, had had only two months of schooling in his entire life. Born after his father's death in 1857 in Tennessee, he had known the extremes of poverty and hardship.

During the Civil War he and his family had struggled to survive on army hardtack found floating down the river. When he was nine he obtained a job in a photograph gallery. There he had charge of the solar camera. There, too, he made his first telescope. From these beginnings he had gone on to gain fame as the greatest astronomical observer of his time. His skill in measuring the diameters of the planets and of Jupiter's satellites, his dramatic discovery of Jupiter's tiny fifth satellite, bore glowing testimony to his observational powers. Unlike Burnham, who was interested only in double stars, Barnard was enthusiastic over everything that appeared in the sky. If there was even a remote chance of observing, nothing could keep him from the telescope. Yet, at Lick that "petty dictator" Holden had rationed Barnard's time with the great 36-inch in a way he considered quite unmerciful. In the isolated observatory on Mount Hamilton, just as in any remote army post or arctic station, small differences grew to mammoth proportions. For many months before his departure, they had communicated only by notes.[25]

Now at Yerkes at last, Barnard, delighted with Hale as director and with his surroundings, plunged into work. As soon as the 40-inch was ready, he would use it at every available moment. He was, however, overjoyed to share it with his old friend Burnham, who had left Lick some time before to return to his job as court reporter in Chicago. Each weekend Burnham would arrive at the observatory, accompanied by his indispensable box of cigars, and preceded by his equally necessary case of Burgundy. Whether he was observing or telling one of his incomparable tales on a cloudy night, he smoked incessantly; wherever he went, his path could be traced in an ashen trail.

The contrast between the two friends was indeed great. As Hale had soon realized, they differed not only in temperament but also in personal habits. Barnard had no "vices." He did not smoke; he did not drink. He was by nature impatient. On overcast nights when Burnham, accepting the clouds philosophically, turned to a book or a game of whist or congenial talk over a glass of wine, Barnard would spend his time dashing in and out to see if there was any hope of clearing or if the wind had changed. After a long spell of cloudy weather, Barnard's mood was one of despair. Even if the thermometer registered 28 degrees below zero, nothing could keep him from work as long as a star shone in the sky.

One freezing night the Irish astronomer Robert Ball arrived at

Yerkes. Afterward he gave a vivid picture of Barnard at work in the frosty dome. "We found what looked like a moving cylinder of fur coats within the axis of which the great Barnard was to be found moving about, running as briskly up and down as if he were playing football. Indeed he had to be well clad, for that night he worked from five in the evening till six in the morning." [26]

Years later, when Barnard died after completing his great photographic atlas of the Milky Way, Hale paid him glowing tribute: "Never since the days of Herschel has there been such an eager observer or such a living storehouse of celestial knowledge. Although his observation of every class of celestial objects, stars, nebulae, comets, planets, satellites, meteors and many others, were numbered by tens of thousands, he could always recall the day and often the hour of one of them, as well as the exact details recorded.

"Barnard was a magnificent example of that high regard for the exact and unqualified truth which the genuine man of science must embody. No finer illustration of the ideals of science could be found than in this simple, sincere and lovable astronomer, whose one and only object was to extend the boundaries of knowledge. He greatly succeeded, and his name is recorded in the rolls of the foremost explorers of the heavens." [27]

In the outside world, meanwhile, the national picture was brightening. The long depression was lifting. Everyone looked forward to a new and prosperous era. In November, 1896, McKinley was elected, and the Hale family, ardent Republicans all, rejoiced. Perhaps now, with money flowing again, Yerkes might feel inclined to do something more for his observatory. "Professor Young," Hale told Harper, "has said that an Observatory equipped with one great telescope is much like a warship with no rapid fire guns." [28]

Hale hoped to dedicate the observatory by the fall of 1897. By then, as he tried to remind Yerkes once more, the place should at least appear respectable. But Yerkes at this point in his stormy career was in no mood to think of the stars. Struggling to push through his franchise bills, he was using all his resources to sway a suddenly reluctant legislature. Moreover, he had built a $2,000,000 mansion in New York, said to be one of the three most splendid houses in America. Here he had his collection of famous paintings, including canvases by Rembrandt, Franz Hals, and Fragonard, and of fabulous Oriental rugs. He explained that he had built the house because he wanted a place where he could get away and think. Actually, it appealed to his love of pomp and grandeur, the same love that had prompted him to give the observatory. During the previous August he had made a tour of inspection of the great domed building at Williams Bay. No doubt he realized even then that it would give him a fame unattainable in any other way. But he wanted his immortality as cheaply as possible. When Harper approached him for additional funds, he refused; and his reply was tinged with bitterness and irony:

"On the same date I received your letter I also read of a meeting of the Civic Federation, at which you were present. It was a notable gathering of the great and good few who represent the great and good part of our city, and I see by the names that these people are the ones who uphold all the charities of our city, who are always fairly throwing away their wealth so that others may be benefitted thereby, who are building up the great institutions of the city in a most daring and reckless manner, while such people as myself, according to the theory of your friends, are doing their best to pull to pieces and destroy what little honor and integrity is left in your community." [29] He concluded with the suggestion that the money needed for the observatory should be provided by "friends" such as these.

As Hale read this new explosion, he sympathized with Yerkes's feeling, but he was in despair at its implications. The road to the magnate's pocketbook appeared closed forever. The Civic Federation was railing at his methods; the press was denouncing them. His life in Chicago was ending. In a way it was hard not to feel sorry for him, but the situation made the observatory's outlook dim. Yerkes's rebuff came in April. In May the 40-inch lenses arrived on a special train, accompanied by their maker, Alvan G. Clark, his two daughters, and the foreman of the Clark works, A. R. Lundin. At the Williams Bay station Hale and Barnard were waiting on the platform.[30]

Hale, intensely relieved to see the lenses safely at Williams Bay, surpervised their loading on the farm wagon that was to carry them to their final destination. The following day they were mounted in the great tube that rose like some prehistoric monster above the rising floor. That night, as fate would have it, was cloudy.

A reporter from the Chicago *Times Herald*, eager to know more about the history of the great instrument, interviewed its maker. Clark described the five years of "arduous labor" that had gone into the "forming and testing" of the lenses. "The work was constant, requiring the utmost patience and care. The deviation of the breadth of a strand of a spider's web would be fatal to the purposes of the glass. The crystal had first to be examined to see if it was without flaw, and, after this, it went to the grinding room. For two years sand, fine steel and emery were used. Then the finest jewelers' rouge and beeswax for another year or more . . . During the last stages the lenses were set in a sixty-foot tube, so I know that the object glass and its mate are perfect." [31]

The reporter was obviously impressed, and even more so when he learned that the lenses were worth $65,000.

In answer to a question on how the monster worked and what were the advantages of its great size, Hale explained that the primary object of a telescope is to bring the light from a distant object to a focus, and to form an image of it. In a refractor like the 40-inch the lenses are used to gather this light. The light is then bent or refracted to the point called the focus at the far end of the tube. The larger the diameter of the telescope objective (that is, the lenses), the more light will be collected, and the brighter the image that will appear in the eyepiece. Thus, Hale noted, "If we double the diameter of the object glass we get four times as much light in the image of a star, for the amount of light collected depends upon the area of the object glass, and this increases as the square of its diameter." Likewise, the larger the diameter of the objective, the wider the original beam of light will be, and the greater the detail the observer will see in the image.

The size of that image is, as Hale was also to explain, dependent not on the size of the objective but on the distance between the object glass at one end of the tube and the focal point (that is, on the focal length of the telescope). "Suppose," he suggested, "we have two telescopes, with object glasses of the same diameter, but of different focal lengths. The one of longer focal length will give a larger image. If the focal length is twice that of other telescopes, the image will be twice as large. With the same eyepiece, therefore, the magnifying power of the longer telescope will be twice that of the larger one." [32]

To the astronomer, however, mere magnification of the size of the image generally means little, unless he can gain a corresponding increase in illumination. Therefore, Hale's eternal cry as he sought to penetrate the depths of the universe would be for "more light."

The day after the lenses were mounted in the telescope, a special train arrived, loaded with trustees, members of the faculty, and friends of the university. That night they gathered in the dome; this time, although a strong wind was blowing, the skies were clear. President Harper was the first to look through the telescope at Jupiter. One by one the other guests looked and marveled at the brilliant planet's beauty.

Later, when all had gone, Hale returned to the dome with Barnard. Together they turned the telescope, first on one object, then on another, testing its powers. As far as they could tell, the lenses were all that Clark had claimed.[33] After some further tests, Hale wrote enthusiastically to Yerkes. He told him that both Barnard and Burnham agreed that the Yerkes telescope was decidedly superior to the Lick telescope. "Such objects as the ring nebula in Lyra, the great cluster in Hercules, and the dumb-bell nebula are shown in a surprising way. The brightness of these objects in the telescope is remarkable, and it is already clear that no other instrument in the world will show nebulae in so perfect a manner.

"But," he exclaimed, "the most satisfactory proof of the superiority of the objective to that of the Lick telescope, at least as regards its light gathering power, is afforded by a discovery made by Professor Barnard." Close to the brilliant blue star Vega, Barnard had found another star "much less brilliant" than the faint companion discovered by Winnecke. It lay in a region where Burnham had searched again and again with the Lick telescope. "On account of Professor Burnham's extraordinary skill as a discoverer of such objects, it may safely be said that this new star would in all probability have remained undiscovered with the Lick telescope." With the Yerkes telescope, however, it was easily seen.

"The discovery of such an object," he explained, "is a very severe test of the performance of the object glass. It shows that its polish must be very perfect in order that the field of view shall not be so brightly illuminated by Vega as to completely conceal any faint objects in its neighborhood. It shows too that the definition of the object glass must be very good; for if the feeble light of the star were not brought sharply to a distinct focus it would be too faint to be visible. And finally it demonstrates most conclusively the great light gathering power of the instrument, and leaves no question that the Yerkes telescope is not only the largest, but also the most powerful in the world." [34]

Yerkes was, of course, immensely pleased and intensely proud of the behavior of his giant infant.

Unfortunately, however, if the lenses were good, the mounting was not. In fact, it was, as Hale reported to Warner and Swasey, the builders of the mounting, "almost useless for observational purposes." It would have to be taken down and corrected. This meant another serious delay.

Meanwhile Yerkes, who feared that some of his enemies might injure the telescope, sent an ultimatum to Harper, asking "that no one be allowed, except those in charge, to approach anywhere near the glass. It should never be lowered so that visitors could look at it, or be on the same plane with it. In fact, every precaution should be taken to keep visitors away from it . . . it is there for scientific research and for that only!" [35]

Harper sent the letter on to Hale, who had been hoping to have at least one open night for the public each week. Hale read it on May 28th. During the first part of that same night he had worked with the 40-inch; Ellerman and Barnard had started work at 12:45 A.M. and got some "superb" observations of the Swan Nebula in Sagittarius. Once, while the floor was rising, Ellerman thought he heard a squeaky grunt "similar to a piece of wood being crushed or pinched." He stopped the motor, but could find no apparent cause for the noise. At 3:15 they closed the dome and went home to bed.

Three hours later, at 6:43 A.M., J. C. McKee, the contractor, was on his way to the observatory. He was about two hundred yards away from it when suddenly he heard a crash. He ran to the dome, climbed up on a windowsill, and peered through the window. The huge rising floor lay in ruins under the great telescope. McKee clambered down and ran to Hale's house. Hale leaped out of bed, threw on a robe, and dashed to the observatory. Soon Barnard arrived. Together they gazed in silence at the ruin in the dome pit, then up at the telescope, which seemed intact. Yet it was impossible to tell whether the lens had withstood the shock or whether flying debris had cracked it. Hale ran to the top balcony and tried to examine the lens through binoculars. He thought he saw some tiny cracks. (Later he discovered that these were actually the fine threads of a spider's web, built in the large tube.)

Afterward McKee wrote a report on the "reck" which Hale sent to Warner and Swasey, builders of the ill-fated floor. He had "found that the ends of the two cables holding Counter Balance weights had given out." News of the "axedend" spread rapidly, and Williams Bay was soon overrun by reporters. "It looked," said a man from the Chicago *Tribune,* "as if a cyclone had slipped through the slit and gone on a rampage." [36]

It was a miracle that no one had been killed and a wonder that the lenses had been saved. If the floor had fallen four hours earlier or fifteen minutes later, fatalities would have been inevitable. If it had fallen on the night of the trustees' visit, the disaster would have been tremendous. Nonetheless Hale was discouraged. Again there must be weeks, possibly months, of delay—and this after waiting five years for the completion of the observatory. Only a week after the accident Alvan G. Clark died of apoplexy.

During these disastrous weeks everything else was ignored—the sun was forgotten, the writing of letters suspended. For months after the crash the dome could not be moved. Only stars visible through the narrow, now immovable slit could be observed. Within that limited range, however, Hale worked to determine the color curve of the objective, and Barnard discovered some fifteen or twenty nebulae as well as four or five new double stars. "The 40″ is a great thing for bagging nebulae," Hale remarked.[37] It was the middle of September before the dome and floor could be used. It took even longer to put all the motors in running condition. Hale chafed at the delay. Finally, however, all was ready. Then, whenever the skies were clear, the telescope was busy. By day it was aimed at the sun; by night at the moon and planets, stars and nebulae. Like "the numerous Chicago saloons of those unregenerate days," it might, as Walter Adams was to comment, "have borne the sign above the entrance to the dome, 'We have thrown away the keys. Open all the time!'" [38]

When, some time later, the exuberant Oxford astronomer H. H. Turner arrived from England, he looked in amazement at the constant whirl of activity in this "virgin forest." "They work too hard at the Yerkes Observatory," he declared in his Oxford Notebook. "Morning, afternoon and night the work seems to go on continuously, with only that hour from five to six as an intermission. The centre of it all is the 40-inch, which never rests. The whole performance is splendid, and strikes awe into the beholder if he happens to come from lands where folk still retain the mistaken idea that one ought to rest every now and then." [39]

Others who arrived to see the new observatory were equally impressed. One of these was a reporter who described Hale as "slight in figure, agile in movement, of high-strung nervous temperament, over-flowing with formulae, technical facts and figures, theoretical speculations, almost ad infinitum. His mind seems made of some stellar substance which radiates astronomical information as a stove sheds heat."

Meanwhile, preparations for the dedication went on. It was now set for October 21, 1897. Still, at times, as Hale forged ahead with his plans, he pondered the prospect of dedicating an observatory consecrated to pure science to a man who was daily being castigated in the press. He concluded that he had to believe it better to have obtained the telescope in this way than not to have had it at all.

The day of dedication finally came. Hale had arranged for several special astronomical meetings before the actual day, and astronomers arrived from all parts of America, even from abroad. They were stowed around the observatory in every available nook. They were enthusiastic about everything—the telescope, the observatory, the grounds, *and* the director, now just twenty-nine years old.

At five-fifty on the evening before the dedication, Hale was at the station to meet the donor. Handsome in an immaculate gray outfit, Yerkes arrived, exuding charm. That night he stayed with the Hales. Evelina Hale considered him the most charming man she had ever met. After supper he made a tour of "*my* observatory." Although the sky was cloudy, he insisted on seeing through the telescope. He professed to be delighted to meet the "learned" astronomers as well as university officials and honored guests who came in hordes to celebrate the great event. He talked amicably with the reporters. "Everything," he exclaimed, "is in excellent condition. I am satisfied with the entire plant. The management, too, could not be improved upon." Here he smiled and patted Hale on the back.[40]

The following morning, a reporter described a scene of "tremendous bustle," with the cook running upstairs and the director down, while the visiting astronomers had their shoes polished by an enterprising waiter, turned bootblack.

Shortly before noon on the 21st, seven hundred people gathered in the dome under the great telescope. From the platform Hale looked down at the distinguished audience, which included many old friends—E. C. Pickering of Harvard, G. W. Hough, whom Hale had known at the Dearborn Observatory; and Simon Newcomb from the Naval Observatory in Washington; the good Brashear, his eyes shining, even though he had not made the great lens; Hale's old teachers, Professor Ira W. Allen from the academy and Dr. H. Belfield from the Chicago Manual Training School. Among the scores of scholars from the faculty of the University of Chicago were James H. Breasted, the Egyptologist, and Robert Millikan, who were destined to become closely associated with Hale in later years.

James Keeler, now director of the Allegheny Observatory, had been chosen to substitute for Huggins, whom Hale had originally invited to give the main address. His subject was "The Importance of Astrophysical Research and the Relation of Astrophysics to Other Physical Sciences." It was a brilliant address, remarkable for its analysis of the possibilities in this new field as well as for its broad view of science.

"There may be some," Keeler declared—and Hale looked around at Newcomb, Burnham, and Hough, all confirmed believers in the old astronomy, all suspicious of this new trend—"who view with disfavor the array of chemical, physical and electric appliances crowded around the modern telescope, and who look back to the observatory of the past as to a classic temple whose severe beauty has not yet been marred by modern trappings. . . ." In contrast, Keeler himself, like Hale, looked forward to a future in which astrophysical research "with respect to its own ends and to its bearing on the advance of knowledge in other fields" would grow from year to year.

After Keeler came the other speakers—Harper, Yerkes, Newcomb, and Hale himself. "If I mistake not the signs of the times," Hale said, "the Yerkes Observatory can render no better service to both astronomy and physics than to contribute in such degree as its resources may allow, towards strengthening the good will and the common interest which are ever tending to draw astronomers and physicists in closer touch. . . ."

The day after the dedication Hale took time to read the papers. In general they were charitable, praising Yerkes and speaking proudly of the honor he had brought to Chicago. But some, as he had anticipated, denounced the whole business as a farce. One ran:

YERKES BREAKS INTO SOCIETY

Street Car Boss Uses a Telescope as a Key
to the Temple Door and It Fits Perfectly.

Another account referred to the sumptuous banquet that was held in Chicago at Kinsley's restaurant the following night. At that dinner Ferdinand Peck, the Chicago capitalist, suggested that Yerkes's earthly activity might now be connected with his heavenly aspirations. Cable and electric cars might be run to the stars and planets. "It does not appear how the franchises for these roads are to be obtained but it is said he anticipates friendly legislatures in these starry worlds, though hostile common councils may prevail there as well as here." [41]

Afterward one of the reporters at the banquet concluded: "If his banquet excursion did it today, Charles Tyson Yerkes will have won the distinction which he has vainly desired for many years—a position in polite society. George Gould, prince of New York's social pretenders, rode into Gotham's select circles on a yacht; Mr. Yerkes is trying his prettiest to ride into Chicago's sacred temples on a telescope."

CHAPTER SIX

Life at Williams Bay

✳ *1897-1903*

THE DEDICATION was over. Five years after Yerkes's original promise, Hale could at last plunge into the work that had been waiting so long. While the staff for such a huge observatory was small, its members made up in energy what they lacked in numbers. Like Mattie and Will, they entered into his plans with the greatest gusto. He was so full of his subject (as well as of numerous other subjects), and his mind buzzed with so many ideas, that they could not help being infected with his enthusiasm.

They soon learned, too, that if he was a dynamo, driven by an overwhelming faith in research, he was also a very human individual, keenly sensitive to the welfare of others. He was the first to help them in time of trouble, the first to offer sympathy, and even, when necessary, financial assistance. Doubtless the concern that had pervaded his childhood during his mother's long years of illness had heightened his awareness and made him unusually conscious of suffering in others and of the troubles it inevitably brought in its wake. In the small red notebooks in which he itemized all his expenditures—even the most minute—lists of flowers and gifts sent to relatives and friends would always be long. From his mother and father he had inherited a strong New England reserve. He was, therefore, not inclined to be demonstrative. But it was not in his nature to be indifferent.

The original staff of the observatory, in addition to Hale himself, included Barnard and Burnham, Ritchey and Ellerman, and F. L. O. Wadsworth, who had been Michelson's assistant at Chicago. For some time, despite the diversity in their talents and personalities, they got along. Accepting life as it came, they laughed at themselves and at each other. It was, in general, an extraordinarily youthful group—most of them were in their twenties and thirties—and they shared a common delight in the days and nights at the observatory, filled with work, but filled, too, with endless excitement and amusement. If occasionally there were those who did not share the general enthusiasm, these were the exceptions. One such was a young man who spent only a short time at the observatory, then departed because he apparently could not comprehend Hale's inability "to suffer fools gladly" or to tolerate sheer laziness or careless work. When Hale reprimanded him, he retorted by blaming the director for the difficulties in the observatory. "The management of the place is decidedly slipshod, possibly due to a superabundance of enthusiasm which has led to attempts to do too much with the means at hand." Hale, piqued by such uncalled-for criticism, replied angrily: "The difficulty is that for some reason, you do not enter into your work with sufficient interest and enthusiasm. No one can ever be a success under such circumstances." [1] With his own boundless enthusiasm for everything he entered into, he was unable, throughout his life, to comprehend the indifferent or the perfunctory.

In general, those who worked with him shared that enthusiasm, and delighted in it. As Philip Fox was to write, "His mind was seething with plans for research and the plans were no sooner conceived than they were put into action. Of him it might also be said that he was 'fertile in resource.' . . . He formulated problems with crystal clarity; he loosed the hold which old and established methods often retain."

Yet, Fox went on, "I would not wish to give the impression that George Hale was always intense, relentlessly driving himself and his associates. He could and did work at fever heat, but those who have seen him at Orchestra Hall or at the opera have sensed his love of music; those who have seen him reenact the puppet show exploits of the brave knight Rinaldo, or heard him recite the tales from the Monasteries of the Levant know his keen dramatic sense and his deep and abiding appreciation of humor. He was a most delightful companion, equally eager as an auditor or raconteur." [2]

To those in the outside world, it is true, the extraordinary dedication of these astronomers to the study of the heavens must often have seemed incomprehensible. Doubtless many of them wondered what impelled these men to spend all their time in extraterrestrial pursuits. Absorbed as they were in their work, earthly events often meant little. In April, 1898, the shadow of the war with Spain hung over the country; yet at Williams Bay there was a feeling of detachment. "This is a very peaceful region of the world," Hale wrote calmly, "where we occasionally hear faint rumors of disturbance, but are not much concerned with such doings ourselves." They were, in a way, like sailors on a remote voyage. Chicago was only eighty miles away, yet it often seemed like another planet, especially in winter, when the observatory was barely visible above the drifts. Like skilled navigators, they knew the skies—the position and color of the stars, the mythical pattern of the constellations, the ways of moon and planets, the characteristics of the nebulae. They had learned all these things in the solitude and serenity of the night, when all the world slept and only they waked and worked. It was not surprising, therefore, that they felt a certain detachment from this small planet spinning around its little star.

In time the original group was expanded. One of the earliest additions, and one of the most valuable, was Walter Sydney Adams, whom Hale soon came to consider his closest astronomical friend. Adams, then a lean and lanky boy of twenty-two, had earned a reputation at the University of Chicago as a mathematical shark. The son of missionaries, he had been born in the small village of Kessab near Antioch, North Syria, and had spent his early years there. Yet, like his parents, he remained a true New Englander, with a shyness that disappeared only when he came to know his fellow astronomers well. From the beginning, however, his quiet, unexpected wit and eagerness to join in the escapades that enlivened their days made him a welcome member of the group. In time Hale found that Adams had wonderful tales to tell of his boyhood in the Near East. These tales were a special delight to one who, as long as he could remember, had been fascinated by anything that had to do with the Orient. Adams's knowledge of ancient history went back to his childhood when, in their isolated Syrian community, he had devoured the contents of his father's library, which, apart from theological books, consisted largely of histories and classical texts and treatises. Then, too, he had lived near Antioch, crossroads of the Crusades and of most of the campaigns of ancient history, and was surrounded by medieval and Greek and Roman remains. In the back yard, he could play in the ruins of a Crusader castle. In the surrounding fields he could pick up coins from the time of the conquests of Alexander the Great.[3]

When Adams was eight, the family returned to Derry, New Hampshire, and there for the first time he attended a regular school and followed a more customary educational routine that led to Dartmouth College, and eventually to graduate work at Chicago. When asked what had led him to the choice of a branch of physical science, and astronomy in particular, he said he thought he could distinguish two main factors. "The first was my innate preference for exact subjects, that is, those in which the fundamental processes of reasoning and application are relatively unchanged, as in mathematics, astronomy, physics and chemistry." As against this, he noted, he had much less interest in the social sciences, "the bases of which are subject to constant change." The "exactness of thought and reasoning" to which this preference led was, as Hale soon realized, one of Adams's outstanding characteristics.

Yet his most outstanding quality remained his modesty. Years later, the astronomer Harlow Shapley commented: "Modesty was instinctive with Walter Adams. He strove to excel in everything he undertook—in endurance at the business end of a telescope, in quality of spectrum plates, in hiking speed up the mountain trail from Sierra Madre, in tennis, golf, billiards, bridge—and he did excel. But I never heard him call attention to his excellence. I remember complimenting him once on his designing the series of powerful and tricky spectrographs that were used in the Mount Wilson stellar and solar work. 'It is a very low form of cunning,' he replied." [4]

Adams was eight years younger than Hale. Yet from the beginning a friendship characterized by an intense and unshakable loyalty developed between them. As the years passed, Hale was to rely increasingly on the younger man for help and advice.

Adams joined the Yerkes staff in 1898. That summer E. F. Nichols arrived from Dartmouth with an instrument he had invented, called a radiometer, that was sensitive to the light of a candle five miles away. With it he was to carry out his spectacular measurements on the heat of a star. By 1900 he was measuring the heat radiation of the brilliant stars Arcturus and Vega—a feat hitherto considered impossible.

One of the next to arrive was the gay and exuberant Philip Fox, who came from Dartmouth to join the "inspiring company," and to help Hale in his contributions that "defined an epoch." Fox, like all the other astronomers at Yerkes, soon became aware of Hale's love of music. One day they were working on a globe on which Hale had projected the sun's image in order to measure the coordinates. "Pausing in thought, Hale started to whistle the Vorspiel to *Lohengrin*, with Ellerman and me joining in, whistling on to the end, then laughing a bit and going on with adjustments. It revealed his complete absorption, and moreover his deep love for music which I saw manifested many times."

At first, Hale had hoped to collect a staff of astronomers from all over the world—then a novel idea—as he was eager for astronomers everywhere to benefit from the magnificent powers of the new telescope and the incomparable facilities of the observatory. Unfortunately, the infinitesimal size of the salaries and the costs of travel made his offers unacceptable. Karl Runge's reply from Germany, "Impossible," was characteristic.[5] Only much later would Hale see his dream realized, to the enormous benefit of international astronomical cooperation.

Yet the list of those who came to the observatory, without pay, eager only to use the facilities offered, is long. Some came only to visit. Others stayed on for weeks, even months. In the list appear such names as A. A. Belopolsky from Russia, Gilbert Walker from India (an expert at the art of the boomerang), Pietro Tacchini from Italy, Arthur Schuster and Hugh Newall from England. All these and others brought a knowledge of their special fields that delighted Hale and broadened the vision of the Yerkes astronomers. They, in their turn, were amazed by the superb conditions they found at Williams Bay. One European astronomer, after using the 40-inch on the planet Jupiter, exclaimed that his years of work with a smaller telescope seemed almost useless. "So much more detail could be perceived at a single glance."

They were amazed, too, by the astronomers' willingness to observe even on the most frigid nights when the thermometer registered as low as 21 degrees below zero, and even 30 degrees below. Nothing daunted them. On one such night, as Storrs Barrett, a member of the staff, who had also been at Kenwood, reported to Hale, his alarm rang at 1:30 A.M. With its ringing the sky cleared. It stayed clear while he hauled on "two sets of underwear and one pair of pants, one shirt, two coats, two pairs of socks, a collar (no tie), a neckscarf, a sealskin cap, a fur overcoat, a pair of shoes, a pair of arctics and two pairs of mittens." It stayed clear until he had plowed through the drifts to the observatory—roused the night assistant, and climbed the heights to the 40-inch—"and then it curdled." [6] Nor was this night unique, as Hale and every other astronomer knew well from long and frustrating experience.

Even when the skies were clear, the intense cold often made observing difficult, if not impossible. On such nights the moisture on their lanterns would freeze on the inside of the globe so that it was almost impossible to see anything in the flickering shadows. Yet, despite the cold, the work of guiding the big telescope required so much physical effort that, as Ellerman reported, even with the thermometer reading 22 degrees below zero, he was hot, though wearing only a sweater and coat. "My fur coat I had laid on the floor could be held out almost like a frozen cloth." [7]

In contrast to the majority of the Yerkes astronomers, Hale was to spend most of his time observing the sun. He did, however, have certain programs that required night observing. Then, rigged out in tall boots, fur coat over another coat and sweater, and fur hat, he would trudge through the snow to the observatory. In the distance he could often hear the gray wolves howling. Often he floundered on the way. He reached the observatory looking more like a snowman than like a human being. His hands were cold, his feet numb. Yet, if it was clear, he would take time only to shake himself to throw off the snow, then head for the dome. There, as always, lest heat waves obscure the seeing, the temperature had to be the same inside and out. Soon he would be eagerly at work at the eye end of the telescope.*

Yet life at Yerkes was not all work. During the day, when clouds obscured the sun, there was always the chance to go on a flying jaunt in an iceboat on the lake. After one such excursion, Hale wrote ecstatically to Goodwin, "It was glorious." Or there was golf. In summer the game was played on a six-hole course in the fields north and west of the big dome. In winter the hard snow was the course, with the golf balls painted a brilliant red. In these games, as Adams noted, the contrast between the players soon became evident. Ellerman, accurate and precise, studied his strokes carefully and executed them well. Hale, eagerness personified, developed a good game but often "pressed and ruined his shot." [8]

So the work, interspersed with play, went on. Yet there was still no prospect of getting the funds needed to run the observatory. The

---

* Today, although observatory domes in northern climes and on mountain peaks remain as frigid as ever, the advent of electrically heated suits has robbed observing under such conditions of some of its discomfort.

university, faced by huge deficits, could do nothing. Ever since their unforgettable sojourn on Pikes Peak, Hale had dreamed of having James Keeler, who was director at Allegheny, join him. No one else in the country shared so fully his devotion to astrophysics or his crusading faith in its future.

Soon after this, at the end of 1897, Catherine Bruce, the generous supporter of astronomy, offered $15,000 to cover Keeler's salary at Yerkes for three years. Ironically, soon afterward, in March, 1898, Keeler was appointed director at Lick. At the last moment Hale urged him to reconsider. He described the disadvantages of a directorship—the need for spending time on administration, on settling rows between members of the staff, on begging money from people "who are far from anxious to give it." He urged him to consider the problem of family life in a place like Lick on Mount Hamilton in California. He noted that he himself would not be inclined to accept a position that would mean living permanently on an isolated peak. If Williams Bay was also "rather out of the world," it was, at least in summer, a place where "one can, if he wishes, have the companionship of many cultivated people." "To you and me this is a consideration of little moment, but it is certainly one that must affect our families." In the end, however, Hale knew, as Keeler knew, that such considerations were secondary. The ultimate decision must "rest largely on the effect it is likely to have on one's subsequent scientific output. For after all the most important thing we can do is to contribute as far as we are able to the advancement of knowledge." [9]

Keeler hesitated once more, but finally decided to go to Lick. Shortly afterward, amusingly enough, he wrote to Hale, "If you succeed in raising the money for completing and running your reflector, why not come *here*. . . . All that would be needed would be the observatory for the reflector and a small home for the observers. Perhaps the Lick Observatory could furnish brick at cost prices, as we shall burn some this summer."

As soon as he arrived at Lick, Keeler had begun work with its Crossley 36-inch reflector—"an ungainly machine," he called it, "but, I hear, better than it looks." [10] In his skillful hands, as he "tinkered" with it to make it usable, it did indeed prove to be better than it looked. Soon he was taking the most magnificent photographs of nebulae ever made up to that time. A beautiful photograph of the great spiral in Canes Venatici, taken on May 16,

1899, was succeeded by others that seemed to indicate that the majority of nebulae are spiral in form. Estimating that there were not less than 100,000 nebulae within range of the Crossley, Keeler looked forward to years of fruitful work. Hale, rejoicing with all the astronomical world in these spectacular results, foresaw a brilliant future for his friend, a future in which he still hoped they might work together. But, like Hale, Keeler, in his eagerness to ferret out the nebular secrets, was driving himself to the limit. A few months later, in July, 1900, he mentioned a cold he was trying to shake. On August 2nd Hale replied, urging him to take a vacation in the form of a trip to Yerkes. Ten days later the unbelievable news came that the cold had turned to pneumonia and that Keeler was dead.[11]

A few days later a letter came from his widow. In it she wrote: "I always felt the friendship between Mr. Keeler and yourself was one of the strongest and most disinterested that ever existed. In the old Allegheny days I used to joke with him about the inexhaustible interest you both found in talking together and I laughed over the rejoicing exchanged over any success accomplished by the other. He looked forward to seeing you again as one of the brightest prospects of the future, never a doubt crossing his mind as to the speedy fulfillment of that expectation." [12]

When he died, Keeler was only forty-two, and Hale found it impossible to realize that never again would they discuss the exciting problems that had absorbed their hearts and minds. He wrote to Brashear: "I find it hard to believe that Keeler is gone beyond hope of recall." Moreover, this death, coming on the top of two overwhelming losses in his own family—the death of his father and mother—seemed more than he could bear.

In October, 1898, he had been called to Chicago. His father had been brought back from a business trip to Mexico City by his wife in a private car rented from Mr. Pullman. He was suffering from Bright's disease. For four more weeks he lived on. During the whole time his son stayed at his side, unable to do anything but watch helplessly as this man whom he loved and counted on more than anyone else lay dying. At 6:30 A.M. on November 16th, the end came.

The newspapers ran long columns of praise for what William Hale in his quiet way, but with his "prodigious energy," had done for Chicago. "By Perfecting the Modern Lifting Apparatus He Made Possible the Skyscraper," ran one headline. Others described the buildings he had erected in Chicago—the Calumet, the Rookery, the Insurance Exchange, the Reliance; and in Kansas City, the Midland Hotel. No one realized the vacuum created by William Hale's death more than Brashear. "Your loss is indeed a great one," he wrote, "for your good father appreciated your interest in your life work as no one else could do. I do not remember in all my life anyone who was so devoted to a work not in his own line as your father was, and my dear George, *you* can have the great consolation and a pleasurable remembrance that your work from the very first was a source of unbounded pleasure to your father." [13]

William Hale had wanted his son to become an engineer. But when he saw that his interests lay along other lines, he had done everything in his power to advance those interests. Still, if George failed to become an engineer, many of his astronomical tasks required engineering skills that he inherited from his father. No doubt they had both recognized this fact. In any case the elder Hale would have been pleased by the comment of the noted astronomer J. C. Kapteyn, over twenty years later, when he wrote of the spectroheliograph and of George Hale's abilities: "I do not think that anybody, not having something of the engineer in him, would ever really have enriched the world with this instrument. It is so with the big telescope, with the great mirror, with everything. I really begin to think that the true recipe for making a first rate astronomer is: Take an engineer and teach him some astronomy, not: take an astronomer and teach him some engineering." [14] And William Hale, who had always encouraged his son to design and build his own instruments, would surely have agreed.

After his father's death Hale returned to the observatory, but he had little heart for work. He went through his daily routine like an automaton. Years later, dreaming that his father was near, he would wake, his eyes wet with tears.

His letters and journals and the statements of his friends bear evidence that during this dark period George Hale, now thirty years old, was profoundly concerned with the ultimate realities. What is the meaning of life and death? What hope is there of a hereafter? Should he, he wondered, go back to the beliefs his mother had taught? He no longer knew. Sorrowfully he wrote to Harry Goodwin: "I hardly know how it happened, but I found myself coming back, as though to solid ground, to many of my old beliefs, and before Father died I could appeal for support to a source which of late

I have known but little if at all. . . . I do not know how I could bear Father's death if I had no hope of seeing him again at some future time. We may not be able to find scientific proof of immortality, but I think we *must* believe in it." [15]

But he felt constrained to add, "Now do not suppose that I have gone completely back to the somewhat hysterical views of our early struggles, or that I intend to go into retirement and renounce the pleasures of the world. . . ."

In the weeks that followed, he needed all the strength he could find. As his mother's health declined, he spent more and more time in Chicago, neglecting his research. Suffering from kidney trouble, she lived on for seven months. Slowly she faded away. The last two weeks he stayed with her continuously. She died on July 11, 1899.

In a way Hale felt it difficult to regret her death. She had suffered so long that death could come only as release. Nevertheless it increased his sense of desolation. Ever since he could remember, his mother had suffered. Yet, in spite of it, she had done everything in her power to broaden her childrens' lives.

In August, 1899, after George Hale's return to Williams Bay, Margaret celebrated her third birthday. A lovely child, she was her father's joy. At night when she lay in her crib, he would sit beside her, telling bedtime stories. One of her favorites had to do with a remarkable tribe of bears who lived in the Great Cave. There were endless chapters, but one she liked especially was called "Dinner in the Great Cave." As he told of the Polar Bear she listened eagerly to his account of "the biggest and best Bear Cook in the whole kitchen" and of all the other Black Bears who sat down at the High Table with "the rhinos and the hippos and the baby camelopards and the pidgwidgeons" to partake of the fabulous dinner that began with rich green soup "tasting of the ten thousand turtles of the Seven Seas." "So they ate and ate, and the Band played and played . . . and everybody got fuller and fuller until they all finally stopped because they couldn't stuff another mouthful. . . ." [16] As each instalment ended, Margaret was usually falling asleep, hugging her stuffed dog. And her father would hurry off to the observatory, back to his desk piled high with letters to answer, articles to edit or translate, and measurements on solar photographs to make.

At other times, Margaret would sit on her father's lap, learning to read from the entrancing pages of *St. Nicholas* or listening to fragments of verses he loved to repeat, like "Dance a jig to the Granny's

Pig" and "Old Gammert Hipple Hopple," which he had learned from Nellie, the Irish maid in his childhood.

In the summer of 1899 Hale, to his surprise, found himself sharing in the formation of an astronomical society. Two years earlier, Simon Newcomb, the leading mathematical astronomer in the country, had proposed the founding of such a society. For twenty years Newcomb had been director of the Nautical Almanac Office, which in 1894 was made a branch of the Naval Observatory, under naval control. In these last years of the nineteenth century, Hale, as a member of the Board of Visitors of the observatory, had had the chance to know Newcomb well, as he had unsuccessfully worked to wrest the observatory from naval control. Now Newcomb was urging that the headquarters of the new society be in Washington. But Hale, unable to forget Newcomb's extraordinary statement in 1888 that "we are probably nearing the limit of all we can know about astronomy," was extremely doubtful of the fate of a society in which Newcomb "would be sure to run the whole thing."

Shortly after Newcomb had made his proposal, the astronomical conference at Yerkes in 1897 had taken place. The following summer Hale, hoping to repeat its success, had invited astronomers, physicists, even geologists who might be interested in astronomical problems, to join him at Yerkes. The response was enthusiatic. G. W. Myers of the University of Illinois wrote, "I hope you may succeed in making Williams Bay, or rather Lake Geneva, the Mecca of Astronomers. You will find me among the pilgrims." [17] Joseph Ames of Johns Hopkins was equally eager.

Soon, however, Hale discovered that the feeling existed that he intended to form an astronomical society with headquarters at Yerkes—a thought that had never occurred to him. "My idea," he wrote hastily to Pickering, "was simply to have an informal gathering of astronomers and physicists from time to time." [18] Indeed, he feared that a formal society might involve questions of politics rather than of science, of fighting for office, "with all the attendant evils." He feared too that in a formal society the presentation of papers would become the chief function.

As opposition to a conference at Yerkes increased, he decided with Pickering that it would be best to hold the meeting at Harvard before the fiftieth-anniversary celebration of the American Association for the Advancement of Science. So it was that Hale found himself in Cambridge on a hot day in August, 1898, one of a large group of astronomers. Throughout the meeting the powerful Newcomb, with his leonine head crowned by a shock of gray hair, dominated the proceedings. At every opportunity he discussed his dream of an astronomical society. Many expressed interest in the idea. Finally Hale reluctantly agreed to form a committee to consider the subject. From the beginning, however, he argued that if any society be formed it should be an astronomical and *physical* society, in which physicists would want to join. Only in this way, he was sure, could progress be made. Immediately [19] Newcomb objected to the name Hale proposed. "I hope," he averred, "the astrophysicists will consider their science as a continuation of the ancient and honorable science of astronomy, and allow the new body to be called the American Astronomical Society." [20]

Hale, sure the infant science of astrophysics would be submerged in an organization that ignored its existence in its title, wrote vehemently to Langley, "The time has come for according astrophysics a distinct place among the sciences." Nevertheless he did admit that "to attempt to separate it altogether from astronomy would be not only illogical but unwise." [21] In the end Newcomb gave in, and the name "American Astronomical and Astrophysical Society" was adopted. If it was cumbersome, it was at least specific.

In March, 1899, the organizing committee (which included Hale) met in Washington. Plans were made to hold the first meeting at Williams Bay in August. At that meeting Newcomb achieved his aim of election to the presidency; Young and Hale were elected first and second vice-presidents. "The prospects for the success of the society seem to be excellent," Hale said, and everyone agreed. In this way the society that is still the leading astronomical society in the United States was born. Long after, when astrophysics had achieved the recognition Hale was fighting for, the name was changed to the simpler "American Astronomical Society."

When the first meeting of the society was held at Yerkes in the summer of 1899, astronomers who had been at the dedication in 1897 were impressed. The 40-inch was working beautifully, producing extraordinary results. With it Hale, observing the moon, could see minute structures "no trace of which could be made out with a 12″ on the same evening." With it, too, Ritchey would take some of the best photographs of the moon and star clusters ever made up to that time, and Adams would take photographs of faint stellar spectra never before visible.[22]

Outside, the grounds still looked like an abandoned farm. But at the telescope it was a different story, as it led the way into the new era in astrophysics for which Hale had planned so long. In this new era, while the western frontiers of America receded, the frontiers of the universe would be pushed farther and farther back. The physical nature of sun and stars would be revealed in a way that few astronomers could then anticipate. In making possible the exploration of these new frontiers, Hale was to lead the way. If he could have lived to see the opening of the Space Age sixty years later, he might well have compared this new revolution with the one in which he had played such a dominant role.

Gradually now he could get back to his solar work so long neglected. Now, at last, he could take advantage of the great focal length and dispersion of his new telescope. As soon as he could turn it on the sun, he discovered hundreds of lines in the solar spectrum which had never before been seen. At Kenwood he had identified the green carbon fluting in the solar spectrum (which much later would be shown to be part of the band spectrum of the carbon molecule and CH). But in what region of the sun the carbon lay no one knew. There were even some who claimed that it lay not in the sun's atmosphere at all, but in the earth's. Now, with the Kenwood spectroscope attached to the 40-inch, the answer shone out crystal clear. Near the small b-line, in the chromosphere he noticed a shaded band. This band, he was sure, must belong to carbon. Day after day he photographed the tiny spectrum. Day after day the band was broken up into its component lines and he was able to measure them. From these measurements he proved the existence of carbon in vaporous form in the chromosphere. He showed, too, that the carbon probably lies within five hundred miles of the sun's surface, forming a relatively thin layer in the lowest part of the chromosphere.[23] "Its rays," as the British historian of astronomy, Agnes Clerke, was to put it, "do not come near the surface; they have to be dredged for. Hence the extreme delicacy of detective operations." [24]

By 1899, too, the gigantic, unwieldy and complex Rumford spectroheliograph, built in the instrument shop at Yerkes and weighing 700 pounds, was finally ready. Because of the continuing lack of funds, its acquisition had been painfully slow. A small grant from the Rumford Fund had been provided; then such essential parts of the instrument as the 6½-inch Voigtlander lenses of the type formerly used by portrait photographers had to be searched for until they were found in a pawnbroker's shop. In the end, however, it proved to be one of the most valuable instruments of its time for the study of the sun.

With this powerful tool the way was now open to tackle some of the most puzzling of solar problems. After the countless delays and frustrations, Hale could not wait to get at them. The problems were not easily solved. The difficulties in managing the Rumford spectroheliograph, that clumsy, heavy attachment to the 40-inch, were manifold. The necessary adjustments included thirteen distinct operations that had to be carried out with great precision before a satisfactory photograph could be taken.

But all these difficulties were overcome, the spectroheliograph went into operation, and Hale was soon observing a "multiplicity of fine details" that had never before been seen. At Kenwood he had photographed the bright solar clouds called faculae that rise above the ordinary level in the photosphere. On his first plates taken with the Rumford spectroheliograph, the faculae shone out. But on those plates he also found extensive calcium clouds that did not (like the faculae) rise high enough to be observed in elevation at the limb, but could be seen in projection against the bright disk. Again and again he discovered these objects on his plates, and the differences between these calcium clouds and the underlying faculae became so marked that he felt a distinctive name for the vaporous clouds was needed. Because of the "flocculent" appearance of his photographs, he decided to call them "flocculi." To the eye at the telescope or in direct photographs of the ordinary kind these flocculi had been invisible. Now, with his spectroheliograph, he could record the images quite clearly. Moreover, he observed that they varied not only with the setting of the slit on different parts of the H or K band but also with a change to the line of a different element —that is, from hydrogen to calcium. These observations led to "the striking achievement," which, as R. G. Aitken noted, "savored of the miraculous." By using different parts of the broad K line, Hale was able to study the distribution of the extensive clouds of brilliant calcium vapor at gradually increasing levels in the sun's atmosphere. When he could combine these results with measurement of the radial motions of the gases in the flocculi, a vivid picture of the circulatory processes in the sun's tumultuous atmosphere emerged.

By May 27, 1903, he could write to Newall, "I am full of the spectroheliograph just at present, and would give a great deal for a long discussion with you on this and other subjects. The new results are turning out even better than I anticipated, and there is evidently work enough ahead to keep any reasonable number of observers busy for years to come."

He went on: "In my last letter I think I told you about our photographs of the calcium vapor taken at different levels, by setting the slit at various distances from the center of the dark K band. We have already had an instance of the value of such photographs, which are now made on every clear day. On two high level plates a bright eruptive phenomenon appears, which is very faint on low level plates taken at about the same time. There seems to be no doubt that this was in the nature of a bomb which went up to a considerable height before it exploded. The simile is not exact, as the low level plates show a much fainter object at the same point. A column of flame, comparatively faint in its lower part and brilliant near its summit, would afford a better description." [25]

After this, he went on to measure the direction and velocity of motion in the flocculi. In general it appeared that the vapor was moving upward at the rate of about one kilometer per second. At times, however, the flocculi were shot out of the sun's interior at tremendous velocity and showed extraordinary changes. He concluded that these brilliant eruptive flocculi probably correspond to the great eruptive prominences sometimes seen at the sun's limb.

Then, too, he was investigating the hydrogen flocculi and comparing them with those of calcium. He obtained his first satisfactory plates in May, 1903.[26] He was surprised to find in them evidence of a mottled structure that covered the sun's disk. It resembled in a general way the structure of the calcium flocculi, but "differed in the important fact that whereas the calcium flocculi are bright, those of hydrogen are dark." (These long, dark flocculi proved, like the bright flocculi, to be prominences projected against the sun, and were subsequently called "filaments" by Deslandres.)

Already, however, Hale had realized that, to push his observations farther, another type of instrument was needed. A much longer spectroscope than the short-focus grating spectrograph was required —one that would be far too long and heavy to attach to a moving telescope like the 40-inch. What was needed was a fixed telescope in which a large and sharply defined image could be formed on the slit of a long fixed spectrograph.

In February, 1901, he had written to Harper of his plans for such a telescope, which "represents a new departure in telescope construction and in my opinion is likely to serve as a type for some of the great telescopes of the future." With it for the first time a star could be brought into a constant-temperature physical laboratory, and there "the latest and most delicate instruments of the physicist can be employed in the examination of celestial objects." [27]

Two weeks later, in a letter to Yerkes, he elaborated his ideas: "The instrument which we are building is of the horizontal type. No lenses will be used in the telescope. The images will be formed through the use of three mirrors, one of which will be driven by clock-work. The length of the instrument will be 175 feet, and images of the Sun, Moon, star clusters etc. will be about 3 times as large as the corresponding image given by the 40-inch. At the north end of the horizontal tube of the telescope several laboratories will be erected and equipped with spectroscopes and other instruments far more powerful than any instrument now employed in astronomy. As the image of the Sun or other heavenly body will be formed at a fixed point in the laboratory where it can be maintained for hours together, it will be possible to undertake many investigations which are not within reach of existing instruments. I confidently expect that the photographs to be taken with this instrument will greatly surpass the best hitherto obtained." [28]

For months work on this new instrument went on. In the final design the light of sun or other star was reflected from a 30-inch plane mirror to a 24-inch plane mirror that could be slid along a cast-iron track nine feet long. After reflection from this second mirror, the light was sent horizontally in a southwesterly direction for a distance of 175 feet to meet a spherical concave mirror with a focal length of 175 feet.

In 1902, this new horizontal telescope was set up in a little house near the observatory. But before there was a chance to test it, disaster struck. The insulation in a high-potential line broke down and the house caught fire. Everyone turned out to fight it. But a high wind was blowing, and almost nothing could be saved. The 30-inch flat and the 20-inch concave mirror, which had cost an immense amount of time and trouble, were shattered into a thousand pieces.

The 24-inch flat was saved, but the new grating and the entire spectroscope were ruined. Even if funds could be found to rebuild the instrument, "nothing," Hale moaned, "can ever make up for the serious delay in time." [29] Once again the Yerkes project seemed to be dogged by ill fate.

"The sun is a typical star. . . . It cannot too often be repeated that the sun is the only star near enough to the earth for intensive study of its surface." This concept ran like a refrain through all Hale's work. It was this that had led to his initial plea for a great telescope that would enable him to study the details of the solar surface.

Yet by the time of the Yerkes dedication, his plans had expanded to include all phases of astronomical research, particularly those that stressed the relationship of astronomy and physics and that might lead to an understanding of stellar evolution. Thus, while the sun had been and remained for him the most important star in the sky, there were countless other stars he wanted to investigate with the 40-inch.

Now, therefore, he turned for a while from the study of our little star, hoping to show its relationship to all the other stars in the universe. With this objective he made plans to observe those classes of stars that, he thought, might show some of the same spectral characteristics as the sun. He selected a special group of red stars, first classified by the great Italian spectroscopist Pietro Angelo Secchi. Up to this time these stars had been observed only visually. The carbon bands and a few prominent lines had been recognized, but in general the nature of their spectra was unknown.

The work progressed slowly, partly because of the personal difficulties that forced Hale to be away from the observatory for weeks on end, and partly because these stars were extremely faint and the observations proved exceedingly difficult. The stars were also red, and red-sensitive photographic plates were still in their infancy. The plates had to be treated with special dyes. Even then they were hard to use because of irregularities in the sensitivity. The taking of a single plate required several nights' exposure; it was "liable to interruption by clouds"; the measurements demanded additional computers, for which there was still no money. Hale often became impatient at the lack of progress and the continuing dearth of time. Yet what he was able to do with the help of Ellerman and J. A. Parkhurst (who joined the Yerkes staff in 1898) would earn the praise not only of contemporaries but also of astronomers for years to come. They obtained photographs of the spectra of over twenty stars; and Hale arranged ten or twelve of these in a series that showed progressive changes.

In her *Problems in Astrophysics* (1903) Agnes Clerke described the "splendid" series of pictures obtained. "The arduousness of the undertaking can be estimated from the fact that, with a train of three prisms, exposures of nine hours were required to secure impressions comparable with those given by Betelgeux in twenty seconds." The results, published as one of the Decennial Publications of the University of Chicago, contained 135 quarto pages of closely printed text and tables.[30]

Fifty years later the astronomer Otto Struve, looking back on these photographs, would point out that they could be duplicated with only a very few of the world's largest telescopes—and this despite the revolutionary developments in photography that had taken place in the interim.[31]

In time, as he identified the lines of numerous elements in the spectra of these stars and observed bright lines in several of them, Hale began to see indications of the relationship between these stars and the sun that he had hoped to find.[32] In his classification of stellar spectra, Secchi had distinguished four principal types: I, spectra of white and bluish-white stars like Sirius, which have broad and strong hydrogen and calcium lines, and only a few narrow and faint lines of other elements; II, spectra of yellowish-white stars, which contain many lines; III, spectra of red stars, containing a characteristic series of bands; IV, spectra of another class of red stars, containing the strongly marked carbon bands. (From the reddest stars, which he likened to drops of blood, to the bright blue stars that appear in the constellation of Orion and elsewhere, Secchi observed a continuous range of color, without a break, which indicated a continuity in the development of the stars.)

Now, using Secchi's classification, Hale decided to compare photographs of the blue and yellow regions in the spectra of the stars of the fourth type with those of the second and third types. Then he compared both groups with the solar spectrum. The third and fourth types seemed to show parallel branches of development that, he felt, might be traced back to the sun.[33] Could they then, he wondered, be stars of the sun's type in an advanced stage of life? If so, what was their place in the sequence of stellar development? To answer these and other puzzling questions on the absorbing problem of stellar evolution, he foresaw the need for more powerful instruments.

Yet one immediate and significant result grew out of these observations. One night, observing as usual with the 40-inch, he aimed the telescope at a small star in the constellation of Pisces (19 Piscium). The plate was a beauty. In it the magnificent lines of calcium stood out just as he had seen them in sunspots. He saw that, in this fourth-type group, he was dealing with a class of low-temperature stars that show certain marked similarities to sunspot spectra. How, he asked himself, could he prove this relationship? How could he gain the greater dispersion needed to identify individual characteristics in these lines? The answer, he knew, lay in a larger telescope, and one of a type different from that of the 40-inch refractor. He had long been making plans for such a telescope. Unfortunately, it was not yet ready for work.

The 40-inch was, of course, a magnificent instrument, unsurpassed for certain kinds of work. For Burnham's observations of double stars, for Barnard's measurements of faint satellites and other difficult objects, for Fox's photographs of solar prominences, for Ritchey's photographs of the moon and the photographic determination of star distances by Frank Schlesinger—for all these and many other observations it would continue to prove invaluable. Its "convenience of manipulation, surprising insensibility to temperature change, and large field of view" could not be shared by a reflecting telescope. But it had limitations that would not apply to a telescope of the reflecting type.

In a reflector, the light is not weakened by passage through glass; after reflection from the silver surface, all the rays, independent of color, are united in a common focus. In other words, a reflector is completely achromatic; it can also make use of all available light. (In a refractor, on the other hand, the lens forms a series of images corresponding to the light of different wavelengths.) As Agnes Clerke put it dramatically, "None of the beams they [reflectors] collect are thrown away in colour fringes, obnoxious in themselves and a waste of the chief object of the astrophysicist's greed—light. Therefore, they are in this respect especially adapted to photographic and spectroscopic work." [34]

Many of these advantages had already been shown by George

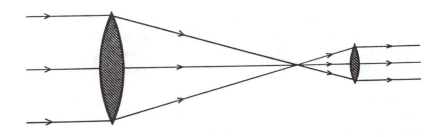

Principle of the refracting telescope.

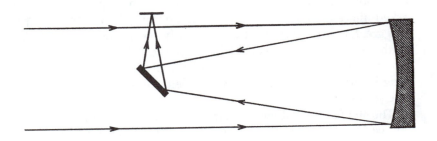

Principle of the reflecting telescope (Newtonian focus).

Ritchey with the 24-inch reflector that he had completed at Yerkes, especially in the photographing of faint stars and of nebulae. With the 24-inch, with an exposure of only 40 minutes, it was possible to photograph stars that lay at the extreme limit of vision of the 40-inch. With longer exposures, innumerable stars, never before photographed, could be recorded.

Long before the dedication of the 40-inch, Hale had been weighing all these advantages with his father, and had discussed ways of obtaining a large reflecting telescope. William Hale, always interested in his son's exploration of the universe, had agreed in 1894 to the ordering of a 60-inch disc from the venerable foundry at Saint-Gobain in France. By 1896, Hale was writing jubilantly, "The great 5 foot disc has arrived and seems to be a beauty." (It had been successfully cast only after numerous failures.) The immense disk, eight inches thick, weighed nearly a ton.

William Hale's gift included the cost of the mirror and the expense of grinding. Later he agreed to give the mirror to the university "on condition that a suitable building, with dome and telescope, mounting and all auxiliary apparatus be provided and that arrangements be made to keep the optical parts of the instrument in good condition." With characteristic modesty he did not want his name attached to the telescope. Instead he had stated that if someone was found who was "willing to give the money required, the telescope (or, if desired and considered necessary, the observatory, which may be regarded as distinct from the Yerkes Observatory) may be called by his name."

In a letter to Harper, Hale again stressed the advantages of a 60-inch reflector over the 40-inch refractor for certain kinds of observation. Again he emphasized the astronomer's eternal need for "more light": "Its first and most obvious advantage will be in the amount of light concentrated in the image of a star; more than twice as much as that given by the 40"." Then he described the type of mounting he proposed to use.[35] This would be built so that the star's image would be brought to a fixed point in a constant-temperature room, where it could be kept on the slit of a spectroscope as long as desired. "It will thus be possible to use with the telescope apparatus that cannot be employed with the 40", which is never at rest and must always be used in an open dome, where the temperature is constantly changing."

In photographing the spectrum of a star with the 40-inch, exposures had to be confined to a single night. With the new telescope, spectroscopes more powerful than any hitherto used would be possible. "Thus," Hale concluded, "the spectra of stars might be photographed on so large a scale as to permit the study of their chemical composition, the temperature and pressure in their atmospheres, and their motions with that high degree of precision which can now be reached only in the case of the sun." [36]

During the previous summer at Yerkes, E. F. Nichols had used a very sensitive radiometer with Ritchey's 24-inch reflector to detect a very small amount of heat from a star—a feat hitherto considered impossible. With the large aperture of the new reflector, these observations could be extended, "and important advances in our knowledge of stellar evolution should result." Finally, the new instrument would be beautifully adapted for photographing stars and nebulae—"work for which the 40" telescope was not designed."

Two days after receiving this glowing account, Harper wrote to accept the telescope for the university on William Hale's conditions.[37] But, as George Hale had soon discovered, the task of finding the money to mount the new telescope again fell on him. He approached two wealthy Californians, H. C. Durand and Griffith Griffith, who many years later would build the Griffith Planetarium in Hollywood. But in 1898 both men turned him down.[38]

In 1901, John D. Rockefeller arrived at Yerkes with the Ryersons and the Hutchinsons and a herd of reporters on their trail. The skies were cloudy. But fortunately the sun shone long enough for Rockefeller to see a small sunspot. "The clouds considerately moved away just as the millionaire placed his eye to the glass," a reporter noted. Afterward Hale led them down to the optical room where Ritchey was grinding the 60-inch mirror. Rockefeller was fascinated by the delicacy of the tests needed to find the accuracy of the mirror's surface. He was amazed to learn that the lump raised by the heat from the accidental touch of a finger to that surface could be detected instantly. He was incredulous when Hale told him that E. F. Nichols had detected the heat from a man's face at a distance of half a mile.

Hale, elated by the evident success of the visit, hoped the millionaire might offer to mount the 60-inch reflector in California. This hope, unfortunately, was not realized. Actually, Hale was destined to wait twenty-seven years for a Rockefeller gift of a great reflecting telescope. By that time the size of Hale's dream had increased to 200 inches and the cost to $6,000,000.

Therefore the 60-inch disk continued to lie unused in the Yerkes basement, while Hale continued to be frustrated in his quest for a solution to its mounting.

Meanwhile, on May 28, 1900, another chance for trapping the sun's corona appeared. A solar eclipse was to pass over the southern United States. Eager to see the corona at time of eclipse if he could not see it at any other time, Hale made plans to go to Wadesboro, North Carolina.

On the morning of May 28th the sun shone brilliantly on a field covered with men and instruments, surrounded by spectators. As the time of eclipse approached, an eerie stillness pervaded the air. Inside a small hut Hale was busy adjusting the bolometer with which he hoped to measure the coronal heat. His chief assistant was Harry Goodwin. On the way to Wadesboro his first bolometer had been broken, but with the help of the physicist C. E. Mendenhall he had rigged up an excellent substitute. All went well until totality. Then, at the very last moment, a stick was knocked over. The spot of light that was to measure the coronal heat went flying off the scale. Frantically, while the moon's shadow traveled slowly but inexorably across the sun's face, Hale tried to balance the bolometer. Though it was almost hopelessly out of kilter, he somehow managed to readjust it, and cried out to Harry, "Expose!" Harry obeyed orders, but it was too late. The sun was reappearing, and the world, which a few moments before had been shadowed in a weird light, shone once more in all its brilliance. The long-awaited eclipse was over. Hale's results were exactly zero—just as they had been whenever he had tried to get the corona before.[39]

Although disappointed, he consoled himself with the thought of the success of the other members of the party. He was especially delighted with the discovery made by Charles Abbot, Langley's assistant in the new Smithsonian Astrophysical Observatory, who found that the corona sends out a "measurable amount of heat"—though a very small measure.[40] Now that Hale had an idea of how much heat he had to deal with, he began to plan for an attack on the corona in full sunlight, using a much more sensitive radiometer. Once more he dreamed of success.

About four months after this eclipse, another event whose time could not be predicted so exactly took place. On the night of November 6th, the night of William McKinley's election, William Ellery Hale, named for his grandfather, was born to George and Evelina Hale.[41] Like Margaret, he was a frail baby. From the beginning he suffered from stomach troubles and other ailments that defied analysis. Soon his father was worrying about him as he had done about Margaret, and in time Evelina became as overconcerned as her husband.

At four, Margaret shared none of these fears. She was enchanted by the baby, and took him as her special charge. One day of her own accord she began to read aloud to him. No one knew where she had learned. Her father listened in amazement.

In 1900, three years had passed since the Yerkes dedication. The house on Drexel Boulevard had been sold, and the Hales had rented an apartment in Chicago. As a result, in winter Hale spent the weekdays at Williams Bay and Sundays with his family in Chicago.

The sight of his slight figure dashing down the hill at the last moment on his way to the train became a familiar one at the observatory. It was an unsatisfactory existence—for months he could never really share a normal family life. He felt constantly torn between his responsibility to his work and his devotion to his family. To visitors the arrangement must also have seemed odd. The Irish astronomer Robert Ball remarked in his diary, "Mrs. Hale was away with her children in Chicago, but a man and his wife, who also belong to the observatory, keep house for Prof. Hale." [42]

Long afterward Evelina Hale said that most of the responsibility of bringing up the children had fallen on her. Under the circumstances it is hard to see how it could have been otherwise. Yet Margaret had happy memories of those days in Chicago. Once they had rented a house belonging to the noted physiologist Jacques Loeb, which stood where the Divinity School would later rise. Here George Hale would put on magic performances when he was in Chicago. The stage was the top floor, and his accomplice was Dr. H. H. Donaldson's twelve-year old son. For the occasion, Hale, feeling like a boy again, arrayed himself in top hat and tails, while his accomplice was decked out in Chinese costume. Together they put on a magnificent show—to the joy of their small audience. [43]

CHAPTER SEVEN

Pioneer Days on Mount Wilson

✳ *1902-1904*

THE SNOW was falling. The sky was dark. The house at Williams Bay was gloomy. Hale was reading the Chicago *Tribune* when he came across a dispatch carrying a Washington dateline. The date was January 10, 1902. It was headed:

CARNEGIE TELLS PURPOSE OF GIFT

Outlines Plan for Institution
to which He Gave $10,000,000.

The story told of the founding of the Carnegie Institution of Washington "to encourage investigation, research and discovery in the broadest and most liberal manner, and the application of knowledge to the improvement of mankind." One of the institution's chief aims, said the article, would be to "discover the exceptional man in every department of study, whenever and wherever found, and enable him, by financial aid, to make the work for which he seemed especially designed his life work." [1]

As Hale read on, his excitement grew. Later he was to comment: "The provision of a large endowment solely for scientific research seemed almost too good to be true. Knowing as I did the difficulties of obtaining money for this purpose and devoted as I was to research rather than teaching, I could appreciate some of the possibilities of such an endowment." [2]

Some weeks after the initial story, appointment of an Executive Committee was announced. Daniel Coit Gilman, president of Johns Hopkins, was appointed president. Charles D. Walcott, director of the United States Geological Survey, was secretary. Other members included Elihu Root, Secretary of War in President Theodore

Roosevelt's cabinet; Dr. John S. Billings, noted surgeon and director of the New York Public Library; Dr. S. Weir Mitchell, doctor and novelist; and Carroll D. Wright and Abram S. Hewitt of New York. It was a powerful board with a pronounced scientific bent. Hale was never sure who had been primarily responsible for inducing Carnegie to found this institution devoted to scientific research. It was often claimed that Gilman and Billings had exerted the greatest influence. Yet, as he came to know them, he came to believe that Elihu Root's influence must also have been powerful. Like his father, Oren Root, mathematician and astronomer at Hamilton College, Root had a profound belief in the value of pure science. In addition, Hale was sure that in Pittsburgh, Carnegie must have been influenced by Brashear, who had once said, "If I could only get him interested in pure science—what a boon he would give it." [3]

From Brashear, too, Carnegie doubtless derived the belief in large telescopes that he expressed in *The Gospel of Wealth.* Referring to the Lick Observatory, he had declared, "If any millionaire be interested in the ennobling study of astronomy—here is an example which could well be followed, for the progress made in astronomical instruments and appliances is so great and continuous that every few years a new telescope might be judiciously given to one of the observatories upon this continent, the last being always the largest and best, and certain to carry further and further the knowledge of the universe and our relation to it here upon earth." [4]

If Carnegie really believed this, here, Hale felt, would be his chance to carry out that belief. Hopefully he sent Dr. Billings a statement on the 60-inch reflector (which, if it could be mounted, would be the largest in the world). In his plea he spoke of the forty-inch refractor, then stated: "It is quite possible that a larger refracting telescope will never be constructed. The limit of size in this direction seems to have been reached. In a refracting telescope the light of a heavenly body passes through the two lenses of the objective at the upper end of the tube, and in the present case traverses a thickness of over three inches of glass. For yellow and green light the loss occasioned by this transmission is not very serious, but blue and violet light, to which photographic plates are most sensitive, is very greatly weakened, while for the no less important rays of the ultra-violet the glass is impervious as so much steel. For bright objects, such as the central parts of the Great Nebula in Orion, this loss may be tolerated, and important advantages indeed result from the great length of the tube and the consequent large scale of the photograph. But for photography of faint objects, whether they be stars or nebulae, such an instrument is not suited. Indeed, even this immense telescope is far outdone by the two-foot reflecting telescope recently constructed in the instrument shop of the Yerkes Observatory." [5]

In order that Billings and the Carnegie trustees might understand clearly why the powers of a much greater telescope of the reflecting type were needed, Hale explained how it would differ from the more conventional refractor: "The star light passes through no glass, but after entering the open tube falls directly upon the silvered upper surface of a mirror which lies at the bottom of the tube. The image formed by this mirror after reflection to one side at the upper end of the tube, falls directly upon a photographic plate. There, through the use of suitable mechanisms, it is maintained in a fixed position for as many hours as may be needful."

To illustrate his point he enclosed a photograph of the Great Nebula in Orion taken by Ritchey with the 24-inch reflector. It was on a much smaller scale than a similar photograph taken with the 40-inch. Yet the fainter regions of the nebula, which could not be seen when photographed with the 40-inch, were beautifully shown. A more striking case appeared in the Great Nebula in Andromeda. "The photograph taken with the refractor shows only the very center of the nebula, while the result obtained with the reflector gives the best representation of an early stage in the evolution of a system inconceivably greater than that of our own Sun." Millions of stars too faint to be seen or photographed with the 40-inch could be easily photographed with the smaller reflector.

If, therefore, the possibilities had been proved great with the 24-inch, the possibilities with the 60-inch reflector appeared illimitable. "It is easy to see," Hale exclaimed, "that a large reflecting telescope would give extraordinary results, and greatly advance our knowledge of the universe." [6]

Together with this ardent plea, he enclosed letters of recommendation from astronomers the world over for this telescope which Huggins called the "telescope of discovery of the future." For many years astronomers had argued over the relative merits of refracting and reflecting telescopes. In the case of the Lick Observatory, the advocates of a large refractor had won out. Now, as a result of Ritchey's and Keeler's successes with reflectors, the majority of astronomers had changed their minds—at least for photographic and spectroscopic observations of stars and nebulae. Enthusiastic endorsements of reflectors came from leading astronomers, physicists, and especially spectroscopists. Robert Ball in Cambridge reflected the general view in his ardent support of Hale's plans: "I am but repeating what every one knows when I say that you and your staff compose the greatest astronomical observers in the world. You have already given such magnificent proof of your energy and capacity that there is not the least doubt that if the field now open for a reflector be occupied by an adequate installation under your control, its fame will shortly rival that of any scientific undertaking ever yet known." A. Belopolsky in Pulkowa, Russia, wrote with equal enthusiasm: "I can only say that after the striking discovery of the variable nebula surrounding Nova Persei, it is certainly to reflecting telescopes that we must look for the future progress of science. I await with impatience the provision of a mounting for your great mirror." Moreover Belopolsky, with the majority of astronomers, urged that the telescope be set up on a mountaintop in a clear atmosphere where the effects of the blanket of air above us would be least felt.

As Hale sent on all this convincing evidence of the need for his 60-inch telescope, he asked Billings if there "would be any reasonable chance of securing the funds from Mr. Carnegie. If so, will you be kind enough to give me the benefit of your advice as to the best manner of getting the question before him?"

Soon after this, while he awaited an answer, a letter arrived from E. C. Pickering of Harvard. As chairman of the Advisory Committee on Astronomy (formed to recommend how the Carnegie Institution was to allocate its funds for astronomical research) he invited Hale to serve as a member. The other members were Samuel P. Langley, now Secretary of the Smithsonian Institution, whom Hale had first met many years before at Allegheny; Lewis Boss, director of the Dudley Observatory in Albany; and the former director of the Nautical Almanac Office, Simon Newcomb. All were leading astronomers; all had made notable contributions in their divergent fields. But, as Hale soon realized, it would be difficult to reconcile their different outlooks. Newcomb was famous for his contributions to theoretical and mathematical astronomy. To him, as to Boss, a knowledge of the positions, distances, and motions of the stars was of paramount importance. To them the chief instruments needed in an observatory were a telescope and meridian circle, a clock and a micrometer. They could never overcome their suspicion of that "new-fangled" thing—the spectroscope, which provides a key to the understanding of the physical nature of the heavenly bodies, and makes it possible to weigh the stars, to find their chemical composition and to discover something of their life histories. These representatives of the "Old Astronomy" would never really understand Hale's overwhelming interest in the "New Astronomy," or astrophysics—in which the spectroscope, once a minor accessory of the telescope, would eventually gain a place as the primary instrument, with the telescope as its indispensable auxiliary. Even E. C. Pickering was primarily concerned with a study of stellar brightness and of stellar spectra on a statistical basis. This meant that Hale was the only member who was vitally interested in the physical nature of sun and stars, and in the development of instruments and methods for their detailed study. The scales were obviously weighted heavily against his outlook. Moreover Langley, who had done valuable work on solar radiation in the past, was now preoccupied with that other "new-fangled device" the airplane, and had almost abandoned the sun. Newcomb had even less use for the airplane than he had for the spectroscope. "Like the mathematician's efforts to square the circle, to trisect the angle or duplicate the cube, this effort to make man fly is doomed to failure," he averred. As usual, when the old and the new, the traditional and the "revolutionary" clash, explosions were inevitable.

The committee met in New York in the middle of a heat wave. Its first task was to survey the field of astronomy and to recommend a number of small grants. It was stated that "large and expensive projects are beyond our scope." This ruled out the mounting of the 60-inch. Yet, as Hale soon learned, the secretary of the Executive Committee, Charles D. Walcott, had larger ideas. On a day in August, when he was back at Yerkes, deep in his work on the sun, the tall and impressive Walcott appeared at the door unannounced. Walcott, a paleontologist with an overwhelming interest in fossil trilobites, was as far from Hale in his scientific interests as possible. Yet, as he showed him around, he sensed that Walcott was as aware of the value of great telescopes for the study of macroscopic realms as he was of the importance of microscopes at the other end of the universe.

It was important, Walcott said that day, to present great as well as minor problems to the institution, no matter "whether the amount involved were a hundred thousand dollars or five hundred dollars a year." He had, he said, discussed the question of a large observatory with Carnegie himself. Carnegie had agreed that, if in the astronomers' opinion, such an observatory could obtain results as important as those gained by the Lick or the Allegheny Observatory, he would provide the necessary funds "even if they amounted to as much as three million dollars." In fact, Walcott noted, "he showed the keenest interest in the observatory and dwelt particularly upon the fact that he had always been especially interested in astronomical work." [7] To Hale this was marvelous news. It was what, remembering Carnegie's *The Gospel of Wealth*, he had hoped.

He was overjoyed, too, by Walcott's evident enthusiasm for the 60-inch disk and his support of the plan to mount it. Yet he knew that some of the more conservative members of the committee (particularly Boss) might not share Walcott's enthusiasm. To Boss, the thought of mounting a great reflector on a distant mountain peak seemed nothing but a flagrant waste of money. When Hale talked of using a reflecting mirror with a large Rowland concave grating to increase the dispersion, the whole business seemed to him not only incomprehensible but also quite mad.

As Hale tried again and again to explain his reasons for wanting to mount the 60-inch reflector, letters flew back and forth between Williams Bay and Albany. At times Hale was amused, often discouraged, sometimes downright disgusted. At the end of August, a revised report arrived from Boss. Hale saw that his efforts had been useless. He wrote back heatedly, "You cannot expect me to join in damning the instrument with the faint praise you have accorded it." [8] Still, in view of this attitude, he decided to eliminate his request for the 60-inch from the report. And here, for a while, the matter rested.

Soon afterward a grant had been made to Boss, Hale, and W. W. Campbell of the Lick Observatory to "investigate the proposal for a southern and solar observatory." [9] This grant, as it turned out, laid the groundwork for future consideration of the various projects. W. J. Hussey of the Lick Observatory was chosen to make an extensive search for sites, first in the United States, then in Australia and New Zealand. He started out in 1903. In the months that followed, Hale traced his journeyings eagerly. He read with enthusiasm his reports on atmospheric conditions in Southern California—from Mount Lowe and Wilson's Peak above Pasadena, down to that vast unknown country to the south that was accessible only on horseback. Years later he would have cause to remember Hussey's poetical account of Palomar Mountain, "a hanging garden above the arid lands." Even Boss was so impressed by Hussey's glowing account of this "paradise" that he was all for investigating it further. But Palomar was isolated. It was eighty miles (as the crow flies) south of Pasadena, and much farther by road. It could be reached only over a very rough and steep trail. There was no regular stage, no telephone connection. As a result, while he was entranced by the surroundings, Hussey concluded that Wilson's Peak remained the best site he had seen.[10] (See map, page 418.)

Hale's interest in this peak had first been roused years before. In January, 1889, when he was working at the Harvard Observatory, an expedition had been sent out from Harvard to explore the possibility of establishing an observatory there. Its object was to find a site for the mounting of the famous 40-inch lenses that had eventually found a home at Yerkes. From Pickering, Hale had learned many of the details of the plan, and the name of Wilson's Peak and its climatic advantages remained stored in his mind. (Another detail that stuck in his memory was the expedition's comment on the prevalence of rattlesnakes.[11]) In 1896, when his father had given him the 60-inch disk, his first thought of a site for that telescope was Wilson's Peak.

This peak is one of many that form the southern boundary of the Sierra Madre range, and this, as he came to know it, is the way Hale described it: "Standing at a distance of thirty miles from the ocean, it rises abruptly from the valley floor, flanked only by a few spurs of lesser elevation, of which Mount Harvard is the highest. Except for a narrow saddle, Mount Wilson is separated from Mount Harvard by a deep cañon, the walls of which are very precipitous. Farther to the west, beyond the saddle leading to Mount Harvard, the ridge of Mount Wilson forms the upper extremity of Eaton Cañon, which leads directly to the San Gabriel Valley. East and north of Mount Wilson lies the deep cañon through which flows the west fork of the San Gabriel River, and beyond this rise a constant succession of mountains, most of them higher than Mount Wilson, which extend in a broken mass to the Mojave Desert. The Sierra Madre range forms the northern boundary of the San Gabriel Valley, which is further protected toward the east from the desert by the high peaks of the San Bernardino range." [12]

Now, on a glorious June day in 1903, he arrived in Pasadena to explore the site for himself. He was joined by W. W. Campbell from Lick. Following Hussey's directions, they sought out Charley Grimes, the proprietor of a restaurant on Fair Oaks Avenue, who was to supply them with burros. (Hussey had written, "Take a Santa Fe railroad train to Santa Anita station; a bus from Santa Anita to the foot of the trail to the summit; the ride on the burro will take about four hours.") [13]

Long before dawn the following morning Hale woke in his room at the Hotel Guirnalda. All he could see in the dim light was a dense bank of fog. There was no sign of the great San Gabriel range. His spirits fell. He doubted if it was worth starting out. Still, he decided to rouse Campbell, and soon they were on their way to catch the train for Santa Anita. There, at the yellow, red-rimmed station, they were met by an imposing individual named Twycross, who sported a black moustache and wore a straw hat of "ancient vintage." He reached out a shirt-sleeved arm in welcome and led them to his extraordinary black bus that resembled a dilapidated butcher's cart. It had two large wheels in back and two smaller ones in front. It was drawn by two horses of "dubious pedigree" and "uncertain age." On the top and along the sides were the alluring words:

SIERRA MADRE AND WILSON'S PEAK

Mr. Twycross climbed up to the roof-covered ledge at the front, Hale and Campbell climbed into the rear, and with a flourish of the whip they were off on the two-mile uphill jaunt to the foot of the Sierra Madre trail. The whip had a stick two feet long and a lash of equal length, with which the driver flicked his ancient steeds. All the way he kept up a running fire of news and rumor for his passengers' benefit. Highly amused, they were almost sorry when they reached the foot of the trail. [14]

There they were diverted by the sight of two sad-faced burros, whom Hale would come to know intimately. Feeling like a boy on his first camping trip, he mounted his burro. Campbell, too, was soon mounted. Up the Little Santa Anita Cañon they plodded through the fog that still hung darkly over the valley. Hale was depressed by the gloomy skies, but Campbell, used to the fog at Lick, tried to reassure him.

For some time the fog persisted—until they reached a ridge above the valley. Then, suddenly, a beautiful blue sky burst out on Hale's delighted gaze. It was a moment he never forgot. Years later he described it to his colleague Harold Babcock, who used the incident in his poem "In 1903."

"Across the miles of transparent air" they could see Old Baldy:

> So sharp, clear-cut, distinct, it seemed to them
> Impossibly far away as told
> By maps.

They climbed on and discovered that the rocks were hot—sure indication that the sun had been shining on the ridge the morning long. Their enthusiasm grew. From Martin's Camp, looking back over the valley, they saw the fog slowly vanishing. Still they climbed:

> Through oaken thickets splashed with clematis
> The trail led on—a short remaining span—
> To reach the summit of the peak at last.[15]

On the peak they found Hussey with his nine-inch telescope. In the distance lay Catalina Island, a silver band on the horizon in the blue Pacific. To the east rose Old Baldy. Beyond towered the twin peaks San Gorgonio and San Jacinto, still guarding their wintry snows. Overhead the sun continued to shine.

It was midafternoon. They scattered to explore the mountaintop. Hale headed east toward Echo Rock, a huge outcropping seemingly poised on the edge of the world. For a long time he stood gazing out. In Babcock's words, "a curious homely feeling of content" came over him. After a while he wandered on. Through a manzanita thicket he pushed his way until he came to a promontory he would later call Monastery Point. He looked back at the plateau covered with live oak and big-coned spruce. He gazed down into the canyons and ravines, filled with cottonwood and maple, alder and spruce. Again he thought what a magnificent place this would be for his observatory. The luxuriant growth would shield the ground, thereby reducing the radiation "so injurious to the definition of the solar image." He could not imagine a more perfect site.

That night they tested the seeing on stars that glowed steadily in the dark night sky. But it had been a long day, and at an early hour Campbell turned in. Hale promised to follow soon. Instead, too excited to sleep, he wandered off among the great ponderosa pines to gaze once more at the brilliant stars overhead and the twinkling lights of Pasadena and Los Angeles scattered over the valley below.

The next morning Campbell roused him early to see the rising sun. He jumped up and dashed over to Hussey's telescope to examine the sun's image. It was just after six. The day was magnificent. With the telescope Hale could see sunspots and the granulation on the sun's surface as vividly etched as in a steel engraving. Many of the finest details he had observed only on the clearest days at Yerkes were visible. He was elated.

That afternoon he set off on another exploratory expedition. First he investigated a broken-down log cabin, called the Casino, with an eye to its use in case the solar observatory plan should go through. Farther afield he discovered the region known as Barley Flats where, as Hussey told them, bands of horse thieves had once hidden their booty. From Hussey, too, they learned that, according to legend, the first white man to climb the mountain was the colorful Don Benito Wilson. It was for him the mountain was named.

The following day, a Sunday, the representatives of the Mount Wilson Toll Road Company, owners of Wilson's Peak, arrived. The party included William R. Staats, who was obviously eager to push the project, and J. H. Holmes, manager of the famous Hotel Green in Pasadena. With them was Hussey's new friend and adviser T. P. Lukens, of the United States Forestry Commission, who knew more about the California mountains than anyone else. Again they tramped over the mountaintop, and Hale jotted down in his diary information on the contours, the vegetation, the possibility of getting materials up to the site, the vital problem of water supply, the question of running an electric power line up the mountain. One notation reads: "Mt. Wilson Toll Road Company owns top of Wilson's Peak. Toll Road is about 8½ miles long—a rise of 4500 feet. This is a trail and would cost about $40,000 to make into a good road about 9½ miles long. Grade would not exceed 10%. This would come out in Eaton Cañon, about 5 miles from center of Pasadena." [16]

At first, Staats and Holmes seemed willing to offer any site on the mountain. Later, difficulties arose. But nothing could quench Hale's ardor. He seemed willing to agree to any proposition, make any concession if he could only build his observatory on Wilson's Peak. Campbell had returned to Lick, but Hussey watched the negotiations in astonishment. "You had not reached the foot of the trail," he told Campbell, "before Hale was ready to give up much that (I am sure) you would have striven to retain, and that without any particular pressure from the owners of the land. Indeed, to give it up appeared to be Hale's notion and not theirs." [17]

The next day Hale reluctantly started down the mountain, following the Toll Road, which zigzagged down into Eaton Cañon, about six miles north of Pasadena, to the west of the old Indian trail. Soon after his arrival in Pasadena, he headed back to Chicago. On the train his mood changed. In his first flare of enthusiasm everything had appeared perfect. As the California Limited moved through the valley below Wilson's Peak, he looked up and began to worry. Had he been overoptimistic? He wrote to Campbell from the train, "In the long hours of reflection last night and again this morning the great responsibility of our position forced itself upon me, and convinced me that our search for sites has not yet been sufficiently complete." [18]

Before mailing the letter, he telegraphed to Campbell. Campbell sent the telegram on to Hussey with Hale's letter. Hussey, perturbed by Hale's vacillation and evident ignorance of mountain conditions, was outspoken in his criticism: "It appears that Hale's characteristic is to seize one idea and forget all the rest; to oscillate from one point of view to another. One day he thinks Mt. Wilson the best site in the United States—the next he wants Flagstaff tested; and a day later after meeting Oatman on the train, he thinks San Bernardino Mountains fulfill all conditions—forgetting that we have discussed these very matters in detail. These considerations make me less moved by his telegrams. When final arrangements down here are to be made, I would suggest the desirability of your being present—the interests involved are so great they require your clear head and steady judgment." [19]

Apparently Campbell shared Hussey's opinion. He wrote confidentially to Boss: "We were on the mountain three and a half days. Every day, we called attention to the dust. . . . But Hale was charmed with the place and constantly said that this must be the best place in the world. While I was packing my grip he wrote the telegram and signed both our names to it, which I sent you from L.A. His telegrams of dissatisfaction came to me two and three days later without warning. It's a good sign to change one's mind occasionally. But I must say the circumstances this time rather affected me." [20]

Meanwhile Hale, unconscious of Hussey's and Campbell's criticism of his mercurial nature, was nearing Chicago. He was again optimistic. In cheerful mood, he sent John Billings his first impressions. He wrote extravagantly of the accessibility, unlimited supplies of water, excellent building stone and firewood on Wilson's Peak. Comparing the night seeing with that on Mount Hamilton, he exclaimed, "There is certainly no observatory site at present known which seems to offer advantages equal to those found at Wilson's Peak." [21]

Still, to guarantee the wisdom of this choice, Hussey continued his journeyings through the west, then set sail for Australia and New Zealand.

On October 3rd, the Executive Committee on Astronomy met in Washington, and the report of the Advisory Committee on Astronomy was submitted. As a result of Hale's urging, it included a recommendation for a high-altitude solar observatory where the 60-inch reflector could be mounted, and also for an observatory in the Southern Hemisphere. However, there were eighteen other advisory committees in as many different fields, including geophysics, archaeology, nutrition, physics, and chemistry. The representative in each field naturally considered his project the most urgent. The committee expressed interest in Hale's solar observatory, but regretted that with the funds available it could not be undertaken. It recommended that the question be referred to the trustees at their meeting in December. Meanwhile Billings asked Hale whether, if the entire amount requested could not be granted, a modification of the plan would be possible. Hale, defending his plans for the 60-inch, said he doubted if any investment the Carnegie Institution could make would result in more important advances or have greater influence on telescope construction and the development of future instruments.

Soon, however, he became worried that he had been too emphatic. Five days later, therefore, he wrote again: "The essential thing is not the establishment of a large and pretentious institution, but the accomplishment of certain definite pieces of work under improved atmospheric and instrumental conditions." [22]

Still Billings hoped that Carnegie might increase the endowment "so that we can carry out both observatory plans." [23] He promised to see Carnegie before the meeting. This, at least, was hopeful. With the powerful support of Billings and Walcott, there appeared to be a good chance of convincing Carnegie of the importance of a solar observatory. Hale, however, was not aware that opposition continued to exist among other members of the Advisory Committee. Boss, still unable to get over what he considered the grandiosity of Hale's plans, was critical not only of him but also of his supporters. He wrote confidentially to Campbell, expressing his view that although they had been able to cut Hale's original estimate, they should have "stuck" for much smaller amounts.

Campbell, too, was skeptical. He confided to Boss that Hale had apparently set his heart on calling this new institution an "astrophysical" observatory. "My impression," he noted, "is that we should be slow in opening up the entire field of astrophysics." [24]

At this point Hale saw there was little he could do to push the project until the trustees met in December. Still, unable to wait, he decided to go ahead to California and gamble on his own. On Mount Wilson he could test the conditions further. If possible, he would take the Snow horizontal telescope with him. His primary concern was, of course, the observatory and his hope of eventually mounting the 60-inch. Yet, for months past, Margaret's bronchial asthma had been so bad that at times he had despaired of her recovery. California again might be the answer. He arranged to leave in November. At the last moment he himself became ill. So, sending his family ahead, he waited in Chicago for news of the trustees' decision.

December 8th came. All day he waited for the doorbell or the telephone to ring. He waited for three more days before a telegram came from Billings. He tore it open anxiously and read it, unbelieving: "No result as regards solar observatory. Will write." [25]

Finally a letter of explanation arrived. Before the meeting Billings had talked with Carnegie, and urged support for the solar observatory. "He listened and appeared interested . . . but gave not the slightest hint as to what he would do in the matter." At the meeting, it became clear that in view of other projects the idea for a large observatory must be discarded unless a special gift were made. Carnegie offered no such gift. At dinner he asked why the work proposed by Hale could not be done at Yerkes. Charles Hutchinson, a trustee of the institution as well as of the University of Chicago, described the need for a clearer atmosphere, then "expressed himself strongly as to the desirability of having the solar observatory under your direction, although they would be sorry to lose you in Chicago."

"I am sorry," Billings exclaimed, "that I have not better news for you, but looking back on the matter I do not see that I could have done any more towards securing what you desire. Instead of $150,-000 for the use of observatories etc., the total sum appropriated, I think, was $15,000."

The outlook was as gloomy as it was unexpected. As Hale read the letter, he realized how totally unprepared he was for this negative decision. In bed that night he pondered his next move. Long afterward he was to write of this turbulent period, "My future plans were far from clear." [26] He decided that the "first obvious step" was to join his family in Pasadena. The lure of California was strong. He could not forget "the great possibilities of research with new and powerful instruments in such a climate." There was always the hope, too, that Carnegie might change his mind.

On December 18, 1903, therefore, he set off for the West. "I am practically on an expedition from the Yerkes Observatory." But, at the moment, there was no financial support for such an expedition, except what was coming out of his own pocket. He left Edwin B. Frost in charge as acting director.

In the weeks of waiting he had made extensive notes on things to take. These included a Casella sunshine recorder, a Richards wet and dry bulb thermograph, an American anemometer, and (recalling Pickering's ominous report on rattlesnakes on Wilson's Peak) an antivenom remedy prescribed by the Pasteur Institute. He also listed "pillow slip in bag, colored shirts, Macintosh, Frock Coat, Stationery, Gun, Fishing Tackle." [27]

On December 20th, just six months after his first visit to the mountain, he arrived in Pasadena. The sky was a brilliant blue, the sun as warm as on a summer's day. On the station platform he found his family waiting, dressed in summer clothes, startlingly different from the wintry garb they had worn when he had seen them off in Chicago. He rushed up, and as he hugged them his spirits bubbled over. The future might be uncertain, but now he was sure he was doing the right thing. No wonder that "depression quickly vanished." [28]

From the station they drove to the charming inn called La Solana. As with everything else in this enchanted place, Hale fell in love with their shingled bungalow, surrounded by a garden rampant with flowers. A few days later, he wrote ecstatically to Goodwin: "Here I am sitting in the shade of the porch of our cottage with the sun too hot to stay in, the birds singing around me, and the flowers so numerous that I can't begin to tell their names. William is out in his cart, and Margaret has just run away to join him. Both are bare-headed and wear no jackets. I haven't been here long enough to be in the least blasé, and I can't say enough of the beauties of the place and the climate. The orange trees in the yards around us are full of fruit. One of them, so heavily loaded that it seems ready to break, is a mass of oranges and roses, a great rose tree beside it thrusting up bunches of flowers in the midst of the fruit. I would give anything if you could be here to enjoy it all." [29]

He wrote, too, that he had renewed hope that the observatory plan might still be carried through. A recent letter from Horace White, editor of the New York *Evening Post*, trustee of the institution and Uncle George Hale's brother-in-law, had been encouraging. During a golf game he had asked Carnegie about the observatory prospects. "He spoke of it with decided approval," White reported, "but said that 'we' couldn't do everything at once. Alexander Agassiz must be helped first. The solar observatory would probably be the next thing on the list. . . ." [30]

About the same time, Charles Walcott, fully aware of Carnegie's belief in the "exceptional man," had written to the magnate: "In Prof. George E. Hale you have an exceptional man to take charge of the work. . . . The productive period of every man's life is limited. Hale has reached a point in his development where his services should be utilized as soon as possible. Will you not take the matter up with a view of providing the endowment for carrying this work on. . . ." [31] Carnegie's reply to Walcott's plea was short, direct, and discouraging. "My impression is that the Carnegie Institution has quite enough on its hands for the present." [32]

Unaware of this exchange, Hale in Pasadena continued to revel in his new surroundings. Soon after his arrival he toured the town with his wife. They rode along the magnificent Orange Grove Avenue, "one of the most stately boulevards in the world," past the sumptuous millionaires' mansions, with their great carriage houses. They gazed at the estates of Adolphus Busch and T. S. C. Lowe, the latter the fabulous balloonist of Civil War days, who had built the Mount Lowe Railway and Lowe's Opera House. They climbed Raymond

Hill to the huge and lavish hotel at the top, built by Walter Raymond, leader of the famous excursions that had transported so many Easterners to this "paradise of the west."

In contrast to Chicago, Pasadena was a young town. A little over twenty-five years earlier, the first band of colonists had arrived from Indiana. Many of the streets were unpaved. Often the dust lay inches deep and had to be sprayed from watering carts whose drivers were paid a dollar a day. Yet, with its 25,000 inhabitants, Pasadena was reputed to be the wealthiest town of its size in the world, and its boosters dreamed of unlimited expansion. Why, they asked, should anyone want to live anywhere else? In this attitude there was a certain suffocating smugness, but of this Hale at the moment was unaware. Nor could he foresee that in this fertile soil he would create a great cultural center and carry out some of his fondest dreams.

Each morning he got out of bed and looked eagerly up at Mount Wilson, eight miles away, capped by sunshine. In December the June fogs that had so distressed him rarely appeared. "Atmospheric conditions," he wrote to H. F. Newall in England, "are so important that I would give up my position at the Yerkes Observatory and commence all over again in a log cabin on this mountain, if I could only reasonably take the time as matters stand." [33]

This letter was sent on a day after he had climbed the mountain armed with a three-inch telescope and accompanied by Seward Simons, a young boy of fourteen who lived with his mother in a cottage at La Solana. As they started out, Hale was in a gay mood, and Seward soon became infected with his spirit. On the first day they reached Martin's Camp on a saddle between Mount Harvard and Mount Wilson, and spent the night there. The next morning they were up before sunrise; one carried the tube, the other the tripod. As the sun rose, Hale swore that "by the seven saints of Tyre" he had never seen such a magnificent morning. When they reached the peak they set the telescope up near the old log cabin—the Casino.[34]

This was the first of many expeditions young Seward was to make. In later years he recalled airs from Schubert's "Unfinished" and Beethoven's Fifth and Seventh symphonies that Hale sang as they rode up and down the mountain trail. He had, Seward noted, "a great deal of music in him." From the beginning he treated the boy as an equal, even asking his opinion on astronomical questions. In

retrospect Seward concluded that Hale was "the most vivid character, the most interesting personality I have ever known. . . ."

Seward's most graphic memory, however, was of that first expedition when Hale was making observations of the sun at various
heights above the ground. Before he knew it he was scaling a yellow
pine, dragging the telescope with him. In his diary Hale recorded
the observations. At 10:55 A.M., with clouds forming to the west and
south, he noted: "Definition poor at tree . . . Tested seeing in Tree
at 32 feet and 68 feet above ground. Apparently some improvement,
but found to be better also at ground just afterwards. Tree meas. 80
feet." [35] Perhaps he already glimpsed the possibilities of a telescope
raised high above the ground where heat waves would not distort the
sun's image.

That first night they stayed in the Casino. Through a gaping hole
in the roof they could gaze up at the shimmering stars. For a while
they talked; then, worn out by the day's activities, they slept—to
dream perhaps that the 60-inch telescope was mounted and the
Mount Wilson Solar Observatory a reality.

The following day, after a hurried trip down the old trail, Hale
stopped to talk with the owner of the "corral" about his "animals."
"Driver," he recorded in his notebook, "said unbroken burros cost
about $15—those broken to trail, $35. Some are kept at foot of trail
at 15 cents per day—10 feet extreme length to pack up." On a little
card marked "Bassett and Son," he scribbled other things about the
indispensable burros:

> Hay about 2¢ a lb. at top.
> Grain about 2¢ a lb. at top.
> Burros need about 100 lbs. a week of hay and grain together.
> Burro cost about $25 with saddle, pack saddle, paniers, rope.
> Bassett & Son have 4 year lease of everything, road etc.

If Hale had any doubts about the relative merits of broken and
unbroken burros, such doubts were soon dispelled. One day they
loaded the burros with equipment as usual. One was transporting a
valuable diffraction grating on his back. He started out, but soon
tiring, lay down and rolled over. The grating was smashed to
smithereens.

Back in Pasadena after this expedition, Hale plunged into plans
for setting up a station on the mountain. If he could get good re-

sults, it might help to convince the Carnegie trustees of the value of building an observatory on Wilson's Peak. He also decided to return to Yerkes for the summer. This news was greeted with joy by his friends there. When he was with them, the work hummed; when he was away, a curious stillness fell over the place. Adams,[36] Barnard and others had all written of the state of gloom that prevailed in his absence. They complained of the miserably cold winter, the heavy snow, the bitter winds—a strong contrast to the conditions Hale was enjoying in California.

"It's pretty lonesome business going over to your house now-a-days," Storrs Barrett wrote, then added, "I found Dodge [Hale's man of all work] wandering around like a lost soul, when I went for the book." [37] In fact, he never did get over the feeling of "lonesomeness." Five years later, when the Hale family had moved permanently to California, Barrett recalled nights at Yerkes when Hale, sitting by the fire, had read aloud from Keats's poems. "We read them to ourselves but it is long since the Reader read them to us, sitting by our fireside. Welcome would he be in our snow-bound cottage with a singing wind blowing outside and the white moon casting long shadows among the trees—the kind of night for St. Agnes Eve." [38]

Meanwhile, in that winter of 1904, on February 29th, Hale climbed Mount Wilson with a carpenter named Britton. They spent the night in the Casino. "Plenty warm enough but *some* drafts." [39] In the morning they set to work on repairs that would make the cabin usable as a shelter until a more permanent structure could be built. That afternoon Hale started down the trail. On the way he "shot the slides" that were cut through the heavy covering of chaparral as shortcuts between zigzags of the trail. He arrived at the foot of the trail covered with dust. There he picked up his bicycle and rode it back to Pasadena.

That evening he donned his frock coat to speak on the work of an astronomer at the home of Mrs. Burdette, the "cultural leader" of the community. The following day he again climbed the mountain, this time with George Jones, a genial stonemason and an enormously powerful man who was to become as legendary as Paul Bunyan for his feats on the mountain. To Jones the huge granite blocks on the peak seemed like mere pebbles. He picked them up with complete nonchalance while Hale looked on in amazement. It soon appeared that he could also do almost anything with his hands, and was remarkably skillful as a designer.[40]

With such expert help, the wrecked log cabin was rebuilt. By the time the first heavy storm struck, it was not only habitable but cozy. "When real winter descended with its furious blasts of snow," the huge fireplace, able to hold logs two feet in diameter,[41] saved their lives. It even had an embryonic library containing Agnes Clerke's *History of Astronomy*, her *Problems in Astrophysics*, and a volume of D'Annunzio's poems.[42]

Soon, as a result of this activity, rumors began to leak out. One day a reporter appeared at the bungalow the Hales had rented on Palmetto Drive. Anxious to keep Carnegie's name out of the papers, Hale said he had no information to give, no report of definite plans on Mount Wilson. This was literally true, as there was still no concrete support for an observatory there. This was still only an expedition from Yerkes. Still, those who watched the procession of animals carrying lumber, tools, and equipment up the trail must have wondered.

During these early days, and far into the future, burros and horses were the only means of transportation on Mount Wilson. Despite the burros' innate stubbornness, Hale was entranced by these perverse animals and developed a curious fondness for them. From the beginning they were as indispensable to the building of the observatory as any astronomer. Their unexpected antics added to the gaiety of existence. "Books," said Walter Adams, who was soon to make their acquaintance, "could be written about the personal characteristics of these sagacious beasts and the infinite variety in their individual behavior. One would deliberately expand his chest when the saddle was placed upon him so that the rider, after a good start, would presently find the saddle rolling beneath him at some awkward point in the trail; another would groan heavily when the grade became steep, but if the rider once dismounted he would be fortunate to overtake his mule within several miles; while still a third would show an irresistible desire to roll over, frequently selecting a stream bed for this purpose." [43]

The load capacity of the animals, varying from 150 to 200 pounds, limited the weight of materials that could be transported. For the same reason, no board could exceed eight feet in length. When the mules were heavily loaded, it was a slow and tedious trek up the trail. The packtrain would start in the early morning; before it arrived at the top the sun had often set. Then they would either unload quickly and descend immediately or spend the night on the mountain. All this required organization and manpower. It also required funds, most of which continued to come out of Hale's own pocket.

From the beginning he entered with relish into this pioneer life. To those who joined him on the mountain later, his ability to adapt to the primitive conditions came as a surprise. "Apparently," Adams writes, "combined with a deep-seated love of nature in every form was the spirit of the pioneer, whose greatest joy is the adventure of starting with little and taking an active personal part in every phase of creation and growth."

In the first week of March, 1904, Hale made three round trips to the mountain. He walked up the trail on Monday, down on Tuesday, up on Wednesday, and down again on Thursday (just in time to miss a gale that nearly blew the Casino off the mountain), then back up on Friday.[44] On the top he worked harder at pure physical labor than he had for years. Yet, if the work was strenuous, it was exciting. As he wrote to Goodwin, "There is a lot of exhilaration in tackling the work on Mt. Wilson with only the few tools I have plus future possibilities."[45]

CHAPTER EIGHT

Founding of the Mount Wilson Solar Observatory

✳ *1904*

IN THE PRIMITIVE surroundings of Mount Wilson, news from the outside world seemed like word from another planet. One such item arrived in January, 1904.[1] It came from W. H. Christie, the Astronomer Royal of Great Britain, and it announced the award of the Gold Medal of the Royal Astronomical Society to George Hale "for his method of photographing the solar surface and other astronomical work."[2] At the time, the recipient was helping to patch the windows and repair the roof in the old Casino. London and the civilized life of Burlington House seemed far away. "I must say," he wrote, "that as my work has just begun, I would have been content to wait for many years before receiving so great an honor. But since my good friends in England have been so kind as to give this official recognition of my work, I can only attempt to justify it by carrying on what I have been doing. . . ."

At the request of the president of the society, H. H. Turner, he took time off from his labors on the mountain to write an account of his life and work that the English astronomer could use in his address when the award was made in February.[3] While he could not be there to receive the award, he read Turner's address with the greatest interest when it appeared in the *Monthly Notices* of the R.A.S.[4]

The address began with a familiar quote from Hale himself. "It cannot too often be repeated that the Sun is the only star whose phenomena can be studied in detail." Turner went on: "We all assent at once not only to the proposition itself, but to the obvious corollary that particular attention should be paid to solar phenomena, and yet it would appear from the history of astronomy, that the statement, however often repeated, has fallen on deaf ears. . . ." As he reviewed the recent history of solar research, he showed vividly how Hale, more than anyone else, had been responsible for making those deaf ears hear again.

He spoke of Hale's success in photographing for the first time, in full sunlight, "not merely the prominences themselves, but their spectra." He traced his achievements from 1889, which, "though not very successful in themselves, contained the germs of the spectro-heliograph, and therefore of all the recent successes." He followed his spectacular results with the spectroheliograph in the photograph-ing of the faculae all over the sun's disc so that they could be seen as plainly as the prominences. "In this achievement of 1892 the key note is struck of the theme which he has recently developed so mag-nificently." This, as Turner pointed out, had led Hale to the more recent discovery of the flocculi with the help of the Rumford spec-troheliograph at Yerkes. "A new degree of freedom has been im-parted to solar research by the discovery of this method. The new development is in its infancy; but we have already had some most beautiful illustrations of its practical working exhibited to us, and from these alone we may confidently predict a great and important future for the study of the 'flocculi.'" Turning then to Hale's work in other fields, Turner described his work on sunspots: "What vast new fields has not Professor Hale prospected by showing that every spot is a region which must be explored from hour to hour, from section to section, and from one chemical element to another!" Finally he showed how Hale had, in this way, been led to a study of the fourth-type stars in the hope of finding a correlation between the spectra of those stars and of sunspots, and so of proving his life-long belief in the importance of the study of the sun as a typical star.

As he came to the end of his address, he spoke of the birth of the *Astrophysical Journal* and of the founding of Yerkes, and exclaimed, "I venture to think that we have seldom had a better example of versatility than on the present occasion, when we find the same man initiating and establishing one of the foremost observatories, inventing a new instrument which has an obviously great future before it, and recently finding an entirely new field of research for that instrument." He concluded his glowing tribute with an ex-planation of the reason for Hale's absence from "our front bench." "He is filling a place of even greater honour; he is initiating a new work of science."

Yet, even while Hale tried to spend most of his time on the "new work" there were frequent interruptions, some of them welcome, others less so. One night, late in January, 1904, he gave a lecture on the "Evolution of the Stars" before a large audience at Throop, a small technical institute in Pasadena. Soon afterward Campbell sent him a letter of introduction to John D. Hooker, who had made a fortune in the hardware and steel-pipe business, and was then president of the Western Union Oil Company and vice-president of the Baker Iron Works. Possibly, said Campbell, Hooker might be willing to contribute to the Yerkes expedition to Mount Wilson.[5] He had been a founder of the Southern California Academy of Sciences in Los Angeles, and was interested in astronomy. He even had a small telescope in his fashionable mansion on West Adams Street in Los Angeles.[6]

On February 3rd, Hale met Hooker at lunch downtown. Hooker was affable, and expressed interest in the observatory plans Hale described so enthusiastically.[7] Yet, as with Yerkes, Hale had no idea then that this new prospect could prove just as difficult as the traction magnate.

Soon after this first meeting, he decided to go out to West Adams Street to call. On his arrival he found a large yellow house, surrounded by elaborate grounds. He was ushered into a large living room, its floors adorned with magnificent Persian rugs. Here he was greeted by Hooker, who then led him on a tour of the house and grounds, of which he was inordinately proud. In the stables behind the house he kept two fabulous trotters; in the garden, which was surrounded by a high spite fence that he had built to keep out the eyes of inquisitive neighbors, he had a fantastic collection of roses that ran to thousands of varieties. He also kept his telescope in this garden.[8]

This last, of course, again brought the talk around to astronomy and so to the project for which Hale hoped to get Hooker's support. He had brought along some of Ritchey's beautiful photographs of nebulae. As he exhibited these, he talked of Barnard and of his long-cherished hope of extending his photographic atlas of the Milky Way to stars invisible at Yerkes. Before the day was over, Hooker, evidently captivated by the charm and the persuasiveness of his visitor, had agreed to pay the cost of bringing Barnard and his Bruce 10-inch photographic telescope to California.

Meanwhile, too, Hale had been in touch with another California tycoon. Actually, though the meeting had little effect on the observatory, it had unforeseen results in quite another direction. On the day after his first meeting with Hooker, he noted simply in his diary: "Met Mr. Huntington at California Club. He was interested in observatory project and agreed to transport instruments for further tests on Lowe." [9] This "Mr. Huntington" was Henry E. Huntington, nephew of the wealthy railroad magnate Collis P. Huntington, and father of the rapidly spreading red-car transportation system in Los Angeles. For a while, under Huntington's influence, Hale considered building an electric line up Mount Wilson, extending it from Mount Lowe. Later, warned by Campbell of the possible dangers to the observatory of such a line, Hale abandoned the idea. But he did not forget Henry Huntington.

At this time, in February, 1904, other developments forced his attention back to Yerkes. For weeks he had waited to hear whether Charles Yerkes would continue to provide the salaries of at least two astronomers. Now the answer came—in the negative. This was not surprising. Yerkes, so rumor ran, had left Chicago for his mansion in New York with $15,000,000 in cash, and had sailed for England to work out his plans for the London underground.[10] The prospect was gloomy indeed. As soon as Adams heard the news, he wrote, urging Hale not to feel obligated to keep him on the staff. He added wistfully: "The prospects of a bohemian year on Mt. Wilson with you appeal to me very strongly indeed, and I should be only too glad to join you for what would cover 'Omar's loaf of bread and jug of wine'!" [11]

Hale had nothing to go on, but he told Adams not to worry. "I feel more and more confident of the outcome of the solar observatory scheme, and even if it fails, I intend to have a station where you and Mr. Ellerman and I can work to advantage on Mt. Wilson. . . ." [12] At the same time he wrote to Frost. Somehow, he said, the money must be found to bring Adams to California. "In any case we of course cannot afford to lose him."

Just at this time, Harper wrote that Rockefeller, after pouring millions into the University of Chicago, refused to contribute further. There was, therefore, little likelihood of any increase in salaries at Yerkes. These factors pointed up the problems of Hale's own future. Therefore, on March 7th, he wrote to Harper, "As several physicians we have consulted say that Margaret must remain here throughout the year, I think it may be necessary for me to give up my work at the Yerkes Observatory and devote my time to solar research on Mt. Wilson where I am now establishing a small station." [13]

It was true that in February, Margaret had come down with another asthma attack and that young Bill, though "cheerful and strenuous," was ailing. But these were rationalizations. The fascination of life on the mountain and the chance of working on the sun in this glorious climate were Hale's real magnets.

In March, Ellerman appeared from Yerkes. He arrived on a day somewhat different from what he had been led to expect in "sunny" California. Hale met him at the foot of the trail. On the way up the mountain they ran first into clouds, then into rain, and finally, above Martin's Camp, into a heavy snowstorm. That night the thermometer dropped to 17 degrees. The next morning it was 26 degrees in the Casino, with an inch and a half of snow on the ground and ice glittering on trees and bushes. Yet Ellerman was as elated to be there as Hale was to have him. With his skill in instrument design and his ingenuity in making equipment out of available materials, however meager, he entered wholeheartedly into the busy, hectic days on the mountain. He entered so much into the spirit of the place, in fact, that, as Hale observed, he soon looked as if he had always lived there. When Adams arrived some months later, this, as he was to describe it afterward, was the Ellerman he found: "He wore a 'ten gallon hat,' high mountain boots, and a full cartridge belt from which hung a revolver on one side and a hunting knife on the other." He commented, "I was greatly impressed and pictured a struggle for existence on the wild mountain top, which bore little resemblance to later actuality." [14]

Hale had sent to Yerkes for the horizontal telescope that had been used at the Wadesboro eclipse. The essential part of this telescope was a 12-inch plane coelostat mirror, which reflected the sun's rays horizontally through a long tube. The plane of the mirror was mounted parallel to the earth's axis, and, by means of a driving clock, the mirror was made to complete a rotation once in twenty-four hours. This motion of the mirror was just adequate to counteract the effect of the earth's rotation so that the sun's rays would continue to be reflected in the same direction. After leaving the coelostat, the rays fell upon a 6-inch photographic lens that formed an image on a photographic plate at its focus, 61½ feet away.

This, in principle, was the way the telescope worked. In practice there were difficulties. It was mounted near the Casino on the hill where the Mount Wilson Hotel would later stand. By April 10, 1904, it was ready for trial. This is Hale's description of what happened: "The sun rose resplendently in a cloudless sky, and its image as first seen on our focussing screen was sharply defined. We prepared to make photographs, but before we could get the first exposure, the image had become blurred and indistinct. I had feared asymmetrical heating of the 60-foot tube of building paper by the sun and had spread canvas before it as a shield. But this was not enough, and a radical change was obviously needed. Ellerman and I were alone on the mountain and must do the work without delay, as I had booked to leave for the National Academy in Washington two days later. So we attacked the tube, ripped off all its upper faces, dismounted the canvas and stretched it up as a flat shield toward the east, no longer enclosing the heated air. By nightfall we were ready. Visitors to the mountain, when asked in Pasadena next day what they had seen, said 'two men working like devils.' " [15]

Early the next morning the "devils" plunged into work. It was clear and calm, "a day typical of the long Mt. Wilson dry season." The sun's image was sharp, the sharpness *held*, and it continued to hold as a result of their feverish work the previous day. They took eight photographs at different foci, and developed them. The results were superb. Immediately, then, Hale started down the mountain, leaving Ellerman to wash and dry the plates. The next morning he left for Washington, carrying his latest results—on his way, too, to receive the Draper Medal from the National Academy of Sciences.

In Chicago he learned that Miss Helen Snow, the donor of the telescope, had withdrawn her objection to the temporary removal to California of the Snow horizontal telescope for studying the sun. This good news was offset by Yerkes's refusal of further support for salaries at Yerkes. Nevertheless Hale was elated. That night he talked until long past midnight with his brother of his hopes and dreams for Mount Wilson. The following morning Will Hale, who was now a lawyer in Chicago, announced that, after consultation with Uncle George, they had decided to invest some capital in the "solar plant" on that mountain. He called George the greatest gambler in the world, but trusted that everything would turn out well in the end. George, who for the last three months had been gambling on the venture and pouring every cent he owned into it, and borrowing on top of that, was jubilant.

On the afternoon of April 16th, Hale reached the Arlington Hotel in Washington. There, waiting to see him, was the tall and powerful Dr. John Billings, a handsome man with straight Napoleonic nose and clear blue eyes. With him was Mrs. Henry Draper, who, Hale was sure, could also have tremendous influence on Carnegie. She was the widow of the noted spectroscopic pioneer Henry Draper, who, in 1872, had built with his own hands a 28-inch reflecting telescope. With it he had pioneered in the photography of celestial objects, especially the moon, and in stellar spectra. Mrs. Draper, an able and beautiful woman with magnificent auburn hair, was the daughter of Courtland Palmer of New York, a wealthy man from whom she had inherited a large fortune. From that fortune she was endowing the Henry Draper catalogue of stellar spectra at Harvard and had donated the Henry Draper Medal to the National Academy. This medal was to be awarded to anyone in the United States or elsewhere "who shall make an original investigation in Astronomical Physics. . . ."

As for Billings, his power as chairman of the Board of Trustees of the Carnegie Institution was incalculable. Soon, therefore, Hale found himself deep in a discussion of Mount Wilson and its prospects. Now, or never, he felt, was the chance to present his case.[16] He described the glittering prospects to anyone who would listen. He took time out only for the meetings of the National Academy of Sciences.

The academy, founded in 1863 during the dark days of the Civil War, was, and would remain, the leading honorary scientific body in the United States. Its charter, signed by Abraham Lincoln, provided that "the Academy shall, whenever called upon by any department of the Government, investigate, examine, experiment and report on any subject of science and art." Its founders, among them scientists of the caliber of Alexander Dallas Bache and Joseph Henry, had in fact done research on military and naval problems for the North. Often at night during the war President Lincoln had climbed with Henry, the Secretary of the Smithsonian Institution and a leading scientist, to the Smithsonian tower to watch the lights being flashed to distant stations, in connection with tests of new methods of signaling. "It was in such researches for military purposes that the National Academy of Sciences had its origin," Hale noted. But since the Civil War, despite the advance in all branches of science, it had been largely moribund. The majority of its members came to the annual meetings, listened smugly to their own speeches, ate heartily at the banquet, then returned home to forget it entirely until the next meeting. The majority, too, as Hale soon realized, considered that the academy's chief function was to confer honor on those sufficiently worthy by electing them to membership.[17]

Hale had been elected to this august body in 1902, just two years previously. At thirty-three, he was one of the youngest men ever so honored. "It was a great piece of luck I got in," he had written to Harry Goodwin. "I must have pulled through by a mere scratch." He had been particularly pleased because, as he told Goodwin, he had "studied the work of Academies of Science so much during the past year, in connection with our Research Fund" (the William E. Hale Fund, founded in his father's memory). He added: "It is a fine thing to meet men in other lines of work, and to find out what you can about their point of view. I am becoming more and more interested in investigation in general, regardless of particular fields of knowledge. Huxley's and Darwin's life and letters have done me more good than anything I have read in years." [18] Before long this interest in the development of academies, coupled with the breadth of interest developed through his reading, was to bear unexpected fruit.

From the beginning he had felt that the academy, in its capacity as adviser to the government, ought to be doing much more for the advancement of science in America. Here, included in a single compact body, were a majority of the country's leading investigators. In such a store of brainpower, he could see unlimited possibilities. Before long he would set out on an ardent campaign to rouse it from its lethargy and widen its influence.

Now, however, in accepting the Draper Medal, he was giving a paper on "The Rumford Spectroheliograph." In the audience were scientists like Henry Rowland and Simon Newcomb who had urged Hale's election, as well as others, such as William H. Welch, the jovial and friendly Johns Hopkins pathologist, with whom Hale was to develop a close friendship.

As soon as the academy meetings were over, the Executive Committee of the Carnegie Institution met. Billings urged Hale to go along and to wait outside for the decision. After what seemed an eternity, Billings appeared, a rare smile on his face. They had, he said, voted $10,000 for Hale's immediate use, and promised that at least $30,000 would be granted in December for the Yerkes expedition to Mount Wilson.

Somehow news of the grant leaked out. On April 25, 1904, Hale was amazed to read this notice in the morning paper:

PLUM FOR MT. WILSON

Fund of $10,000,000 to be devoted to Prof. Hale's research.

The reporter, obviously carried away, had added three zeros to the actual sum. Hale read on with amusement mixed with consternation. "Mt. Wilson is in a fair way to become known wherever two or three scientists are gathered together the world over." Soon reports sprang up all over the country that made the solar observatory seem an assured thing. This, of course, was far from true. The $10,000 would help. But the 60-inch mirror must still lie fallow in the Yerkes basement.

Now, however, Hale decided to gamble everything on the chance that in December the larger plan, providing for the 60-inch, would go through. He made up his mind to "plunge," using all the money his uncle and brother could or would lend him. After that he would borrow and spend at his personal risk. If things in California could hum in the months ahead, Carnegie might be convinced of the importance of the work on Mount Wilson.

At this point, however, he was still officially on an expedition from Yerkes. With "a mixture of hope and fear" he returned to Chicago, knowing he must come to terms on the future with President Harper. They had a long talk, then drew up an agreement in which the terms were quite specific. When, months later, Harper questioned the legality of his actions, this agreement would become Hale's Magna Carta.[19] Work was to be continued on Mount Wilson under the title "University of Chicago Expedition of Solar Research." Yet, significantly, they agreed "that in case the Carnegie Institution decides, as a result of a new gift from Mr. Carnegie, to establish a solar observatory of their own, this work will naturally be replaced by such an observatory." [20]

With this agreement settled, with the university paying the salaries of Hale and Ellerman, with Ritchey's salary paid out of Carnegie funds, with Hale paying Adams's salary himself, and the rest of the Yerkes staff provided for, Hale left Chicago in a joyful mood on the "Yerkes Expedition." He was joined by Ritchey and by Adams, who, in his vivid account of the "Early Days on Mt. Wilson," described their journey across the country. He tells of Hale's excitement when, toward their journey's end, he realized that "the decision had been made and that a new life with new responsibilities and opportunities lay before us all."

Immediately on their arrival Hale set to work. On his own initiative he met with the directors of the Mount Wilson Toll Road Company to discuss a lease. On June 13, 1904, the lease was signed—in the name of George Hale; it was for ninety-nine years, free of rent. Privacy was assured—the land at the east end of the mountain was some distance from the trail's end and from any point where an electric road might reach the summit.

After this, one of the first tasks was the choice of a site for living quarters. Years before, Hale had become fascinated by Curzon's tale about the monasteries of the Levant, perched on rocky promontories, looking out on distant peaks.[21] He had dreamed of building such a monastery, where the male astronomers could live while observing on the mountain. (He had not forgotten the difficulties at Yerkes, or his resolve made then that, if he should ever found another observatory, the astronomers and their families would not live on the observatory grounds.)

Soon after their arrival, therefore, he set out with Adams in search of a "monastery" site. To explore the ridge they had to hack their way with small hatchets through the dense underbrush. Finally, about a quarter of a mile away, at the end of the ridge, they came out on a small opening. The ground was nearly level; on three sides the land fell away in sheer precipices, revealing a magnificent view of valleys, canyons, and distant peaks. "This is where we must build it," Hale exclaimed. So the future "Monastery" was born.

Six months later, designed by Myron Hunt, it was finished. It had a large living room and a huge fireplace built of granite from the mountain. On a beautiful, mild still night in December, Hale, Adams, and Ellerman moved from the old Casino. They walked down the hill carrying lighted candles, and that night, sitting late around the embers of a fire, they talked of their hopes for a great observatory.[22]

Meanwhile news of the activity on the mountain had spread—over the United States, even to Europe. Before the Monastery was finished a flood of visitors had arrived. One of the first to come was the genial English astronomer H. H. Turner, with his wife. The only available accommodation was the powerhouse, a small wooden building. There they took up residence. They insisted that they enjoyed themselves "hugely" in the company of the gas engine. Nearly fifty years later Daisy Turner was to write, "We both adored that whole experience, and though Herbert had always been fond of Prof. Hale it sealed our friendship forever; and I have always felt proud to share the friendship of such a truly great (and most *lovable*) man." [23]

Soon after the Turners, Harry Goodwin and Arthur Noyes, Hale's old college friends, arrived. They stayed on until he had to leave for the East. Often then they talked again of working together, if possible in California. But many years were to pass before their hopes were to be realized even in part with the arrival of Noyes in Pasadena.

On September 12th, Hale left for Washington. On the way he stopped in St. Louis for the scientific congress that was being held in connection with the great exposition there. He was chairman of the astrophysics section and chief sponsor of plans for the "International Union for Cooperation in Solar Research"—under the significant auspices of the National Academy of Sciences. For many years he had believed that to advance solar research, indeed any astronomical research, it must be organized on an international scale. Eleven years earlier he had sponsored the international congress at the Chicago Exposition. Now, in preparation for the St. Louis Fair, he had written to astronomers the world over suggesting another such meeting. He hoped, too, that it would lead to the formation of an international organization, "so planned as to interfere in no wise with individual liberty, but rather to aid and suggest, wherever it could be of service, and to collect information for discussion."

The response was enthusiastic. Astronomers as well as physicists arrived from England, France, Germany, Italy, Russia, Hungary, Sweden, Switzerland, and even from far-away India, as well as from the United States. The attendance exceeded Hale's wildest hopes.

On September 23, 1904, he called the meeting to order in the Hall of Congresses that had been provided by the far-seeing organizers of the St. Louis Fair. He was acting as Chairman of the Committee on Solar Research of the National Academy of Sciences, which, at his request, had approved his plans and issued the call for the conference. His opening address was eloquent. He pointed out that with new instruments and methods, the possibilities of solar research had barely been tapped. He described the importance of laboratory spectroscopy and stressed the need for applying spectroscopic methods to the problems of solar physics. Over and over he emphasized the importance of individual initiative. "I believe that one of the principal objects of establishing committees on solar research should be the encouragement of little known investigators to bring forward and develop ideas that may have originated with them." But in order to give individual initiative full rein, he urged his belief that it was necessary to work out an international scheme for the collection of information and the planning of cooperative research programs.

As an example of such a cooperative program, he suggested that certain specific regions of the solar spectrum be assigned for study to different observers who in turn might develop "new and original ideas" on the nature of that spectrum. The opportunities, he stressed, were unlimited.[24]

Afterward, as he received congratulations, he expressed the hope that this would be only the first of many international astronomical meetings. Years later, when the International Astronomical Union had grown out of these early Solar Union meetings, the astronomer M. Minnaert would look back on Hale's "remarkable address" in St. Louis and would describe that first meeting as "the inspiration for the constitution of a much more far-reaching and systematic organization of astronomical research."[25]

At the close of the meeting temporary officers were selected. Hale was made president; the great Henri Poincaré of France vice-president. With Arthur Schuster and Svante Arrhenius, Hale was also appointed to a committee for setting up a provisional program of observations.

For Hale, however, the event of immediate significance came in the form of a rumor that, he knew, might mean the difference between success or failure for his observatory scheme. Alexander Agassiz, so the rumor ran, had decided not to accept a large grant from the Carnegie Institution for the study of coral atolls. If this were true—but Hale hardly dared to hope!

That night he left for Chicago to stay with his uncle. And there, for the first time in these last strenuous months, he suddenly realized how weary he was. His head and body ached. His mood was depressed. It was some time before he felt able to go on to New York, where he was to meet again with John Billings.

They met at the Park Avenue Hotel, and from the first Hale felt that everything was wrong. The pessimistic Billings seemed quite unwilling to hold out any hope for success in Washington. He would not even discuss the subject. "I discretely [sic] awaited a better mood," Hale noted, "saying nothing of our work in California."

They had dinner at the hotel. All through the meal Billings remained glum and impassive. Afterward he suggested going to his house. Reluctantly Hale agreed. No one else was there. They sat down, and Billings lighted his cigar. There was a long silence. Then the surgeon turned his close-set, gray-blue eyes on Hale and demanded a report of the work on Mount Wilson. "It was not easy," Hale recalled, "to explain why I had not only spent the ten thousand dollars but much more than this, though I had fully protected the institution, in a manner approved by my lawyer, from any responsibility beyond the actual appropriation."

When he finished, he looked diffidently up at his host. Billings, looking like his ancestral Norse Vikings, continued to puff grimly on his cigar. At last, "in sepulchral tones," he remarked slowly, "During the Civil War I expended on one occasion seventy-five thousand dollars, without authority of law. I was ultimately sustained, but in such cases no one can foresee the outcome." In the same lugubrious tone he continued, "The only thing for you to do is to report fully, in person, to the Executive Committee at its next meeting." Hale meekly agreed. "His cynical manner did not seem propitious for my future hopes. So I did not mention them and soon took my leave." But just as he was putting on his coat, Billings asked, to his surprise, "Why don't you go up with me to see Mrs. Draper tomorrow?" [26]

The next afternoon they met at Grand Central Station. Billings's mood had not improved. They boarded the train for Dobbs Ferry. Billings sat down, said nothing. A cloud hung over his handsome face. Hale gazed out of the window at the Hudson Palisades, and remained quiet too.

After dinner at Mrs. Draper's they went into the drawing room for coffee. Then, while Billings retired to a corner with a secondhand-book catalogue, Hale talked amiably with his hostess of trivial things. Suddenly Billings looked up from his corner to ask, "Why don't you show Mrs. Draper your pictures?"

Fortunately, Hale had brought his latest solar photographs, taken just before he left the mountain. Mrs. Draper looked in wonder at the beautiful images, and praised the work that had produced them. This, naturally, was all the encouragement Hale needed. He launched into an account of their experiences on the mountain, and described his plans in detail, "thinking this the best way to inform Dr. Billings."

After the tense hours on the train and his reception in New York, it was an immense relief to have an appreciative audience. He took full advantage of it. For some time Billings continued to study his catalogue. Then suddenly he looked up to remark caustically, "That's right. All you've got to do is to convince Mrs. Draper!" Then again he buried his head. Not again during their entire visit did he show the slightest interest. The two men returned to New York as grimly as they had come. The next morning, after a night of uneasy sleep, Hale left for Washington.[27]

There he immediately got in touch with Charles Walcott, secretary of the Carnegie Institution Executive Committee. The geologist was most friendly, and invited him to dinner. He confirmed the rumor Hale had heard in St. Louis. Agassiz had decided to conduct his expedition for the study of coral reefs independently. He was to have received $65,000 per year for two years. Now that this money was available, Walcott advised Hale to revive the original project with the 60-inch reflector, and to present it in person to the Executive Committee the following day.[28]

As Hale listened, he felt inclined to follow Walcott's advice. Still, the uncertain shadow of Billings loomed in the background. Hale remembered, too, that the last time he had followed Walcott's advice it had resulted in failure. That night he went back to his bleak room in the Arlington Hotel "still revolving two alternatives" in his mind. The original plan, though revised in the light of the year's experience on Mount Wilson, still called for a large expenditure. The second, providing for work with the Snow telescope on an expedition from Yerkes, involved much less expense. But this left the 60-inch in the Yerkes basement. Which project should he suggest? As he lay in bed the two plans whirled in his head. Sleep was impossible. All night long he kept turning on the light and getting out of bed to make notes for arguments for and against both plans. At last, unable to make up his mind, he decided to present both plans at the meeting the next morning.[29]

Before the committee, which was meeting at the New Willard Hotel, Hale spoke first of the smaller plan, which required the $30,000 practically promised by the trustees at their April meeting. He knew, of course, that this amount would just about cover what he personally had already spent on Mount Wilson. Then he turned to the "big scheme." Thinking of the Agassiz grant, he noted that this plan could be financed by a grant of $65,000 a year for five years. The committee members questioned him in detail, and again he described all that could be accomplished if only the 60-inch could be mounted. Afterward he wrote to Goodwin, "I am not sorry I mentioned the larger plan. They seemed to take great interest in it, and asked me many questions. There was not a suggestion of criticism for spending the $27,000 in advance of the appropriation." [30] Prospects for success, he felt, were great.

Later, Walcott telephoned to say that the committee was favorably inclined to give him enough for the "plant" ($225,000 at least) in three years instead of five, plus running expenses. Afterward there would probably be $40,000 a year for at least ten years. In two weeks the committee would decide definitely what recommendation to make to the trustees. Hale, elated, told Goodwin, "It certainly looks as though it had been worth while to stay over. . . ." [31]

It was late in October, 1904, when Hale arrived back in Pasadena. After these harrowing weeks he was delighted to be home and again at work on the mountain. There he anxiously awaited the decision of the Executive Committee. On November 2nd, it came. Relaying the news to his brother, Will, he wrote, "Walcott just telegraphed that the Executive Committee will recommend a grant of $150,000 for the Solar Observatory for next year. This would be followed by the same amount the second year. Don't say anything about this now, as it must be confirmed by the Trustees in December." [32]

This time Hale was sure the recommendation would be accepted by the trustees. He was surprised, therefore, when his more cautious brother wrote back, "Your telegram from Walcott is very interesting. But I do not regard it in as sanguine a light as you do. . . . For my part, I think that the chance that the Trustees will confirm this are just about even. . . . The Trustees have been known to turn down the Executive Committee very hard."

When at this point George proposed buying a house in Pasadena, Will was appalled. He could not understand his headstrong, impulsive brother. He was willing to do anything within reason to push

George's observatory plans, but he could not see his way to gambling further on an uncertain future. "You can't afford to load up with a house and lot at present," he exclaimed, then added: "Dr. Isham is not altogether wrong when he says you are the biggest gambler he knows. The answer is that sometimes it is necessary to gamble. But the present proposition looks to me really to be a speculation in land." Still, despite his doubts, he concluded, "I would regard a loan to you as perfectly safe, of course; and will ask Martha's consent if you desire." [33]

Meanwhile Hale had received a letter from Horace White; in it he said that the enigmatic Billings, who was actually an excellent judge of character, had expressed himself very favorably with regard to the solar observatory. "Why, of course, we can do it," he had said, "Why else are we here?" [34]

As a result of this unexpected news, Hale decided to send Carnegie some of their "illustrated papers," together with a few of Ritchey's photographs taken with the 24-inch reflector. He wrote of all that these promised for the "extension of our knowledge" with the 60-inch, with its capacity for collecting six times more light than the 24-inch. He concluded, "I think the present opportunity for an important advance has not been equalled in astronomy during the past fifty years." [35]

At last the fateful day arrived. On December 13th the trustees met in Washington. Later, Hale learned from Walcott something of the proceedings. When the question of the solar observatory was raised, the majority favored it. But there were dissenters. One, curiously, was Charles Hutchinson of Chicago, usually one of Hale's most ardent supporters. Now he was hesitant. The suggested appropriation, he felt, was too large for the conduct of the work. "I would not commit myself to $40,000 a year for equipment," he declared, "because if he has it, he is pretty sure to spend it, and if he has not that much he will probably get along on less." [36]

Other trustees echoed Hutchinson's views. "What is the necessity," asked Cleveland Dodge, "after you have the plant thoroughly installed, and the instruments provided in spending so large a sum for new instruments each year?"

Here Billings, as chairman, came to Hale's defense. "It is very necessary," he urged, "to have an appropriation to buy new instruments. No matter how well you are equipped with instruments you will see how to improve them, to get something a little finer." (Hale himself might have been speaking.)

"But if you cannot buy them," Hutchinson protested, "you will do the best you can with what you have! Mr. Hale is an admirable man and he is entitled to support in every way, but of course he will take all he can get, and the entire income if he can get it."

At this point, apparently, Walcott decided the discussion had gone far enough. He moved that the appropriation, amounting to $310,000, be made and that the question be referred back to the Executive Committee for approval. There was some further discussion. Then, to Walcott's delight, his motion was adopted. Immediately after the meeting he dashed out to telegraph to Hale. Nevertheless, confirmation had to await the Executive Committee meeting on December 20th. "So," he said in a letter that followed, "I shall lie low until then." Hale, of course, was overjoyed. This time he could not see how anything could go wrong. In reply he wrote hesitantly: "Perhaps I may trouble you with the request to telegraph yet a third time, in view of the importance of the decision to us? I am still borrowing money perforce, to carry things along, so it will be an immense relief if the final action is favorable." [37]

Somehow that week passed. On December 20th, the day of the momentous Executive Committee meeting, Hale was on his way up the mountain, riding a mule. He had reached Martin's Camp. There, unexpectedly, he was called to the old single-wire telephone. The operator said she had a telegram from Washington. Hale, trembling in spite of himself, asked her to read it. From her voice, which was difficult to hear, he gathered the wonderful news that the Executive Committee had appropriated $150,000 a year for two years and had authorized the immediate execution of the larger plan! [38]

Now, at long last, after endless months of uncertainty, Hale knew, and soon all the world would know, that the Mount Wilson Solar Observatory was a reality. At thirty-six Hale was director of the second great observatory he had founded—one that, when it was completed, would again be the largest in the world.

On that memorable day, December 20, 1904 (the actual date of the founding of the observatory), the new Carnegie president, the noted mathematical physicist Robert S. Woodward (who had been chosen to succeed Gilman), sent Hale the official notice of the grant. He wrote, "I beg to assure you of my warm interest in the great enterprise you have undertaken, and also to assure you that it will be my earnest endeavor to cooperate with you in all the work that may come in the years that are before you." [39]

It was a year to the day since Hale had arrived in California with the Yerkes expedition. Now Walcott sent a check for $30,000, the first installment on the larger sum, which would barely cover the amount Hale had borrowed to keep the project going!

CHAPTER NINE

Life in California

✳ *1905-1906*

February, 1905

On a warm spring day George Hale sat in his garden on Bellefontaine Street near the Arroyo Seco in Pasadena. After renting two other houses, he was delighted to be in a home of his own. The longer he lived in California, the better he liked it. The climate was superb, the beauty of his surroundings ideal. The garden next door, where a house had once stood, was a mass of radiant color, with a magnificent Gold of Ophir rose tree covered with thousands of blossoms, and an orange tree in full bloom. "Just now," he wrote exuberantly, "the odor of orange blossoms can be detected in the mountains more than a thousand feet above the valley. The whole town is a mass of blossoms of every kind, and the green fields are thickly covered with flowers. . . . Every day seems better than the preceding one, and I do not believe it possible that I could ever tire of such glorious conditions." [1]

Moreover, he was having a wonderfully gay and amusing time with a new group of friends whom he had met soon after his arrival in California. In 1903, when Campbell had suggested a meeting with John D. Hooker, the hardware magnate, he had also written: "Mrs. Hooker is exceedingly pleasant, and for that matter, exceedingly talented. Her book *Wayfarers in Italy* is one of extraordinary charm and interest." [2] Soon after his first visit to the house on West Adams Street in Los Angeles, Hale had met the beautiful, highly cultured Katherine Hooker. From the first he was charmed.

Mrs. Hooker, the former Katherine Putnam, member of a famous New England family, was a niece of the noted philologist William Dwight Whitney, and had come to California as a little girl. Behind the yellow house on West Adams Street she had a charming Italian garden, and here, as time passed, George Hale was to spend some of his happiest hours. Here she would read some of her favorite Italian poems aloud, and while he did not always understand the meaning of the words, he was captivated by the sound. Afterward he learned long passages by heart, and recited them on the rough Mount Wilson trail.

At the Hookers, too, he met another delightful woman, named Alicia Mosgrove. Her keen wit, her humor, her insatiable curiosity, her whimsical imagination attracted him strongly, and soon he was calling her, as did everyone else, "Ellie." A unique friendship developed between them, born in a gaiety of spirit that delighted him always. In it the problems of his life seemed to vanish.

"Ellie," too, was fascinated by this man who had suddenly "come down into the heart of Los Angeles" and who, at thirty-six, was the founder and director of the greatest observatory in the world. Later, in colorful phrases, she gave her first impressions of Hale and of the magnetic quality that attracted all who knew him. "You could never think of him without laughing way down inside," she said. Later she compared him with other friends, John Muir, David Starr Jordan, and William James, who came to stay with the Hookers. He had, she reflected, "something that none of them had—an inner excitement—a higher degree of interest—a higher degree of suffering."

In answer to a question about Hale's physical appearance or unique characteristics, she noted that he was slight of build. "His shoulders, in contrast to his large head, dominated by his high forehead, appeared small. His full lips were highly expressive; his finely chiseled nose was good; his eye teeth needed straightening. But his most outstanding characteristic were his eyes. . . . He had the most twinkling eyes I have ever seen." She spoke, too, of his habit of looking down. "Then suddenly he would look up, his eyes alight, and everything around him would shine."

Often, she recalled, he would sit, his hands held curiously, tensely in front of him, at right angles to his body, gazing into the distance. Those who did not know him well, seeing him as a dignified pundit, would expect him to drop learned words about the stars. Instead, to their surprise, he would come out with some unexpected, entirely irrelevant whimsy.

Drawn out by Ellie's ebullience, Hale soon found himself acting quite differently with her than with anyone else. She appealed to a fanciful, imaginative side of his nature that others barely glimpsed and of which he himself had perhaps never been fully aware. Soon after their first meeting he invented a saloon on the Embarcadero in San Francisco in which the fascinating Ellie became the "gal" with diamond heels and a red dress who sold drinks at the bar. To strangers he would often say quite seriously, "By the way, did you know that Ellie once kept a saloon in the Seraglio in Constantinople, and there shanghaied sailors into the bar?" Such visitors, expecting to meet the renowned astronomer George Hale, were astonished. And the more astonished they became, the more he embroidered the tale.

Over the years he would continue to rejoice in Ellie's company and to share with her many of his fantastic schemes as well as his more serious plans. He was interested in everything this versatile, irrepressible woman did in the long years of their friendship—from dealing with labor problems to creating the City Park and Recreation Commission in San Francisco; from founding a children's museum and a hospital there to acting as director of the California Institution for Women in Tehachapi; from climbing Mount Whitney with John Muir to sailing to Tahiti or traveling a thousand miles down the Nile in a native craft.[3]

Soon after he came to know her, Ellie decided to open a school in the Westlake district of Los Angeles. When she asked her good friend, Maude Thomas to help in running it, another beautiful, learned, creative woman, with a charming, responsive smile, was added to the group that gathered in the Hooker house on West Adams Street. From the first, Hale was enchanted, and the house took on an added glamour. In these stimulating surroundings he became, as a younger woman who met him there recalls, "the charming center of a great deal of admiration." When he talked of his work on the sun, of his great telescopes and the universe they revealed, his listeners sat entranced. "He had an extraordinary gift for making astronomy thrilling to those who knew nothing about it," one of the group said. For many years, when asked whom he would like to have for Saturday dinner, his reply would be, "Ellie and Maude." For many years, he sought their companionship as a welcome source of relaxation. As Margaret Hale recalled long afterward, her father would drive over to get them, and they would come and often stay overnight. On Sunday he would drive them home. Even after Mrs. Hooker and Ellie had moved to San Francisco, Maude Thomas came alone.[4]

Occasionally, too, they would go up to stay in the "Hooker cottage" that Mr. Hooker had built on Mount Wilson. One such visit, in particular, stood out in Ellie's memory. Will Hale, who happened to be in Pasadena, had accompanied her up the mountain. The day was as "cold as Greenland." The cook, like other cooks before and after, had become disgusted by the primitive conditions, and left. Ellie, soon after her arrival, discovered that she was expected to substitute. She found, too, that another task, even more arduous, was expected of her. The small spectroscopic laboratory and darkroom had just been completed. It was so primitive that it contained no running water. Ellie was elected "water boy." All morning she lugged pails of water from Strain's Camp until she was exhausted. Yet she was elated by the chance to share in Hale's "spectroscopic research." And he was delighted by her spirit of adventure. At lunch the astronomers, led by Hale, met for a "council" and elected their hard-working slave "a Nubian"—a rare honor on a mountaintop where women were rarely accepted by the monkish inhabitants of the Monastery. At eighty-six, Ellie Mosgrove still considered this one of the signal honors of her life.[5]

With all these new interests and extending social activities joined to the excitement of work on the mountain, life in Pasadena would have been wonderful. However, even in this "Eden" clouds soon appeared. The first came from Chicago. Two weeks after the actual founding of the new observatory, Hale had regretfully sent in his resignation as director of Yerkes.[6] Harper had implored him to defer the move. Yet, in a discussion of the problem, Martin Ryerson, chairman of the university trustees, agreed that Hale's only course was to resign. Letters flew back and forth between Chicago and Pasadena. Then, "like lightning from the blue," a letter arrived in which Harper questioned the legality of Hale's stand.[7] He had apparently forgotten their original agreement, in which Hale had emphasized that, if the grant for a large observatory should go through, the Mount Wilson Solar Observatory would become a separate institution, not connected with the University of Chicago. Hale felt he had carried out this agreement in good faith. But Harper refused to accept the facts, despite legal support of Hale's view or the opinion of Charles Hutchinson, trustee of the university: "Harper's agreement with you regarding the venture is clear, and I see in it no right, legal or moral, for his claim to a refund from the Carnegie Institution." [8] (Harper was asking the return of Hale's and Ellerman's salary from the time the Yerkes expedition had set out for California.)

In the end Harper conceded that Hale had probably intended to do the right thing. This was some consolation. But Hale was wounded by Harper's imputation, and unhappy at the thought of leaving Yerkes on a discordant note. To the end of his life he would regret the misunderstanding, particularly as Harper soon died of the cancer from which he had long been suffering. Then, too, he would always retain an especially warm spot in his heart for the great 40-inch refractor as his first great brainchild.

Years later, in 1932, Hale returned to Yerkes. As George Van Biesbroeck recalls, "He came in, in a hurry as always." He rushed in to see Frost, (his successor as director) in his office, stayed there for about twenty minutes, then came out. In the hall he spoke to Dr. Van Biesbroeck, then started out the front door. Suddenly he turned. "I must have a look at the 40-inch," he said. He dashed up the stairs to the dome, opened the iron door and entered. He took off his hat and stood there looking up at the venerable telescope. Then slowly he said, "Noble instrument." This was the last time he ever saw it.[9]

If Hale could have been in two places at once, he would, of course, have been glad to continue work with the 40-inch. But since this was impossible, he now had to turn all his attention to his latest infant. A week after the Mount Wilson founding, Michelson wrote, "Sorry to have you leave us and will miss you more than you can imagine." [10]

Meanwhile other difficulties had arisen in "Eden." At first the Hale children improved under the California sun, and their father rejoiced. But when the rainy season came, they were forced to stay indoors, and soon became pale and listless. As usual he became worried. In February, 1905, he began to plan for the solar conference to be held at Oxford the following September. Yet he confided to Goodwin: "I don't know how I *can* stay away from the family so long—I always have a dread that the children will be sick while I am away, as William is now. He has had nothing to eat for two days, and it makes my heart ache to see his distress. I hardly think he is critically ill, and look for decided improvement; but the uncertainty of the thing is enough to turn one's hair gray. . . . Margaret is wheezing *badly* every night since the wet weather came." [11]

In the end he went to Oxford, and was gone two months. When he reached home in October, he found that Bill had been ill. His wife had said nothing "because," as he told Goodwin, "she knew I would return immediately and thus lose the Oxford meeting." He discovered, too, that she herself was "a nervous wreck." One day soon after his return she collapsed, and Dr. James H. McBride, the family doctor, insisted that she enter his sanitarium, Las Encinas, at once. Hale felt she could "hardly hope to leave there for several months." "Nervous prostration," he wrote to Barnard, "as you know, is a very hard thing to escape from." [12] He added: "The children are now very well. It is a curious thing that they are almost always ill when I am away." Deeply upset, he blamed himself for his absence. Gradually, Evelina Hale improved, but the effect on Hale himself was disastrous. The continual worry took its toll. "I am caught in the toils as never before," he told Goodwin. Whenever he could he escaped to the mountain.

"Hale was never so happy," Adams was to write, "as when, like a boy on vacation, he could pack a knapsack and start on the eight mile climb over the old trail to the summit." [13] When fog submerged the valley and the peak was bathed in sun, he would often take Margaret along. Under the sun her asthma disappeared, her bubbling, elfin gaiety returned, and her father's spirits soared.

It was on one such expedition that they acquired the famous burro Pinto from Mr. Bassett, the proprietor of Martin's Camp, for ten dollars. This gentle, submissive creature of "unknown age and origin," notable for his long, melancholy face and slow-moving pace, became Margaret's pet. Like the other burros, he was kept in the corral in Sierra Madre, and in winter was put out to pasture on Barley Flats. It was there, as Hale liked to relate, that a mountain lion chewed off half of Pinto's tail, giving him forever an incomplete appearance from the rear. Before the pumping system was installed, Pinto, escorted by an Armenian driver who walked in step beside him, packed ten-gallon cans of water from the springs at Strain's Camp. Water transported in this way was used by Barnard to wash all his plates of the Southern Milky Way. Yet Pinto had one bad habit. His taste in food was omnivorous. He ate newspapers, magazines, and even the somewhat indigestible *Astrophysical Journal*.[14]

Sometimes then, when the day's work on the sun was over and they had watched it setting behind the distant peaks, Margaret would sit with her father under the tall dark pines that, stretching up and up, seemed almost to reach the sky. A rare feeling of peace and understanding would come over them then—a sense of being above the earth, a part of the vast universe. Margaret would sense his love of the mountain's beauty, his fascination with a wandering cloud that resembled some weird, prehistoric bird. Sometimes he would talk animatedly of the things they were doing and planned to do. She listened eagerly; young as she was, she felt she was sharing in her father's work. Again he would talk of art and music or of Egypt, that wondrous land he had long wanted to visit. Soon she shared his longing, and the understanding between father and daughter grew. As the years passed, he was to continue to pour out his thoughts and ideas to her in a never-ending stream.

In these and other ways, the nights and days on the mountain passed swiftly. For Hale, able to see so much happening, they were wonderful. For Margaret, with a thousand fascinating things to watch, they were equally enthralling. Often she would sit on a rock at the trail's edge, watching the burro train carrying supplies or parts of the Snow horizontal telescope for studying the sun up the mountain. One day when she was on the mountain an extraordinary monster appeared on the trail. It was the truck George Hale had designed to transport the heaviest loads up the mountain. Some of the pieces for the Snow telescope weighed as much as 350 pounds. Under such a load even the most powerful mule would have collapsed. For some weeks Hale had puzzled over the problem. One morning, when he was shaving, the solution came to him. In his mind's eye he pictured a unique steel truck that could negotiate the sharp turns in the nine miles of twisting trail, and yet be strong enough to carry the required loads. When the contraption was built, it was indeed a startling sight. First came a man leading a mule, then the truck itself, with a low-slung body about 20 feet long and 20 inches wide on small rubber-tire wheels. Then another mule appeared, led by another man who, like the man in front, could help to steer the "monster." The mule at the back also acted as a brake and, when necessary, could drag the truck back on the trail.[15]

Hale reveled in such challenges to his ingenuity. But there were other difficulties of a less welcome sort. One constant source of exasperation was the telephone line, a single iron wire strung over trees and bushes. Soon after they moved into the Monastery a wild snowstorm swept the mountain. The line went dead. Adams and Hale started through the driving snow to locate the trouble. They finally discovered a broken wire on a live oak. Hale clambered up the tree to repair it. "As he worked," Adams recalls, "his enthusiasm which neither weather nor numb fingers could affect in the least, broke out in the words of his favorite poet, and I heard, 'It was a storm from fairyland . . .' coming down from the tree above the roar of the wind." [16]

The astronomers came to know the mountain in all its moods. As they explored the peak and the surrounding canyons, they learned the countless trails over the San Gabriel Range. As Hale's enthusiasm for these mountains grew, he climbed the trail countless times—on foot, by burro, and on horseback—until he knew every inch of the way in every kind of weather. Often he would walk down the nine-mile trail, then ride his bicycle into Pasadena for needed supplies, which he would then load on his back for the return trek up the mountain. Once, in March, 1906, after climbing the mountain in a storm, he reported to R. J. Wallace at Yerkes, "We have just gone through a little 'shower' on the mountain, which lasted from Sunday morning until Saturday noon. During a portion of this I plunged my way through raging mountain streams and rain which came down in chunks, carrying a pack weighing several pounds, more or less, on my back. When I arrived at the summit, the only part of me that was dry was my throat, and I had some good old Scotch to wet it, so I did not suffer any. We had something over 15 inches of rain during this 'shower.'" [17]

On one occasion he attempted another method of transportation: he tried to climb the mountain on the three-wheeled Indian motorcycle he had just bought. But even with Ellerman's heroic help it proved unequal to the rough Mount Wilson trail.

One day he was riding merrily up Lake Avenue toward his office on Santa Barbara Street. With his thoughts apparently on higher things, he suddenly crashed into the rear of a trolley. His precious machine slid under the trolley, and somehow he was left standing in bewilderment in the middle of the street. Afterward he found it hard to explain what had happened or why. In time he gave up the motorcycle and bought an automobile instead. As Margaret recalls, he drove this vehicle with similar "characteristic push and abandon."

This, however, was not the end of the motorcycle. One day, in a gay mood, he was riding along Columbia Street, near his house, Hermosa Vista. The machine, recently greased and oiled, was running beautifully. Out of the corner of his eye he saw two other motorcyclists approaching. Somehow he got the impression that they called, "Want to race?" Needing no second invitation, he opened his motor wide, only to be startled by a wailing siren. His "competitors" proved to be motorcycle police. He was arrested, and escaped appearing in court only through representation by a friend.[18]

On the mountain the tales of these and other exploits added to the hilarity of the astronomers who gathered in the evening in the Monastery around the great stone fireplace heaped with blazing logs. While the room filled with the smell of burning pine mingled with tobacco smoke, their faces reflected the glow of the fire. There were laughter and good humor in the air, and Hale was the first to laugh at his own folly. But there was also, as Harold Babcock was to note, "a sense of great events in the making." Hale was so full of his subject that those around him could not help being infected by his enthusiasm. He was, of course, absorbed in his instrumental plans and ideas for new equipment. These, as Adams was to point out, were designed to fit the problems he had in mind instead of "seeking the problems to fit existing or preconceived instruments." Yet if he was absorbed in such dreams and plans, he was equally eager to have his colleagues work out their own ideas. He was always receptive to any scheme that might conceivably fit into his overall plans.

In those nights, whether the talk was of the sun or other stars and nebulae or of lighter things, the sense of comradeship was intense. Often Hale listened to tales told by others; often he was the storyteller. With his quick wit, his gift for repartee, he was the match for anyone there. In later years, Adams and others would comment on his "amazing breadth of interests, his great personal charm," and would recall his "stories of important figures in science and national and international affairs." [19]

Occasionally he would recite passages from his favorite poets—Keats and Shelley. And once, at least, on a stormy night when all within felt safe, he composed some lines of his own. They exist to-day in the familiar penciled script on a yellow pad:

What joy to know the swirl of snow
The howling of the gale!
What sport to hear, twixt joy and fear
The thunder and the hail,
For they're without, when storm-gods shout
And grapple in the gloom,
While here beside the hearthstone wide
The flames dance through the room. . . .[20]

This period, as Adams was to recall, was an extraordinarily active one for Hale, both physically and mentally. "In spite of anxiety about the future his health was excellent; he took an active part in the minor construction in progress, and made many long walks along the mountain trails and even into canyons where no trails existed. He combined to an interesting extent the joys of the nature lover with those of the pioneer."

Gradually, as time passed, the group on the mountain grew. In addition to Adams and Ellerman, it included Barnard, who arrived on a temporary basis to photograph the Milky Way with the Bruce telescope; Francis Pease, who had been at Yerkes; and the colorful, humorous Charles G. Abbot, who came on a Smithsonian expedition soon after the observatory's founding. For many years Abbot was to return each summer to the laboratory he had set up below the Monastery to conduct his classic experiments on the determination of the solar constant. Here, too, he worked on his solar cooker, which he felt sure would revolutionize methods of cooking and heating. He also developed his own theory of long-range weather prediction based on sunspot cycles. While Abbot was a hard worker, he was always willing to take time off to spin a yarn. He also shared with Hale a love for *Visits to Monasteries in the Levant,* and was especially fond of the dramatic story of the Jew of Constantinople.

Influenced by their mutual love of the Orient, Abbot began to call Hale "Your Eminence," and referred to himself as "The Sheik of the Southwest." Their conversation, and soon their letters, glittered with Oriental allusions. "In the name of Allah, the Compassionate, and by the beard of his Prophet (on whom be peace)," Hale wrote in a congratulatory note, "I rejoice, Oh Aga, in thy elevation to the supreme seat in the Mosque of our Fathers, of blessed memory." And Abbot replied in similar vein, "Thy words are like pearls of wisdom and shall ever be as lamps in my path. . . ."[21]

Soon after the completion of the small spectroscopic laboratory on the mountain, a young physicist arrived from Chicago to take charge of it. Tall and strikingly handsome, with a powerful physique, Henry Gale had been a member of the first football team at Chicago under the famous coach Alonzo Stagg. But, fortunately for the progress of the spectroscopic work on the mountain, his skill as a physicist surpassed his talents as a football player. A student of Michelson, he soon entered with an enthusiasm comparable to Hale's own into the study of sunspot spectra and into all the other work in this small, primitive laboratory where comparisons could be made of laboratory and solar spectra to learn more about conditions in the sun. Soon he was helping to build the electric furnace that would be used in the long and productive series of observations on the effect of temperature on sunspots.[22]

While the spectroscopic laboratory was being built, Hale was also pushing his plans for the construction of the Snow telescope, with which he had first experimented at Yerkes. It consisted of a coelostat, or slowly revolving mirror, on a clock-driven mounting that received the light from the sun and reflected it to a plane mirror 30 inches in diameter. From this mirror the beam was reflected nearly horizontally to a point 100 feet north, where it fell on a telescope with a 24-inch concave mirror of 60 feet focal length. In this way, a solar image about 6½ inches in diameter was formed on the slit of the spectrograph or spectroheliograph. This, at least, was the principle of the instrument. In practice, however, numerous difficulties appeared. The path over which the beam of sunlight had to pass was very long. Therefore means had to be found to protect that beam from the heated air rising from the hot soil on the mountain summit. To alleviate this effect, Hale raised the telescope high above the ground on huge stone piers. Even then he was to find that, during the hotter hours of the day, the sharpness of the image decreased as the sun's direct rays warped the mirrors, changing the focus and blurring the image after a few minutes of exposure. In the end the worst of these difficulties would be overcome by observing in the early morning or late afternoon, and by shielding the mirrors from the sun between exposures and cooling them with blasts from electric fans.

The site for the new telescope was not far from Echo Rock, and soon after the design was finished, George Jones and his powerful crew set to work. To construct the piers they used the huge boulders

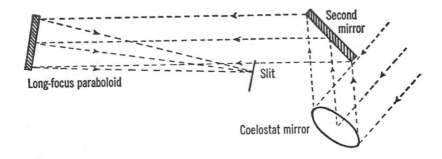

Second mirror

Long-focus paraboloid

Slit

Coelostat mirror

Diagram of the Snow Telescope.

from the mountain. To transport the boulders they built rock sleds to be drawn by mules. The friction was terrific, and the work was dangerous. A Japanese workman lost his life in a plunge down the mountain. One of the mules committed suicide over a precipice. Still, by the summer of 1905 the Snow telescope was ready for work.[23]

In Pasadena, with the help of the Board of Trade, land had been bought for a laboratory, instrument and machine shop. It was located in open country on a short, sandy road called Santa Barbara Street, just off North Lake Avenue. In the shop most of the instruments and other equipment to be used on the mountain were built. One of the first instruments built there was the spectrograph for the Snow telescope. With it Hale was able to analyze the sun's light and the behavior of the thousands of spectral lines in its atmosphere. With the new spectroheliograph that he had designed especially for this telescope he was able to photograph the distribution of the white-hot clouds of individual gases that float above the sun's surface. By means of a high-speed photographic shutter he could take direct photographs of the sun: these showed spot structure and other surface features beautifully. The results surpassed his "greatest expectations." They were superior to the best he had ever attained with the 40-inch at Yerkes.

Moreover, in contrast to the huge movable refractor at Yerkes, this new telescope was infinitely more tractable. At Yerkes it was necessary to attach each heavy instrument, one by one, to the end of a moving telescope tube. Now, instead, the instruments were permanently set up on a pier, where the adjustments need never be

George E. Hale on Wilson's Peak, 1903.

George E. Hale in the Monastery office.

Group of astronomers on Mount Wilson, 1905. *Left to right:* Construction Superintendent Miller, C. G. Abbot, G. E. Hale, Ingersoll, F. Ellerman, W. S. Adams, E. E. Barnard, Backus. *(Photos courtesy of Mount Wilson and Palomar Observatories)*

George Hale and Ferdinand Ellerman on Mount Wilson.

Truck for hauling heavy instruments up the "New Trail," Mount Wilson.

Pack train on Mount Wilson. *(Photos courtesy of Mount Wilson and Palomar Observatories)*

Snow Telescope for observing the sun. The 60-foot tower telescope in the background.

The 150-foot tower telescope.

Sectional drawing of 150-foot tower telescope. *(Photos and diagram courtesy of Mount Wilson and Palomar Observatories)*

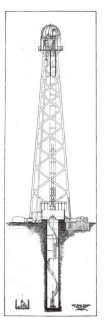

Langley's drawing of sunspot, March 5, 1873, from his book, *The New Astronomy*. Scale is indicated by inset of the Americas at upper left.

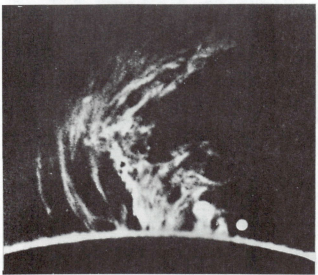

Large active prominence, 140,000 miles high, photographed in light of calcium. *(Courtesy of Mount Wilson and Palomar Observatories)*

Solar corona photographed during the total eclipse of June 8, 1918, at Green River, Wyoming. *(Courtesy of Mount Wilson and Palomar Observatories)*

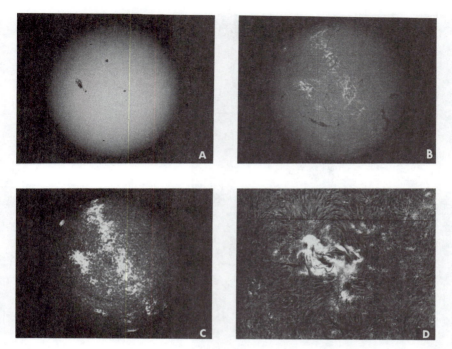

The Sun: (a) ordinary photograph, (b) hydrogen (Hα), (c) calcium spectroheliogram, (d) enlarged hydrogen spectroheliogram.

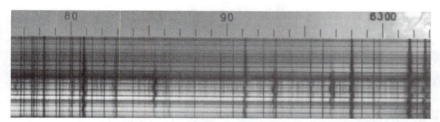

Sunspot spectrum. Narrow horizontal strips divide the spectrum from top to bottom. The strips represent parts of the polarizing apparatus. They show alternately right-handed and left-handed circular polarization. Strips 3, 4, 5, 6, from top, represent penumbra of spot. 8, 9, 10, 11, 12 represent the umbra (having maximum field strength). Bottom two strips represent penumbra. Horizontal separation in vertical lines (as in the pattern to the left) measures strength of magnetic field. The direction of the field (whether toward the observer or away from him) is found from one key strip which has been determined in the laboratory. *(Photos courtesy of Mount Wilson and Palomar Observatories)*

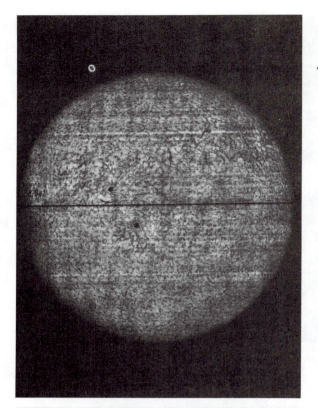

The sun in Hα hydrogen,
October 7, 1908.

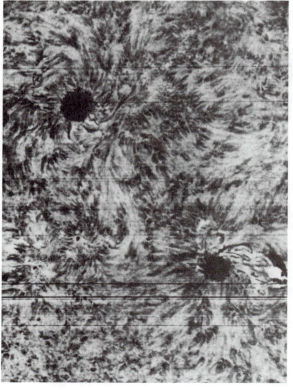

North and south sunspots,
September 9, 1908, in Hα
hydrogen. *(Photos courtesy of
Mount Wilson and Palomar
Observatories)*

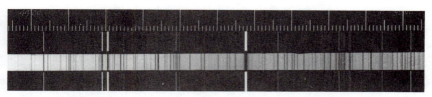

Comparison of solar and iron arc spectra, wavelengths 4900-5000. *(Courtesy of Mount Wilson and Palomar Observatories)*

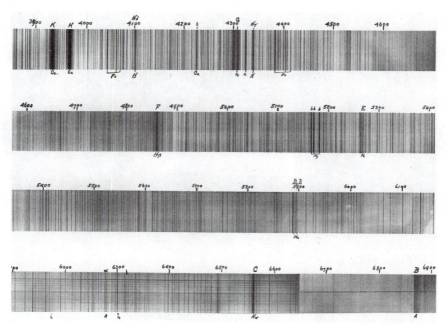

Spectrum of the sun taken with the 13-foot spectrograph from the violet (3900 A.) to red (6900 A.). *(Courtesy of Mount Wilson and Palomar Observatories)*

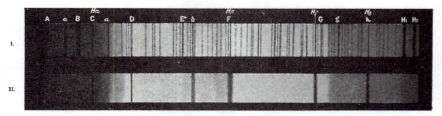

Comparison of the spectra of the sun (I) and a star, Sirius (II). *(From Robert Ball, The Story of the Heavens)*

Andrew Carnegie and George E. Hale in front of the 60-inch telescope dome at Mount Wilson.

The 60-inch telescope.

Group photograph taken at the time of Carnegie's visit to Mount Wilson in 1910. *Front row, left to right:* A. Davidson, G. E. Hale, John Muir, J. D. Hooker, A. Carnegie. *Back row:* Dr. J. H. McBride, H. F. Osborn, J. A. B. Scherer. *(Photos courtesy of Mount Wilson and Palomar Observatories)*

disturbed. "It is thus possible to pass rapidly from one instrument to another, photographing the forms of the calcium flocculi, for example, with the spectroheliograph, and their spectra, only a moment later, with the powerful Littrow spectrograph." [24] Thus, for the first time in the history of the study of the sun, it was possible to examine diverse solar phenomena nearly simultaneously by different methods and to correlate the observations closely. The possibilities were tremendous.

One of the first fields Hale decided to explore was one that had fascinated him ever since he could remember. Sunspots had, of course, been known and studied for three hundred years (ever since Galileo first observed them). As a boy Hale had projected them on a piece of white cardboard attached to his 4-inch telescope, and watched their passage across the sun. At Allegheny he had seen Langley's beautiful sunspot drawings. In them he had observed in vivid detail the umbrae, the central cores in spots, which can attain the fantastic size of 55,000 miles, seven times that of the earth. He had learned that while these umbrae appear dark in contrast to the brilliant photosphere, they are actually brilliant, luminous objects. Around them the lighter surrounding penumbral areas spread to even greater distances and seemed to be composed of radial filaments. (These spots, he would later find, sometimes attain the extraordinary area of 3,000 million square miles, and in one case at least, the gigantic area of 5,000 million!)

Nor had he forgotten Langley's vivid description of sunspots. As he compared his visual observations with a photograph taken at the same time, Langley had observed: "See how full of intricate forms that void, black umbral space in the photograph has become! The penumbra is filled with detail of the strangest kind, and there are two great 'bridges' as they are called, which are almost wholly invisible in the photograph. Notice the line in one of the bridges which follows its sinuosities through its whole length of twelve thousand miles, making us suspect that it is made up of smaller parts as a rope is made up of cords (as, in fact, it is); and look at the end, where the cords themselves are unravelled into threads fine as threads of silk, and these again resolved into finer threads, till in more and more web-like fineness it passes beyond the reach of sight!"

The spot here described by Langley had been observed on September 21, 1870. Three years later, in December, 1873, another even more remarkable spot appeared when, as Langley wrote, "the rare coincidence happened of a fine spot and fine terrestrial weather to observe it in." In this spot, he observed, "the pores which cover the sun's surface by millions may be noted. The luminous dots which divide them are what Nasmyth imperfectly saw, but we are hardly more able than he to say what they really are. Each of these countless 'dots' is larger than England, Scotland and Ireland together." Then, after comparing the resemblance of the spot to frost crystals on a windowpane, he noted, "There were wonderful fern-like forms in this spot, too, and an appearance like that of pine boughs covered with snow. . . . The salient feature here is one very difficult to see, even in good telescopes, but one which is of very great interest. Everywhere in the spot are long white threads, or filaments, lying upon one another, tending in a general sense toward the center, and each of which grows brighter toward its inner extremity." [25]

At Kenwood, Hale had first photographed spot spectra and had discerned some of the principal widened lines in those spectra. But he could not analyze them in detail: they were too faint and too indistinct. With the 40-inch, additional widened lines and bands had appeared. But he was still not satisfied. He wanted to photograph the spectrum on a much larger scale so that he could interpret the details in the spot lines and try to find their origin. More powerful instrumental means were obviously needed. Now, with the Snow telescope with its long-focus grating spectrograph, he hoped to be able to answer such elusive questions as these: Why are spots darker than the photosphere? Are they regions of high or low temperature?

By means of slow-motion electric motors, he was now able to bring a spot exactly onto the spectrograph slit. He was even able to set exactly on the umbra; all other light could be excluded by covering the entire slit except a small region at the center. He could then compare this spectrum with that of the solar surface.

From the beginning the work was promising. He never forgot the day when, after taking his first plates, he dashed down to develop them, or his excitement when he discerned at least three hundred lines in the yellow and green regions of the spot spectrum that had never before been visible. The definition was superb, the detail extraordinary, and the possibilities were immense.

Perhaps now at last he was on his way to discovering the cause of these dark areas that are so vast that the earth could make "but a moment's scant mouthful for them." Now, with adequate instrumental means, a really accurate knowledge of the spot spectrum might be gained for the first time in history. "No archaeologist," he was to write, "whether Young or Champollion deciphering the Rosetta Stone, or Rawlinson copying the cuneiform inscription on the cliff of Behistun, was ever faced by a more fascinating problem than that which confronts the solar physicist engaged in the interpretation of the hieroglyphic lines of sunspot spectra." [26]

The year 1905 ended: 1906 began. Day after day he continued his observation of these extraordinary "hieroglyphic lines." Day after day he continued to measure and catalogue them according to their behavior. But, as so often happens in the history of science, interpretation of his observations proved to be anything but simple. He examined the lines of a single element in a sunspot and saw that they were not all affected alike. "Some are greatly strengthened, or perhaps attended by broad, faint wings." The strengthening was, in fact, so pronounced that lines entirely invisible in the solar spectrum were among the most conspicuous of spot lines. On the other hand, some of the solar lines were greatly weakened or entirely absent in the spot spectrum. In addition there were spot lines that remained unchanged. What, he asked himself, is the cause of these peculiarities?

Just by looking at the sun itself, it was impossible to tell. He turned therefore to the laboratory on the mountain which he had designed with just this sort of problem in mind.[27] If he could reproduce or simulate solar conditions here, he might then discover their celestial cause. Only in this way could he test his hypothesis that spot vapors are actually lower in temperature than their surroundings.

The little laboratory on the mountain contained a number of light sources that could be studied under varying conditions of temperature, pressure, etc. "The immediate imitation in the laboratory, under experimental conditions subject to easy trial, of solar and stellar phenomena, not only tends to clear up obscure points, but prepares the way for the development along logical lines of the train of reasoning started by the astronomical work." In this laboratory, therefore, he had set up arrangements "for studying the spark spectra of metals in air and in liquids; arc spectra in gases at high or low pressure; flame spectra, for which a Bunsen burner and an oxyhydrogen blow-pipe are required; vacuum tube spectra, etc."

There was a small electric furnace for studying the phenomena in the vapors of sodium and other metals which melt at low temperatures. In addition there was a pump capable of compressing gases up to pressures of three thousand pounds to the square inch; an induction coil, giving a 16-inch spark; X-ray apparatus for the study of the effect of X-rays on the radiation of gases and vapors; and finally a small heliostat to supply sunlight. In all these directions he hoped to be ready to perform any experiment that might arise, and so to find a solution of the solar mysteries.

Years before, the presence of iron in the sun had been proved by vaporizing iron between the poles of an electric arc and photographing its spectrum beside the solar spectrum. This, in fact, had been one of Hale's earliest experiments as a boy. Now, with the help of Walter Adams and Henry Gale, he photographed the spectrum of iron vapor in an electric arc; at the same time he varied the amount of current, and thus the temperature, as much as possible. The results were conclusive. Many of the lines were strengthened in the cooler parts of the arc; others were most intense in the hot central core. The similarity to sunspot spectra was striking. The lines that were strengthened in spots were likewise strengthened at low temperatures in the arc; those that were unaffected or weakened in spots were similarly affected at high temperatures in the arc. Temperatures in spots, Hale concluded, must be lower than on the sun's surface.

After this, experiments continued with chromium, nickel, manganese, titanium, and other metallic elements that had already been identified in sunspots. Again the results in the laboratory and on the sun appeared similar. "Sun spots," Hale concluded, "are actually regions of reduced temperature in the solar atmosphere." Before long he was elated to be able to confirm this conclusion in another way. If spot vapors are indeed cooler than the rest of the sun's atmosphere, then some of them, he felt, should unite chemically when chemical combinations impossible at higher temperatures take place. With Gale he measured thousands of faint lines, and identified them as bands due to chemical compounds. In England, Alfred Fowler had discovered magnesium hydride in the laboratory (later identifying it as due to the presence of molecular compounds). Now, on Mount Wilson, Hale and his colleagues had been able to identify the spectra of titanium oxide and calcium hydride, and to compare them with their respective sunspot bands. "In this way we began to form a new picture of these regions of the solar atmosphere and to recognize the chemical changes at work in the spot vapors."

So through observation and experiment Hale had proved his theory brilliantly. By July 24, 1906, he could report to Dr. Woodward, the president of the Carnegie Institution, a success that far exceeded his fondest expectation and promised wide application.[28]

From the sun he now turned to other stars. The sun, as he had pointed out again and again, is a typical star. Now he hoped to prove it. In a series of historic observations of spectra of the brilliant red stars he had first studied at Yerkes, he found characteristics similar to those in sunspots. This, in turn, led him to the formulation of another hypothesis of far-reaching significance. Perhaps, he reflected, all stellar spectra might be classified on a temperature basis, "thus providing something better than a purely empirical foundation for a study of stellar evolution." Here, at last, he thought he could see grounds for a "truly philosophical conception of the variations in stellar spectra." Enthusiastically he wrote to H. H. Turner in England, "I must say that I am much pleased with the outcome of this investigation." [29]

As Walter Adams was to point out, this discovery of the temperature classification of spectral lines meant much more than a means of interpreting the behavior of lines in the sunspot spectrum: "It formed essentially a new method of attack upon the physical problems of sun and stars, especially as it was soon supplemented by other results showing the behavior of spectral lines under varying conditions of density and pressure. A partial breach had been made in the hitherto impenetrable wall surrounding the interpretation of the spectrum." [30]

Long afterward, when more was known of the behavior of atoms and the ionization theory was developed, this temperature classification would assume unexpected importance as a starting point in the analysis of spectral lines according to energy levels in the atom.

Significantly, too, these pioneer experiments on sunspots would serve as the initial step in Adams's classical measurement of stellar distances by the method of spectroscopic parallaxes—a tale Adams himself has told in his brilliant article "Sunspots and Stellar Distances." [31] He wrote, "It forms an interesting illustration of the ramifications of researches in physical sciences and of the unfore-seen consequences which may follow them."

About this same time, also, in 1907, Hale devised a special instru-ment that he called a heliomicrometer. It, too, was built in the Mount Wilson instrument shop. With this ingenious instrument it was possible to read off the latitudes and longitudes of the flocculi directly, without extensive computation. Hale described the results afterward: "Their change of position from day to day yielded a new determination of the law of the solar rotation, which was found to differ at the calcium and hydrogen levels. At the lower level of the calcium flocculi, the period of rotation at the sun's equator is 24.8 days, increasing gradually to 26.8 days at 45° latitude. In other words, the gaseous sun does not rotate like the solid earth, on which points in all latitudes complete a rotation in 24 hours. It turns more and more slowly as the poles are approached, points in high latitude lagging behind those nearer the equator. If this could happen on the earth, Jacksonville, which is almost due south of Cleveland, would be far to the east of it 24 hours hence. In the higher levels of the solar atmosphere, where the hydrogen flocculi float, the period of rotation for any latitude is less than the levels below, but the differ-ence in rotation time between pole and equator is less marked than in the lower atmosphere."

These observations, made on plates taken with the Snow tele-scope, were interesting. But, from the beginning, Hale realized that the constantly changing flocculi were not very "satisfactory objects for rotation measurements." He was delighted, therefore, when Adams developed a much more accurate method that depended on measuring, with a powerful spectrograph, the velocity of approach and recession of the east and west limbs (or edges) of the sun. Later, Hale described this method and the results: "The east edge is moving toward the earth on account of the sun's rotation; this causes a displacement of the spectrum lines toward the violet. At the west edge, which is moving away, the lines are equally displaced toward the red. The double displacement, measured at different latitudes, gives the velocity of approach and recession in kilometers per sec-ond. An investigation of this kind threw much new light on the peculiar law of the solar rotation, giving with high precision the rotation period at different levels and the change of its value from equator to pole." [32]

On March 9, 1908, Hale wrote to Robert Woodward, to describe the results of this important series of observations: "I have just sent to the *Astrophysical Journal* a paper by Mr. Adams giving the results of his spectrographic observations of the rotation of hydrogen in the sun, and another by myself on the rotation of the sun as determined from the motions of the calcium and hydrogen flocculi. It is extremely interesting to find, by these independent methods, that the hydrogen in the upper chromosphere shows no evidence of the equatorial acceleration observed in the case of sun-spots. The motion of the hydrogen is also more rapid than that of the calcium flocculi which are in the lower chromosphere where the equatorial acceleration is still in evidence."

He concluded, "The future development of this work promises to be of great importance."

Astronomers the world over were astounded by these and other results that continued to pour from Mount Wilson in these years. Astronomers, like J. Halm, wrote with justifiable envy from the Cape of Good Hope: "With the greatest interest I have read Adams's account of his observations on the solar rotation, though while reading it I could not suppress a feeling of sadness at the astounding fact that he would do on one plate what took me a whole year's troublesome visual observations. Let me offer you my heartiest congratulations on the far-reaching results of his investigation and the marvelous accuracy of his measurements." [33]

Thus, again and again, Hale's belief in the tremendous advantage to be derived from the use of photographic methods with fixed spectrographs of the highest dispersion in studies where visual observations had previously been used was demonstrated. In the case of sunspot spectra, as he was to point out, the advantages were even more striking "since the catalogue of spot lines we are now publishing (the results of work with the tower telescope) will contain at least ten times as many lines as any previous catalogue."

Sir William Huggins, commenting on these latest results, reflected the general view when he exclaimed, "You are leading the world and I am looking forward to the triumphal progress of your results. It is in your power to turn over hitherto unread pages of Nature's secrets!" [34]

If, therefore, this period at the beginning of the twentieth century was often a puzzling time in which to work, it was also one of intense excitement and glittering promise. In the rapidly advancing field of atomic physics, Hale continued to follow the latest developments eagerly, hoping always that some new advance might help to solve his own perplexing problems in the sun and stars.

CHAPTER TEN

A Time of Great Discovery

✳ *1908-1915*

IN THE FALL of 1905 the spectroheliograph that Hale had designed for work with the Snow telescope was completed. As soon as he could get it up the mountain, he returned to work with this remarkable tool that had first appeared to him "out of the blue" sixteen years earlier.

At Kenwood, with his first spectroheliograph, he had succeeded in cutting out all light from the photographic plate except that of one element at a given wavelength, and had photographed prominences for the first time. At Yerkes, with the Rumford spectroheliograph, using the light of calcium, he had photographed the great whirling gaseous clouds that do not (like prominences) rise high enough to be observed in elevation at the limb. He had called them "flocculi." Moreover, like a miner on the earth, delving under its surface to explore different levels, he had shown that he could sort out the different levels in the sun's atmosphere, to discover what is going on at one level alone.

Now, at Mount Wilson, working with the Snow telescope, he returned to his delving. At Yerkes he had used the bright H and K lines to gain his spectacular results on the white-hot calcium clouds that often overlie and obliterate sunspots (doubtless as a result of the high level of the calcium gas in the sun's atmosphere). But he had long been eager to test other spectral lines, particularly those of hydrogen, the most abundant element in sun and stars, which, he thought, must lie at a higher level in the solar atmosphere than the calcium flocculi.

At this time the process of sensitizing plates for the study of the red hydrogen line remained highly experimental, and the results were problematical. The photographic plates then on the market were insensitive to red light, and the strongest line of hydrogen—the line Hale was most eager to use—lies in the red region of the spectrum. To overcome this difficulty R. J. Wallace at Yerkes had been experimenting with red-sensitive organic dyes. Emulsions treated with such dyes, if dried quickly and exposed immediately, could be used with some success. Hale decided now to experiment with plates treated in this way.

On the morning of April 30, 1908, he set the spectroheliograph slit on the α-line of hydrogen, and turned it across a spot group near the sun's center. He photographed it, then rushed down to the darkroom. As soon as he had developed the plate, he held it up to the light. It was not much to look at, but it proved to be "by far the most interesting photograph" he had ever taken. First he noticed that the hydrogen clouds, unlike the calcium clouds, were dark against the solar background. (Later he would find that some of the hydrogen clouds near sunspots were bright and could also affect earthly phenomena.) Next he made an even more extraordinary observation. On the plate he could see a hydrogen filament that was distinctly vortical in structure, with spiral arms spreading around the spot center. Looking at this amazing "solar storm," he was reminded of a terrestrial tornado. As he wrote to the aging Huggins, he could see with surprising clarity "the radial and spiral structure in regions surrounding spots and other disturbed areas." To which he added significantly, "As all of the spots on the plate show similar structure, I think we are undoubtedly observing for the first time, great whirls in the sun's atmosphere which are intimately associated with spot formation." [1] Before long he observed a strong similarity in the distribution of the hydrogen flocculi to that of iron filings in a magnetic field.

As eager as always to share his latest success with old friends, he sent Brashear a sunspot photograph that showed a great cyclonic storm. Brashear wrote back: "Don't know when I was so much elated over anything as when I received your splendid—glorious—magnificent (wish I had a large fund of adjectives) picture of the sun. McDowell says I've gone daffy over it. Let it be so. It's a fitting climax to your great work, and if the good Lord were to call you away from your work now—you would certainly have enough to your credit to get along in the eons on the other side. . . .

"Bless you, my dear fellow, how you must have felt when that picture was developed. When I think of the first spectroheliograph with its tiny image—and this marvelous picture with that furious monster-cyclone—and that you have made it and I have lived to see it—well, I feel almost (but not quite) like good old Simeon—ready to depart in peace." [2]

Nor was this the end. In June, 1908, an astronomer arrived from Oberlin to join the Solar Department. His name was Charles St. John, and he had spent several summers at Yerkes helping E. F. Nichols with his experiments on the heat radiation from the stars. At fifty-one he was eleven years older than Hale, and therefore older than most of the other Mount Wilson astronomers. But in spirit he was as young as the youngest. His enthusiasm for life, and especially for life on the mountain, was boundless. Soon he became widely known as St. John of the beaming countenance. This sunny trait was shared with notable powers of concentration and absentmindedness. This last characteristic was to become the source of endless stories at the Monastery. Like Hale's motorcycle exploits, some of St. John's most daring escapades were in mechanical vehicles. It was related that he once drove down the railroad tracks in Pasadena, having mistaken them for the regular highway. At another time he mistook the back of his garage for the front and drove right on through. Even on the mountain, life was precarious. One night, warming his hands by the fire, he was busily talking when, without thinking, he sat on the object nearest at hand. The object proved to be a roasting-hot distillate stove! Yet, despite his sometimes bumbling ways, St. John was to prove himself one of the ablest members of the staff—especially in the field of solar spectroscopy. [3]

Soon after St. John's arrival on June 3, 1908, he took a photograph of the sun that showed a dark hydrogen flocculus in the very act of flowing into a large spot. Hale, delighted, suggested that his new method "would ultimately lead to an explanation of the cause of sunspots and the nature of the circulation in the sun's atmosphere." The following day he reported to Edwin Frost at Yerkes: "We are having such an exciting time with the solar whirlpools that I find it difficult to attend to my other work." [4]

As a result of this "exciting time," he was soon led to the answer to a problem that had long puzzled him. In the closing paragraph of his paper on "Solar Vortices," [5] which appeared in the *Astrophysical Journal*, he described his observation of widened and doubled lines in spot spectra. Could these, he asked, be due to the presence of intense magnetic fields?

Recently J. J. Thomson in England had shown that when bodies become hot they emit electrons—and the sun was of course intensely hot. Thinking of Thomson's discovery, Hale suggested the hypothesis: "If positive or negative electrons were caught and whirled in a vortex they would produce a magnetic field, such as we obtain by passing an electric current through a coil of wire."

Long before, Henry Rowland had observed that a rapidly rotating, electrically charged plate would produce a magnetic field; then in 1896 Pieter Zeeman of Leiden had discovered that a magnetic field splits and polarizes the spectral lines emitted by a source placed in the field. Hale never forgot the day he had read of Zeeman's discovery in his early days at Yerkes. "While I then had no clearly defined hope that we should ever be able to apply it directly in astrophysics, I could not help feeling that such a fundamental advance in spectroscopy might ultimately prove significant to astronomers." [6]

With Zeeman's discovery in mind, Hale had provided a Du Bois magnet and "suitable" polarizing apparatus as part of the first equipment on Mount Wilson. As early as 1905, he had written prophetically to Woodward, "I am taking up some work on the effect of a magnetic field on lines that are widened in sunspots, as no investigations have hitherto been made in this direction." [7] In 1906, Hale had tried unsuccessfully to detect evidence of magnetic separation by visual observation of the widened lines in spot spectra. Now, two years later, he asked himself whether this vortical motion might indicate a revolution of electrically charged particles in spots, similar to what Rowland had observed, and which, therefore, would produce a magnetic field. If so, he should be able to detect it through the Zeeman effect and the polarization of the sunspot lines.

The answers he was seeking could not, however, be found without the aid of the physical laboratory in which he had mounted a powerful electromagnet for the study of the Zeeman effect. Here, by placing an iron arc or spark between the poles of the magnet and photographing its spectrum, he was able to study the effect of magnetism on light. He discovered that "the lines behave in the most diverse way, some splitting into triplets, others into quadruplets, quintuplets, sextuplets, etc." One chromium line was even resolved by the magnet into twenty-one components. Now, if a magnetic field were really at work in sunspots, "we should," as he was to write, "anticipate a close correspondence between the behavior of each solar line and its laboratory equivalent."

To test his hypothesis, a powerful spectrograph that could spread out and separate the lines in the spot spectrum far more than that in the Snow telescope was needed. Fortunately, owing to his foresight, the 60-foot tower telescope was ready on the mountain. This instrument, born of his tree-climbing expeditions and of the wooden towers he had built to study the "seeing" at different elevations, was mounted straight up in the air. At the top of the tower, the sun's rays were reflected from two plane mirrors, through a 12-inch objective (of 60 feet focal length) that formed an image of the sun on the slit of the 30-foot spectrograph in a constant-temperature well 30 feet underground. The mirrors, thicker than those in the Snow telescope, mounted 60 feet in the air, were little affected by the sun's heat; they also escaped some of the warm ground currents. Therefore the images were sharper, and the potential periods of observation were longer than with the low-lying Snow telescope.[8] By the end of November he could report to Carnegie, "After a brief hospital experience in September, I find work with this instrument, involving many trips daily up and down the ladder, a most valuable means of exercise."[9] About the same time he wrote to Gale that he had been going up and down the tower telescope so much "that my muscle, appetite and sleeping ability have reached a high pitch."

For two weeks, from the middle to the end of June, 1908, he spent all his time on the mountain. The weather was good, the solar "seeing" superb. Some of his photographs were "simply exquisite." On June 8, 1908, he wrote to Brashear—unable to contain his joy: "I have been carried away by the solar whirlwinds and have been unable to get out of their influence. We have an extraordinary series of pictures showing how the whirls operate, and one day last week a large prominence was drawn into a sunspot within a few minutes. The whole matter is so interesting that I find time for nothing else."[10] Because Ellerman was away, Hale had been making all the observations himself. Day after day he spent every available moment on the mountain.

In the end the result he had been seeking came dramatically. One day, as he was on his way to the Monastery for lunch, he saw Charles Abbot ahead. He hurried to catch up. Then, "with mischief in his eye, and almost in a stage whisper," he exclaimed breathlessly, "I think I've got it!" "And," Abbot recalls, "he surely had!" This was on June 26, 1908.[11] Using the 4-inch Rowland grating that he had first employed in 1889, he had observed a doubling of the lines that he was sure showed a true Zeeman effect. When he compared these lines with those he had observed in the laboratory, he found an amazing correspondence. "So far as I can see," he wrote to Harry Goodwin, "this *must* be a true Zeeman effect, as we know of no other doublets having such properties."[12] And so it proved to be. For the first time in history, an extraterrestrial field had been detected and measured and related to phenomena previously observed only in an earthly laboratory.

When he wrote to Woodward of his latest results, Woodward's reply was vivid, "This is surely the greatest advance that has been made since Galileo's discovery of those blemishes on the sun."[13]

To make sure of the reality of his discovery, Hale continued to observe. By July 6, 1908, he could report to Huggins that there no longer seemed any doubt of the existence of strong magnetic fields in spots. Not only the double lines in spot spectra were true Zeeman doublets; all or nearly all the lines appeared to be widened by the effect of a magnetic field. "Of course," he added, looking ahead, "the great question is whether these fields will be found capable of producing magnetic storms on the Earth. This cannot be answered satisfactorily, I think, until we can systematically determine the areas, intensities and polarities of the magnetic fields on the sun, for comparison with simultaneous records of terrestrial magnetism. . . ."[14]

Now even Huggins (at the age of eighty-four), who had been skeptical about magnetic fields on the sun, was convinced. "A master observation," he called it. Immediately, too, the great physicist P. Zeeman telegraphed congratulations, and J. C. Kapteyn wrote, "You are opening up a new field and nobody can tell at present to what unexpected and far-reaching results it may lead. . . ."[15] Even Hale could not foresee the truth of Kapteyn's comment.

After considering Hale's results further, Zeeman wrote again: "After balancing the evidence for and against the probable existence of magnetic fields in sun spots to be deduced from the nature of their double lines I very decidedly come to a favourable opinion. It can hardly be supposed that a false theory would explain in so satisfactory a manner, as the magnetic separation does, the very different groups of phenomena to be observed in the sunspot lines."

He concluded warmly: "We live in a beautiful period of physics and astrophysics. You may look back upon the past year with immense satisfaction. I hope you may demonstrate one day the magnetic field in a spiral nebula." [16]

The praises echoed the world over, while closer to home Hale's own family came to realize the dramatic nature of his latest discovery and to sense its far-reaching implications. One day, when he was working on his definitive paper on sunspots and magnetic fields, Evelina Hale wrote proudly to Margaret: "Father has been up the mountain, but comes down today. He is writing his *monumental paper*. It is the biggest thing he has ever done and will put him in the same class with Newton and other discoveries."

Years later, Walter Adams, looking back on this, the greatest discovery of Hale's life, and all the intensive work that had led up to it, would comment: "It is probably not too much to say that this investigation of sunspot and laboratory spectra, originating with Hale, with all its ramifications and applications, has been one of the most fruitful in its results of any in the field of astrophysics and spectroscopy." [17]

Harold Babcock, too, who had made many of the pioneer laboratory measurements of the Zeeman effect, looking back in 1965, writes: "For centuries astronomers recognized in the motions of celestial bodies the operation of mechanical laws which derived from terrestrial experience. Also, the chemical elements familiar on earth had more lately become accepted as the building materials of stars. Hale's discovery of magnetic fields in sunspots was a further immense extension of knowledge since it placed magnetism, like matter, in the cosmic framework." [18]

Today, it should be pointed out, it is known that the vortical motion, observed by Hale above sunspots, has nothing to do with the generation of the magnetic field in these spots. It is also held by the majority of astronomers that a clearly marked vortex is indicated in only a rather small percentage of spots. Nevertheless it was his hypothesis on the cause of vortices above spots that led Hale to explore the possibility of finding magnetic fields in those spots. Therefore the methods he used and the way he reached his result are invaluable as a part of the historical record. As so often happens in the history of science, while the path that led him to his result is no longer held to be the correct one, that path was vital to his discovery.

In an article, "Solar Magnetic Fields," published in September, 1965, V. Bumba and Robert Howard examine current views on sunspots. "Earlier theories," they note, "usually attempted to explain the presence of magnetic fields in the cool spot by some means, such as generation of the fields by vortex motion of ionized particles around a region of low pressure. In more recent times more has been learned about the behavior of plasmas in a magnetic field, and recent theories have dealt with explaining the low temperature as a consequence of the presence of strong magnetic fields. It is generally believed now that convective motions, which account for almost all the outward flow of radiant energy for some distance below the visible surface of the sun, are effectively impeded in the strong vertical magnetic field of a sunspot (or at least that the modes of motion which are most efficient in the transport of energy are impeded). With no energy received from below by convection, the temperature of the sunspot decreases until the energy lost by radiation from its surface is in equilibrium with the energy gained by radiation transfer from below and from the side, or by other means.

"This," they conclude, "is not intended to imply that sunspots are thoroughly understood. A complete theoretical understanding of the sunspot phenomenon may not come for some time." [19]

Thus, as Horace Babcock, a leading specialist in this field, points out, "On at least one current theory of sunspots and the solar cycle, vortical motion is thought to be related to twisting of the submerged magnetic flux tubes below sunspots. Thus, while the vortical motion is not of great importance, it may yet have some significance." [20]

Over thirty years ago, in February, 1934, Hale wrote to Sir Joseph Larmor in England, "The fact is that we don't seem to *know* much about the sun, try as we may." [21] Doubtless many experts in the solar field would still agree with this view.

After Hale's discovery of the magnetic fields in spots in 1908, others followed. Zeeman had shown that the distance between the components of a spectrum line is directly proportional to the strength of the magnetic field. Now, "by determining the separation corresponding to a magnetic field whose strength could be measured in the laboratory," Hale found he could "easily" derive the strength of the field in spots. "The strength of the magnetic field produced, which is measured by the degree of separation of the triple lines, increases with the diameter of the spot. The field is strongest near the center of the spot, where the lines of the triplet are most widely separated, and decreases to very low intensity at points just outside the edge of the penumbra. The spectrograph, when equipped with suitable polarizing apparatus, serves as an extraordinarily delicate means of measuring these fields, which can be observed in regions where they are not much more intense than the magnetic field of the earth." [22]

Before long, this measurement of the strength of field in spots was to lead Hale to the intensely difficult and elusive question of whether the sun itself has a magnetic field. If such a field existed, it must, he was sure, be exceedingly weak. The quest in search of that field, begun at this time, proved to be one of the most difficult in his entire life. It was one that even in the end would leave him uncertain, though he would return to it again and again at intervals throughout the rest of his life.

In these years, as these and other discoveries poured in during an astonishingly short time, Hale continued to spend every available moment on the mountain. "I can see him still," said Harold Babcock (who had come from the National Bureau of Standards to join the staff in February, 1909), "hustling from the laboratory on the mountain to the Snow telescope or the 60-foot tower. Along the steep little paths—just dusty little paths with perhaps a step here and there. He didn't lose any time on the trip. He could make it in an amazingly short interval—with a particular specialized gait. He'd arrive out of breath, but that didn't matter. . . ." As soon as he arrived he plunged into work.

In these years, too, the results of that work continued to arouse enthusiasm among astronomers and physicists the world over. In July, 1908, Albert Michelson, the physicist, wrote, asking for a list of Hale's papers to present to the Nobel Committee in Sweden, with the hope of convincing them of his qualifications for the Nobel prize in physics. Hale replied, "As Hasselberg has all of my papers, it would probably be sufficient to suggest that the award be made for researches with the spectroheliograph. The discovery of the Zeeman effect in sunspots is a natural consequence of the other work." [23]

Interestingly, and somewhat ironically, Hale had just at this time recommended Michelson for the same prize—specifically for his determination of the length of the standard meter. This was a purely physical problem, and Michelson received the Nobel prize—to Hale's great satisfaction. In his own case, as in that of the few other astronomers who have made revolutionary advances in *astro*physics, the committee apparently considered his discoveries too astronomical in character. Therefore the prize was denied him, although his discoveries were of far greater importance than those of many of the winning physicists. (Nobel, it was told, had borne a grudge against an astronomer, and had written his will in such a way that it would be almost impossible for any astrophysicist to qualify for his prize.) The fact, however, that Hale's case was given long consideration, was in itself rewarding.

Nevertheless Hale continued to receive recognition for his achievements, particularly from other astrophysicists who could weigh their value. In March, 1909, Arthur Eddington, an outstanding British astronomer, spoke at the Royal Institution on "Recent Results of Astronomical Research." He described the magnificent results on Mount Wilson, and his praise for Hale's discovery of magnetic fields in sunspots was glowing. He called it "the greatest result of any that recent years have afforded to astronomy." [24]

Soon after this, in the spring of 1909, Hale himself received an invitation to lecture at the Royal Institution. In the preceding months he had not been well, and had gone abroad for the "rest" prescribed by Dr. McBride. He hesitated over the invitation, but finally accepted. He worked on his speech until he was exhausted—crossing out lines, writing in words that would convey his ideas more vividly. On the appointed day he went to the Royal Institution and took his place at the table once owned by his idol, Michael Faraday, who here, sixty-three years earlier, had discovered the effect of magnetism on light.

The hour for the lecture came. Sir James Dewar gave the introduction. Then Hale launched into what was afterward called "the most remarkable address of his career." "It is customary," he said, "to distinguish sharply between the observational and experimental sciences, including astronomy in the former. In physics or chemistry the investigator has the immense advantage of being able to control the conditions under which his observations are made. The astronomer, on the other hand, must be content to observe the phenomena presented to him by the heavenly bodies and interpret them as best he may." He pointed out, however, "that the distinction between these two methods of research is not so fundamental as it may at first sight appear." [25] In his own work, experiment had been as important as observation; the two had, in fact, been inextricable.

As he ranged over the history of astronomy from the Chinese and Galileo to the astronomers of his own day, he showed in vivid detail how the work of Faraday and Crookes, of Thomson and Zeeman had paved the way for his discovery of magnetic fields. This brought him to that subject so close to his heart and to the central theme of his talk—the importance of applying the physicist's methods to astronomical research. As he described his latest discoveries he showed how they would have been impossible without preliminary experiment in the laboratory; how, too, astronomical theories could be proved or disproved by experimental methods as conditions in the sun and other stars are imitated in the laboratory. At the end he received a tremendous ovation.

This talk was given on May 14, 1909. By this time the great 60-inch reflector had finally been set up after years of uncertainty and struggle. The founding of Mount Wilson in 1904 had guaranteed its mounting. Yet progress had seemed painfully slow. As usual there had been countless snags. Some were expected, some totally unexpected; some were caused by the sheer cussedness of men, others by "acts of God." Before the mirror or mounting or any parts of the building could be transported up the mountain, it was first necessary to widen the so-called new trail. Hale tried to get the Toll Road Company to share the $10,000 cost. At the beginning, its members had been eager to help with observatory plans. Now, even though they would inevitably profit from the road, they refused to pay a cent. In the end the Carnegie Institution underwrote the expense.[26]

The construction job proved both difficult and dangerous. The road, nine miles long, had to be built up steep grades over "loose and friable ground." There were deep holes to fill, stream beds to cross. Hale engaged Godfrey Sykes, who had been at the Carnegie Institution's Desert Laboratory in Tucson and had helped Percival Lowell build the dome at Flagstaff, to supervise the construction. Later, Sykes described the work in A Westerly Trend. As he pointed out, the source of power could be either mules or men. The ground was better suited for pick and shovel than for plows and scrapers. Therefore he chose the men. "Time," he writes, "was all important because astronomers are always in a hurry except on cloudy nights."[27]

Before work could be begun, however, winter in all its fury flung itself on the mountain. On January 10, 1907, a violent gale roared out of the north. That morning Hale had started from Pasadena in a pouring rain. As he climbed, the rain turned to snow. At the Half-

Way House, he finally had to give up. It was only the middle of the afternoon, but it was so dark it might have been the middle of the night.[28]

For several days no one could reach the top. Telephone and power lines were down. Landslides carried hundreds of tons of rock over the trail. For weeks no packtrain could get through; no equipment could be carried to the top. Finally, in March, a packtrain started out. Just before it reached its goal, one of the most heavily loaded mules lost his footing and rolled over, down to the bottom of the canyon. He was saved by the deep snow, but until it melted there was no possible way of getting him out. So, as Adams recalls, "for some weeks the position of the mule and the driver were reversed, the driver packing hay on his back to feed the mule, who apparently enjoyed his vacation." To which he adds, "The language of the driver was notable even among the remarks of his highly competent fellows." [29]

Thus a winter was lost. Nor was this the only natural disaster that delayed the 60-inch reflector. In 1906, when San Francisco was devastated by earthquake and fire, Hale was in the East. Not knowing how far the earthquake belt extended, he tried anxiously to get news. A telegram arrived from his wife, intended to relieve his anxiety. "Nothing fell on me," it read. Hale, his imagination working at peak capacity, had visions of his wife saved, but the house crushed and the children lost in the ruins. He rushed back to California, only to learn that the telegram was intended to read, "Nothing felt on Mt. Wilson." Still, he had other cause for worry. The 60-inch mounting was being cast at the Union Iron Works in San Francisco, and it was some days before he learned that it had barely been saved from the fire. Finally it was shipped south. But once more there were endless delays, as widespread strikes stopped work on the dome.

Still, by November 27, 1907, all the 150 tons for the building and dome were finally on the mountain, and the skeleton of the dome was being riveted. By this time, too, Ritchey had finished the long and incredibly difficult task of polishing the 60-inch mirror "with no error in form exceeding two millionths of an inch." [30]

In the optical shop in which he had worked the windows were double and carefully sealed; outer air was admitted to the room through a cheesecloth filter; the temperature was kept constant within two or three degrees by a hot-air furnace, controlled by a thermostat. The walls and ceilings were varnished; the cement floor was painted, and kept wet whenever the polishing was going on.[31] A canvas screen was hung above the glass, to keep any particles from falling on it. Every precaution was taken so that even the most microscopic scratch could not mar the mirror surface. Into this room, for four years, ever since the disk had come from Yerkes, Ritchey had entered, dressed in surgeon's gown and cap. Here, hour after hour, week after week, year after year, the work had gone on. Each day he entered his sanctum, cloaked in an aura of mystery, eternally grinding the mirror to its final perfect parabolic form.

Then came the ticklish job of getting the mirror up the mountain. While the road had been widened, it was still narrow and precipitous, filled with zigzag turns that gave the uninitiated nightmares. At any moment, a slight error of judgment could send the mirror hurtling down the mountainside. Reporters who arrived to survey the scene were appalled by the prospect. A story in the Los Angeles *Examiner* called the delicate task of lifting a ton of deadweight, "which the touch of a baby's hand would mar," to the height of a mile up a steep, rough mountainside, the most tedious and nerve-racking task of the kind ever undertaken.[32]

But finally, with the aid of the special truck bought to transport the heavy equipment and a powerful mule team, the mirror was carried to the peak without harm. By June 15, 1908, Hale could report to Woodward that the telescope tube, the 5-ton fork, and the great steel float, 10 feet in diameter, were safely on the mountain. Over fourteen years had passed since William Hale had given the 60-inch mirror to his son. In the intervening years George Hale had often wondered if it would ever be mounted. How he must have wished that his father could have been with him at the top of the trail to watch its last triumphal journey!

It was December 7, 1908, when the 60-inch mirror was finally cradled in the telescope, which, as a result of Hale's planning, was to prove a most versatile instrument.[33] As shown in the diagrams on page 231, its mounting was so arranged that it could be used in four different ways. In diagram 1, the light is reflected from the silvered mirror to a focal point at the upper end of the tube. The photographic plate may then be fixed in the axis of the tube, or it can be mounted at one side, where it receives the image after reflection from a plane mirror diagonally mounted. In arrangements 2, 3 and 4 the diagonal plane mirror is replaced by a convex mirror, and the light rays are sent back to the base of the tube. Here they are again reflected by a plane mirror, either to a point at one side of the tube, where they may be photographed as in 2, or where a spectrogram may be taken, as in 3. Or the rays can travel down through the hollow polar axis "to unite in an image on the slit of a powerful spectrograph mounted in a constant temperature chamber." [34]

On December 20, 1908, just four years after the founding of the observatory, the first photographs were taken. The results, Hale reported to Huggins, were "admirable." The star images were small and perfect, the details of the Orion Nebula beautifully defined.[35]

Now, at last, night work on the mountain could start. With that beginning, the 60-inch reflector, the largest in the world, became the center of the most intense activity. Like the 40-inch refractor at Yerkes, it was never idle so long as a star was shining. With it, Hale and his colleagues turned to observations that, they hoped, might help to solve the eternally puzzling but ever-fascinating problem of stellar evolution, which had been Hale's chief reason for the building of the 60-inch telescope and the founding of Mount Wilson.

In 1908, the year the 60-inch was mounted, Hale's book *The Study of Stellar Evolution* was published by the University of Chicago. For many months he had spent every available moment on it. "It cost me a lot of trouble and I am delighted to have it off my hands," he told Frost. "I can never get such a thing into satisfactory form, and finally had to let it go as it stands." [36] In its pages he ranged over many astronomical problems—from telescopes and spectroscopes to the study of sunspots and stellar temperatures; from the possibilities of new instruments to opportunities for amateur observers, always a subject close to his heart. Still, at the core of the book lay the problem of stellar evolution, which, he hoped, this new great telescope might finally solve.

The investigation of evolution in the universe, as he showed in his opening paragraph, had been given a tremendous boost by the publication of Darwin's *The Origin of Species*. "It is not too much to say that the attitude of scientific investigators has undergone a radical change since the publication of the *Origin of Species*. This is true not only of biological research, but to some degree in the domain of the physical sciences. Investigators who were formerly content to study isolated phenomena, with little regard to their larger relationships, have been led to take a wider view. As a consequence, the attractive qualities of scientific research have been greatly multiplied. Many a student, who could see in a museum only a wilderness of

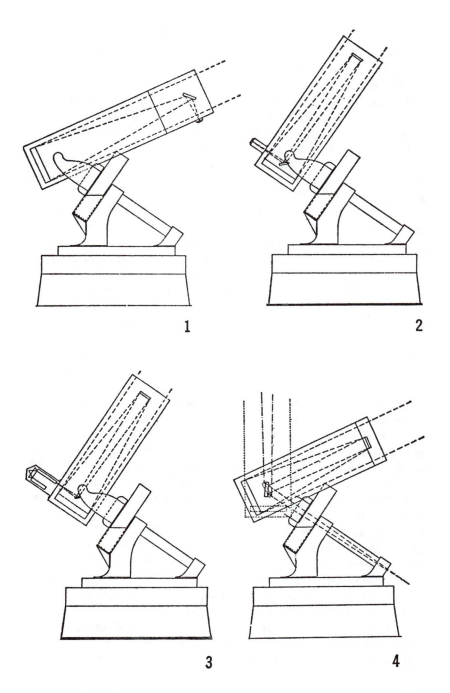

1

2

3

4

Drawing by Hale showing various mirror combinations in the 60-inch reflecting telescope.

dry bones, now finds each fragment of profound interest if the part it plays in a general scheme of evolution can be made clear." And what is true on earth is true also in the skies.

"Our problem," Hale wrote graphically, "is like that of one who enters a forest of oaks, and desires to learn through what stages the trees have passed in reaching their present condition. He cannot wait long enough to see any single tree go through its long cycle of change. But on the ground he may find acorns, some unbroken and some sprouting. Others have given rise to rapidly growing shoots, and saplings are at hand to show the next stage of growth. From saplings to trees is an easy step. Then may be found in the form of dead limbs and branches, the first evidences of decay, reaching its full in fallen trunks, where the hard wood is wasting to powder." [37]

So, scattered over the heavens, are millions of stars (including our sun), each representing a certain stage of development. In the nebulae and in these stars—white, yellow, red, blue—he hoped to read the story of the rise and decline of life in the stars, of "cosmic evolution." This story proved to be more complex than either Hale or most of his contemporaries dreamed. Yet it formed a solid basis on which to found his plans—and dreams—for great telescopes. "Little progress," he repeated in a familiar refrain, "can be made without powerful means."

The 60-inch was, of course, a tremendous leap beyond any telescope then in existence, and no one could be sure of its success in advance. Now as it was launched on its long and distinguished career, it did not take long to be assured of its superior powers. For nearly ten years it would remain the largest telescope in the world. To this day it is still the favorite of many Mount Wilson astronomers. Even after five years, Hale could write that those years had brought out "all its admirable qualities and provided a rich store of photographs for the study of stellar evolution." [38] These photographs would be of the most varied character—star clusters (showing in one case some 30,000 stars), nebulae, stellar spectra, as well as the moon, planets, and comets.

To carry out all this work, as well as numerous other schemes, Hale had foreseen that additional astronomers would be needed. One of the first to arrive after the mounting of the 60-inch, and one of the most ideal, had been Harold Babcock. From the beginning he had charmed everyone with his poetic nature, his enthusiasm, his genial spirit. He became indispensable. "I wish," Hale exclaimed some years later, "we could have half a dozen Babcocks."

Soon after Babcock's arrival Hale invited him to go up Mount Wilson to make some tests with red sensitive plates on the 60-inch. They reached the peak and managed to get a few plates before the fog settled in. Then, for a long time, with the great telescope looming above them in the darkling dome, Hale talked, outlining his plans for the new telescope. As Babcock listened, he realized that he had entered a new and stimulating world, unlike any he had ever known. "It became clear even to a novice like myself that Mr. Hale was thinking years ahead of the actual current date—a tremendous thrill to a young fellow like myself. Every time I turned around apparently I got ideas that were new and in the making." [39]

Babcock was delighted to find a young Italian student on the mountain, who was equally enthusiastic about Hale and the possibilities of work with the 60-inch. His name was Giorgio Abetti, and he was the son of Hale's old friend Antonio Abetti, director of the Arcetri Observatory above Florence. As a physics student at the University of Padua, he had read "with deep admiration and interest about the work Hale was doing in the field of Solar Physics at Mt. Wilson." He had come to Yerkes on a scholarship. But his greatest desire had been to meet the Maestro (as he called Hale). In answer to a letter from the young man, Hale had written back, urging him to hurry on to California, as time with the 60-inch would be available. He added, "I shall, of course, be delighted to welcome you here, and to give you every possible opportunity to become familiar with our work."

"So," Abetti was to record over fifty years later, "wrote the Maestro to an obscure future disciple, and nothing else was required to make him wing-footed. It seemed to him he was reaching Paradise, when immediately after his arrival he was able to accompany Hale to the mountain. Climbing Mt. Wilson along the 'old trail,' my admiration for the enchanting place joined the emotion I felt on hearing Hale quoting with a perfect accent, some verses from the *Divina Commedia:* 'Lo commo er'alto che vincea la vista.'" [40]

To Abetti, like Babcock, the work on the mountain seemed to move at an extraordinarily rapid pace. Yet, to Hale, with all his dreams of what he wanted to do, progress often seemed painfully slow. Conditions were primitive; the spectrographs were rough, wobbly wooden affairs, improvised from materials at hand and necessarily experimental in nature. It was still impossible to buy red-sensitive plates. Therefore it was necessary to take the blue-sensitive ones and sensitize them to the red. Even then red filters were needed. All this was new; it was slow and difficult. All the work had to be done on the mountain. Nevertheless, some superb and revolutionary results were obtained. Babcock, working with Adams, (using a rigid instrument of 18 foot focus in a pit at the 150 foot focus of the 60-inch) was able to get some magnificent spectra of six or eight of the brightest stars in the sky. One of these, Antares, because of its red light and the insensitivity of the plates and also because of its low position in the sky, required five nights' exposure. Today, as a result of photographic and instrumental advances, the same spectrum can be photographed in twenty minutes.

These results with the spectroscope, as well as countless others made by direct photography, soon confirmed Hale's long-held belief in the potentialities of a powerful reflector. For over fourteen years he had looked forward to using its great light-gathering power. Now at last his hopes were being realized. With its huge area (57,600 times that of the human eye), it could reveal stars of the 18th magnitude; with long exposures of four or five hours, stars as faint as the twentieth magnitude could be detected. (To an astronomer "magnitude" denotes the brightness of a star. Two stars differing by one magnitude differ in brightness by a factor of 2.5; an interval of five magnitudes corresponds to an intensity ratio of 1 to 100.) With each gain in magnitude, myriads of stars never before seen became visible. The faintest star visible with the 60-inch was about 100 million times less intense than the brightest star in the sky. By 1915 Hale could write: "Hundreds of millions of stars have been photographed and the boundaries of the stellar universe have been pushed far into space, but have not been attained." [41] As those boundaries were pushed farther and farther out, more spiral nebulae, more globular clusters were found. And so, in time, the views of our own Milky Way, as well as of the rest of the universe, would be radically changed. Nevertheless, astronomers continued to ask those fundamental questions that have puzzled man for centuries. Is the universe infinite, or do these stars and nebulae decrease in number in the farthest reaches of space?

One astronomer, in particular, who arrived on Mount Wilson in the summer of 1909, was especially concerned with the structure of the universe. This was the noted Dutch scientist whom Hale had first met at the St. Louis Congress in 1904. At that meeting Kapteyn had first presented his hypothesis of two star streams in our Milky Way system, which, he hoped, would provide a key to that structure. Now the 60-inch might help him prove or disprove his hypothesis. He came as a research associate, one of a long and illustrious line that would contribute inestimably to the growth of astronomical knowledge as a result of their work at Mount Wilson. For five summers, until the outbreak of war in 1914, he came to the mountain. And there, while carrying on his profound investigations into the nature of the universe, he lived happily with his wife in a tent. "His observatory," as H. D. Curtis was to write, "has been his desk and a room fitted with instruments for measuring photographic plates, his subject matter photographic plates and observations of star positions made by other astronomers; his tools the methods of mathematical analysis." [42]

For many years he had been measuring the positions of hundreds of thousands of stars on plates taken at Cape Town by Sir David Gill. His results had led him to believe that he had found two great intersecting streams of motion in the sky. Gradually, too, he felt he was gaining evidence that the Milky Way in which we live is a spiral nebula. "But," as his good friend Harold Babcock points out, "it wasn't such a straightforward conclusion in those days as one might suppose. The whole discussion of whether the well-known spirals were so-called 'island universes' or just members of our own system was still quite unresolved." [43] In the hope of solving these problems, Kapteyn had formulated his plan for studying "selected areas of the sky" in which the positions, motions, and distances of numerous stars uniformly scattered over the sky could be measured and analyzed. The 60-inch, with its great light-gathering and space-penetrating power, as well as telescopes in other observatories around the world, were to take part in the program. It was the kind of worldwide, even universewide, cooperation that Hale believed in most strongly.

At first, however, Hale had hesitated before taking on this large program. It was not exactly in line with his own plans for the use of the 60-inch. Kapteyn was "not directly concerned with those studies of physical condition and of evolutional progress in which we hoped to aid in tracing the life-history of stars and stellar systems from their birth to their decay." Therefore Hale asked, "Could we afford to postpone some of our own investigations, at first sight very different from his, even for the very important purpose of advancing his great undertaking?" [44]

However, he did not hesitate long. He foresaw that the physical development of stars might depend "upon just such associations as the discovery of star-streams had disclosed." He knew, too, that many questions might remain unsolved if attacked from one angle only. Therefore an intensive investigation of the magnitudes of stars and their velocities was launched with the aid of the large spectrograph attached to the 60-inch telescope. These measurements, based on observation, proved invaluable. And yet, as so often happens in the history of science, the main features of Kapteyn's hypothesis of two star streams are considered quite differently by modern astronomers. His view of our galaxy as one in which the stars are concentrated at the center and thin out toward the edges is no longer held in its original form. Moreover, the observational evidence that would alter his picture would come from work with the 60-inch and later with the 100-inch telescope.

CHAPTER ELEVEN

The Origin of the California Institute of Technology

MEANWHILE, as the plant on the mountain grew, Hale's own surroundings in Pasadena had changed. In the fall of 1907 Dr. James McBride suggested that they swap houses, that Hale buy "Hermosa Vista," and that in turn the McBrides take over the house on Bellefontaine Street. "Hermosa Vista," a large brown-shingled house, overgrown with vines, was said to have been the first hostelry in South Pasadena; part of it had been used as the post office. Old pictures show horses and carriages at its door while their occupants collect their mail and the latest gossip. It was surrounded by five acres of wide-spreading grounds with magnificent trees and gardens.

Hale, charmed by the place, decided to buy it. He was particularly entranced by an L-shaped room upstairs that he decided to turn into a library. Here, surrounded by his books, he was to spend much of his time in the years to come. Here were the tales of adventure that had delighted him since boyhood, and other books that had helped to shape his life—like *Cassell's Book of Sports and Pastimes,* in which he had first discovered the magnificent world of the spectroscope. Here, too, were detective stories like Wilkie Collins's *The Woman in White,* which were his greatest source of relaxation; biographies and autobiographies of such fascinating figures as Gertrude Bell and Richard Burton; and still others on art and music, poetry and archaelogy—an extraordinary range of titles. In addition there were piles of book catalogues, for further exploration into "unknown realms," which he marked for possible purchase and with which he often read himself to sleep. Here he had the best in reproduced music in the records he loved to play. Here, in his chair in front of the fire, he worked, wrote, read, and welcomed his friends.

As time passed, this library was to become a sanctum into which he would retreat more and more.

The house had another advantage. From the front veranda he had an unobstructed view of the observatory and a large part of the new mountain road. With his 8½-inch telescope he could see the Japanese laborers near the top, nine miles away, and even discern such minute details as the kind of tools they were using and the clothes they were wearing. He felt he was as close to paradise as he could ever come on earth.

Mrs. Hale, however, found the residence big and burdensome, for it had never been designed as a house, but was adapted from its earlier days. A month after they moved, Hale confided to Harry Goodwin that she had had a return of her nervousness—caused by the children's illness and the trouble of getting settled. He added philosophically, "I hope it will soon disappear."

In time Evelina Hale became absorbed in the cultivation of the rose garden that was her pride and joy. Her husband, sharing her delight, would often walk along the brick paths and point out to friends the plantings she had made during the week. Mrs. Hale also became an ardent member of the garden club, and took great pleasure in entertaining intimate friends at Sunday dinners. These came to include the members of the "club" and their wives—the McBrides, the Holders, and in time the James Scherers. On Sundays there were also "tea hours" when it was known that the Hale family was at home. Once a year there would be a large garden party for all the Mount Wilson staff. All these things Margaret would recall, together with the memory of "occasional dinners that were beautifully put on."

When they moved to Hermosa Vista, Bill was seven and Margaret eleven. From the first the children were enchanted by this rambling place with its "dark, sheltering feeling" and sense of mystery. They liked the great barn out back that offered untold possibilities, and the rear porch covered with wisteria vines. They also enjoyed the walks on Sunday mornings with their father into the unspoiled hills bordering on the Arroyo Canyon nearby. One ravine had been dammed to form a pool; they called it Pirate Lake, and this was often their goal. Along the way their father would point out wildflowers and rock formations, and would tell something of their history.[1]

Yet, as the years passed and Bill grew, George Hale found it hard to understand a boy who never showed the slightest interest in anything scientific. He gave him tools and engines, all the things that had fascinated him as a boy—but without success—apparently forgetting that he himself had been driven by a passionate desire to own all these things, and had obtained them only when he had proved to his father that he was ready for them. Bill was a charming boy, with an engaging personality and an interest in other people. But for a long time he showed no driving interest in anything. His father was baffled. "I wonder what is the matter with Bill," he would say. Bill had, of course, been a sickly child, and was kept out of school until he was eight. But this, he felt, could not be the entire explanation.

In his old house on Bellefontaine Street, Hale's neighbor had been Charles Frederick Holder, a distinguished zoologist who had arrived in Pasadena some twenty years earlier, and the two men soon became warm friends. Holder had been assistant curator of the American Museum of Natural History in New York, and later had become the editor of the *Californian* in San Francisco. About 1885, in answer to an invitation from Walter Raymond, builder of the great hotel on Raymond Hill, he had come to Pasadena to live. His little book *Picturesque Pasadena* soon became the official guide for all who took the famous Raymond tours. No one knew the region so well or could tell about it more delightfully. An ardent fisherman, Holder had been the first to hook a hundred-pound tuna with rod and reel. Ever after, the rod hung in the Tuna Club on Catalina Island, which, like the Valley Hunt Club in Pasadena and the far-famed Tournament of Roses, he had helped to found. Under his tutelage, Hale was soon tying his own flies, and wished he had time to fish in the High Sierras, another of Holder's haunts.

Holder was also a trustee of Throop Polytechnic Institute, a small coeducational school in Pasadena. In addition he was the curator of its museum and a member of its faculty. Amos G. Throop, a former member of the Chicago Board of Aldermen, had decided in 1891, at the age of eighty, to equip and endow a polytechnic institute "to provide for all who may wish an inexpensive, but liberal, thorough and practical education." He called it Throop University. Later the name was changed, and a large red-brick building was erected across from the Universalist Church at Raymond and Chestnut streets. In time the school was expanded to include elementary grades, an art school, domestic-science and manual-training courses. In addition, there were courses in electrical engineering, chemistry, and natural science, and two or three students each year received the B.S. degree. A few, like Frank B. Jewett, president of the Bell Telephone Laboratories, became famous. But with uneven standards, "meagre funds and many other handicaps," Throop was in general little more than an advanced secondary school.[2]

Soon after his arrival in Pasadena, Hale had become interested in the small institute.[3] Even before the founding of Mount Wilson he had spoken there. Often he had discussed its possibilities with Holder. Gradually these possibilities formed themselves into a definite plan in his mind. One evening, as they sat in Holder's den, he suggested that Throop might be made into a scientific institute of the first rank. To compete with German science, technical schools were needed, "capable of turning out men of the best education (not merely training)." In such schools the emphasis should be "upon the encouragement and development of *research*." To accomplish this would mean changing Throop's entire character, dropping all elementary grades and superfluous courses, and concentrating on fundamental science and a few branches of engineering. Moreover, "adequate instruction [should be] given to all students in the humanities." If such a goal could be carried out, if Throop's wealthy friends would help financially, the possibilities were limitless.[4]

Holder quickly caught Hale's enthusiasm, and in the weeks and months that followed he discussed the suggestions with the other Throop trustees. They, too, glimpsed the vision behind the idea. In a letter dated April 29, 1907, S. Hazard Halsted, president of the Pasadena Ice Company and a Throop trustee, wrote to Hale to tell him that at a meeting of the trustees two resolutions had been adopted (both in line with Hale's proposals). One committed the trustees to make of Throop Institute a high-class technical school; the other suggested that the first step in this direction should be the establishing of a course of electrical engineering, if possible second to none. "All of the Trustees," Halsted added, "are greatly pleased at the interest in this subject which you have expressed to several of the members and at your willingness to aid us in this matter."[5]

This letter indicates the confidence Hale had already inspired. At this point, however, in the spring of 1907, he was planning a trip to Europe. He had not intended, as Halsted now proposed, to include a search for a Throop president in the East. Still, he felt he must accept the responsibility.

His first candidate was his former professor at M.I.T., Henry Clifford who, he felt, should also be able to teach the course in electrical engineering. He decided to discuss the question, first with his old college friends Arthur Noyes and Harry Goodwin, then with Clifford himself in Boston. There was, however, considerable irony in this situation. A year before, in the spring of 1906, Hale (then thirty-seven years old) had received offers of two positions, both great honors. Simon Newcomb, the noted astronomer, and the redoubtable Alexander Graham Bell had urged him to accept the secretaryship of the Smithsonian Institution in Washington, perhaps the leading scientific position in the country. He had turned down the offer, telling Newcomb he was "absolutely contented" at Mount Wilson.[6] The second offer was more difficult to reject. This was no less than the presidency of his alma mater, M.I.T. At first he refused quite firmly, but Noyes would not take no for an answer. Nor would the M.I.T. Board of Trustees. Moreover, Noyes offered what he knew would be the most irresistible part of the plan—the hope that the three college friends might at last work together. If necessary, he said, they could wait a year. By that time Hale should have the observatory in good running order "so that your assistants could carry on the work satisfactorily." Possibly he could spend two or three months of each summer in California, which, Noyes noted, "would be, I imagine, of the character of a vacation."[7] (At this, Hale must have smiled.)

Hale had finally made his choice in favor of Mount Wilson. Still, the question of what could be done at M.I.T. continued to haunt him. Sometimes, especially early in the morning, he could think of little else. On one such morning, he wrote down some of his ideas, and a week later sent a draft to Noyes. "I am still regretting," he wrote, "that I could not see my way clear to telegraph you favorably. New possibilities for introducing important work at the Institute strike me daily, especially about five-thirty A.M. when I am likely to be thinking over matters of this kind."[8]

Later he embodied his ideas (which pertained to Throop also) in an article entitled "A Plea for the Imaginative Element in a Technical Education," which appeared in the *Technology Quarterly* (an M.I.T. publication.)[9] "It is probable," he wrote, "that the average

member of a technological school is in more danger of a narrow outlook than any other class of students. In a large percentage of cases
he has rejoiced from boyhood in a mechanical turn of mind, which
has concentrated his attention on engines and machinery and the
splendid achievements of modern engineering.

"Happy is the boy whose career is thus plainly foreshadowed. For
him, life is sure to be worth living, and the dangers of idleness may
be ignored. But this very interest, in direct proportion to its intensity, is almost certain to lead to a neglect of other opportunities. The
absorbing beauties of machine construction and design so completely occupy the boy's mind that they hinder a view of the greater
world. He cannot be expected to perceive that a knowledge of the
details of his chosen profession should not suffice to satisfy his ambition. He does not yet know that to become a great engineer, he
should cultivate not merely his acquaintance with the details of construction, but in no less degree his breadth of view and the highest
powers of his imagination."

Then, thinking especially of "Tech" and his rebellion against routine work in college, he stated his belief that any good technological
institute should "contribute to the world the largest possible proportion of men capable of conceiving great projects and the smallest
possible proportion of men whose ambition can be completely satisfied by the work of executing them."

As one way of cultivating the imagination he proposed a course in
evolution that, in its broadest scope, might be better able to
"awaken the imaginative powers and to develop an understanding
than any other single conception." As he developed his ideas in vivid
detail, he showed how it could be brought into the study of every
aspect of life, even into that of the growth and origin of society. In
this way, too, he emphasized his belief in the importance of the humanities in any technical education that lay at the heart of his plans
for Throop.

In 1907, when most scientific schools emphasized technical training almost exclusively, such ideas were almost unheard of. It was extraordinary, as a friend has remarked, that Hale, constantly absorbed in astronomical and personal problems, could have arrived at
so broad a philosophical view.[10]

The day he arrived in Boston to look for a president for Throop,
all these things formed a background to his visit. The next day he

went to see Henry Clifford in the familiar old buildings on Copley Square. He was delighted to find his old teacher interested, though a bit wary of Throop's wobbly financial state. He promised to come out to Pasadena in July, and with this promise in his pocket Hale hurried on to see Noyes. They went out to lunch with Goodwin, and over the table they argued the respective merits of the East and West. As Hale soon saw, Noyes had still not given up hope of enticing him to Boston.[11]

The following fall, however, Noyes himself accepted the acting presidency of M.I.T., and soon afterward Richard C. Maclaurin became president. Hale was delighted and relieved, "After passing through a period of temptation, during which I was almost ready to believe that administrative work would prove as attractive as scientific, and perhaps more effective, and after having escaped in safety, I came back to my research with a pleasure such as I never experienced before in my life." [12]

Later, in an illuminating letter to Frost, he reviewed the reasons that had led him to decline this and other attractive offers: "I am convinced that I can accomplish comparatively little in research without devoting my *entire* thought to it. New ideas come to me very slowly, and only as the result of continual thinking in and out of working hours. If I have other things on my mind, especially journal work, which must be completed by a certain date, I am so much disturbed that I make no headway. As I look back upon my record, I find that I accomplished nothing whatever in solar research during the entire period of the organization of the Yerkes Observatory which was not altogether completed when I came to California." [13]

In May, 1907, shortly after his lunch with Noyes, Hale sailed with his wife on the *Baltic* on his way to the Solar Union meeting in Paris. Only just before sailing did he learn that Andrew Carnegie would be aboard. On the passenger list there was also the name of James A. B. Scherer, president of Newberry College in South Carolina—a name Hale did not know but with which he would soon become well acquainted—a name, too, that would have unexpected implications for the future of Throop.

As the *Baltic* sailed out of the harbor, it struck on a sandbar. While a tug tried to pull her off the bar, Scherer, as he told Hale long afterward, found the opportunity for which he had booked passage on the *Baltic*. Carnegie was standing on the deck "all swathed in Scotch plaids, a red haired gillie in attendance." Screwing up his courage, Scherer sidled up and murmured, "Is this Mr. 'Carnaygie'?" As Carnegie wheeled, Scherer saw "a slightly oversized netsuke, his skin the color of old ivory, his face fringed with the white whiskers of an Oriental sage and lit with the canniest pair of blue eyes I have ever seen, set wide apart."

"Well," Carnegie exclaimed, "here's a young man so intelligent that he knows how to pronounce my name!"

"Not to know how to pronounce your name, Mr. Car-nay-gie, would have proved me unintelligent," Scherer replied "on a chance." The chance proved a happy one, for the "Laird of Skibo" was vain. "Not a bit of it," Carnegie blustered, "I'm so little known that very few people ever call my name correctly! Where're you from?"

"Oh," Scherer answered, "a little college down south to which you lately gave a building, but of course you wouldn't remember, you give so many."

"What's its name, and what kind of a college?"

"Newberry, and Lutheran."

This last, Scherer saw to his dismay, was the wrong track. Immediately Carnegie plunged into an attack on orthodox religion. Scherer counterattacked. All through the voyage, whenever the two met, the argument continued. Scherer was sure the purpose of his voyage—further support for his college—was doomed.[14]

Meanwhile Hale, unaware of these maneuverings, was pleased when the Carnegies invited Evelina and himself to join them at their table in the saloon. Here was his chance to talk of his own plans for Mount Wilson. Here was his chance, too, to learn more about Carnegie himself. Often, then, he found that Carnegie could be highly entertaining, relating endless anecdotes, reciting long passages from Shakespeare and his beloved Robert Burns, even discussing the philosophy of Herbert Spencer. At other times, however, he could be completely uncommunicative.

Some days out, Scherer, after learning Hale's identity, peered over the "great astronomer's shoulder," saw he was reading a Marryat novel, and introduced himself. Hale threw his book aside, and "proved himself one of the most charming conversationalists" Scherer had ever met.

As the end of the voyage neared, Scherer kept out of Carnegie's sight. As a result of their altercation, he was feeling "bluer than indigo." Hale finally found him in the shadowed stem of the ship after they had reached Liverpool. "I've been looking for you everywhere," he cried.

"Whatever for?" Scherer demanded grumpily.

"The Laird wants to see you," said Hale.

"Well, I don't want to see him!" was Scherer's reply. "We've done nothing but quarrel all the way across the Atlantic, and I don't want to make bad matters worse."

"Nonsense," Hale exclaimed laughing. "He's really anxious to see you!" With that he pushed Scherer along the deck to the gangplank "where the Laird of Skibo stood with his entourage."

Then the incredible happened. Carnegie reached up and affectionately circled his adversary's shoulder and cried: "See here, you! Where have you been hiding yourself? Lou and I are arranging for a private railway carriage to bring Dr. and Mrs. Hale up to Skibo after they've finished their London rounds, and you're to come with them!"

Scherer, forgetting entirely the Sunday-school convention he was supposed to be going to in Rome, blurted out, "But I've engaged return passage on the *Kaiser Wilhelm II.*" Carnegie shouted, "An invitation from me is a command, young man!" to which Scherer replied, "Certainly, Sir! I'll cancel the ship."

A few weeks later the shipboard companions found themselves at Skibo Castle at Dornoch. On its topmost tower floated Carnegie's extraordinary flag—the Union Jack on one side, the Stars and Stripes on the other. The carriage drew up at the door, and the "Laird" rushed bareheaded to greet them. With that introduction, one of the most remarkable visits of Hale's life began. He was both amused and amazed by life at the castle.

The morning after their arrival he got up early to write to Robert Woodward, the Carnegie president. While he wrote, the bagpiper, in full Highland costume, was making his rounds, heralding breakfast. Before the day's end, Carnegie had driven them over his huge estate, taken them trout fishing, and challenged Hale to a round of golf. Unfortunately, Hale caught fish when Carnegie could not. He also won at golf, though he did his best to lose.

That night, after a sumptuous dinner in the baronial hall, the Hales excused themselves. Scherer soon followed. Unable to restrain himself any longer, he invited Hale to join him in his room in their "fortunately isolated wing." And there, locking the door, he burst

Characteristic letter from Hale to Walter Adams.

out with the account of all that had happened on their Atlantic crossing.[15]

In Pasadena, meanwhile, plans for Throop were growing, and soon after his return from Europe, Hale spoke to the trustees on the institute's future. The chairman of the board was Dr. Norman Bridge, a man of "impressive and elegant appearance" who had migrated to California from Chicago after discovering his own tuberculi in his microscope. Other members included Dr. James McBride, a close friend of Holder; Arthur Fleming, a wealthy lumberman, the most important of Throop's early benefactors; James Culbertson, who had also made his money in lumber; C. D. Daggett, and John Wadsworth, "the biggest person on the board."

It was to these men that Hale outlined his ideas of Throop's possibilities. He again pointed out that if American science were to advance, greater attention to German research methods and to research in industry was needed. And there was no first rate technical school west of the Rockies. Then, thinking of the proposed course in electrical engineering, he spoke of the abundant waterpower in the nearby mountains that would make experimental transmission lines, invaluable for teaching and research, easy to install. The opportunities for Throop, together with the enticement of climate, were unsurpassed.

"In developing such a school," he urged, "we must provide the best of instruction and the most perfect equipment that modern engineering offers. But in laying stress upon the practical aspects of the problem, we must not forget that the greatest engineer is not the man who is trained merely to understand machines and to apply formulae, but is the man, who, while knowing these things, has not failed to develop his breadth of view and the highest qualities of his imagination. No great creative work, whether in engineering or in art, in literature, or in science, has ever been the work of a man devoid of the imaginative faculty."

The Throop trustees were practical men of affairs. Yet they were impressed by Hale's vision. They listened closely as he exclaimed enthusiastically, "In most technical schools, the problem has not been solved, and the opportunity stands open for Throop to devise and carry into effect a broad scheme of education which may give proper recognition to all sides of the engineer's life." [16]

In August, 1907, Hale was unanimously elected to fill the vacancy on the Board of Trustees, and Henry Clifford was appointed president to succeed the acting president, Arthur Chamberlain.[17]

Now, however, a snag arose. Clifford, who had visited Pasadena in July, had obviously been impressed by Hale's plans. But California was far from eastern centers of "civilization." There were only two main rail lines across the continent. The roads over which covered wagons had traveled west were rutted still. There was no main highway on which those willing to undertake the hazardous journey could drive. Therefore Clifford had apparently compared the prospects of Hale's "daring venture" with the settled, cultured life of Boston, and regretfully declined the offer. As John D. Rockefeller was said to have asked, "Who would want to go to college west of the Rockies?" [18]

In January, 1908, Hale wrote to his second candidate, George V. Wendell, professor of physics at the Stevens Institute in Hoboken: "How would you like to forsake the east and seek your fortune in California? I have just returned from a short trip to Chicago, and after experiencing in my throat and elsewhere the effects of the weather conditions there, I am more fully convinced that no one should dwell outside of Paradise." [19] Apparently Wendell thought he could continue to dwell elsewhere. He, too, refused.

So the search continued. For some weeks nothing happened. But in April, 1908, when Hale was on his way to the National Academy meeting in Washington, the name of James Scherer leaped into his mind. Up to this time he had felt that the Throop president must be a scientist. Suddenly the idea struck him, "Why wouldn't Scherer be a good choice?" He had shown his ability to extract money from a very wealthy man under the most discouraging conditions. (After the Skibo visit, Carnegie had given Newberry College twice the amount Scherer had hoped for.) He was evidently a good administrator, a scholar, a writer, and a man of vision with charm and wit. His experience of the world at large was, moreover, wide.

He had, as Hale had learned during their days on the *Baltic* and at Skibo Castle, been born in Salisbury, North Carolina, thirty-eight years earlier. He had graduated from Roanoke College in Virginia. After receiving his Ph.D. in English literature from Pennsylvania College, he had gone to Japan as an English instructor in the Imperial Government School at Saga. During his five-year stay there he met and married his wife. On his return to America, he continued to retain his interest in Japan, and later wrote several books on the subject. He became president of Newberry in 1904.

So, on May 9, 1908, Hale wrote to offer Scherer the Throop presidency. It seemed to him now that Scherer was the ideal choice, and he was desperately eager to have him accept the offer.[20] Finally a telegram arrived in which Scherer offered to come out to Pasadena on June 20th to look the ground over. The day of his arrival was hot, and Southern California, after a long drought, was parched. Later Scherer confessed his disappointment at his first view of "Paradise." "My politeness had to be stretched to its utmost to meet the claims of the Board of Trade boosters." But soon, fired by Hale's enthusiasm, these first impressions vanished. The thought of working with this "extraordinary" man was compelling. "The core of its magnetism," he was to write, "lay for me in the laboratories Hale had already developed, after only four years, in connection with his observatory up on Mt. Wilson. . . . Now he aspired to see Pasadena, which he dearly loved, become the center of a new cultural empire, already rapidly expanding. This aim he never let elude him." It was this aim Scherer dreamed of sharing. It was the kind of challenge he liked best. "Logic on fire," he commented, "is a good definition of the higher forms of eloquence, and no phrase could better describe Hale's enthusiastic reasoning." [21]

During his visit, Scherer spent hours poring over the architectural plan for the campus that Myron Hunt and Elmer Grey, the architects, had laid out from Hale's first rough sketches. Arthur Fleming, the Throop trustee and lumber magnate who had been born in Ontario, Canada, had already given the site, a twenty-two acre plot. The terrain was on a slant. A large part of the acreage was planted in orange groves. There was a stream running through it, which in the rainy season formed a sea of mud. Yet, as he showed Scherer the site, Hale spoke glowingly of the structures he envisioned there—the main building, the physics, chemistry, and electrical-engineering laboratories, a center for the humanities. It was as exciting as any project he had ever launched.

After a meeting with the trustees Scherer left for the East. On his way he stopped in Chicago to see Dr. D. K. Pearsons, the donor of the science building at Beloit (which William Hale had endowed anonymously) and a friend of old Amos Throop. When Scherer mentioned the possibility of the Throop presidency, Pearsons "punched me in the ribs" and said, "Go! It's wonderful! It's the greatest educational opportunity in the country! Go! Go soon!"

When Scherer reported all this, Hale was delighted. He was also pleased by Scherer's praise of his book *The Study of Stellar Evolution*. It read, Scherer wrote, like a novel. "I never thought I could get as clear a grasp upon some of the deep things of astronomy as you give me with your lucid and often picturesque style." He sent his love to "Margaret and Willum," and added, "I wish my two were with you tonight as you read of pirates and things. My boy is at my feet as I write." [22] Like Scherer, Hale looked forward to an "extended friendship." Yet both men would have been astonished if they could have foreseen that one day the boy at Scherer's feet would marry Margaret Hale.

Meanwhile, in New York, Scherer had a meeting with Dr. Norman Bridge. While he had some reservations, Dr. Bridge concluded that Scherer was "a man of great power and possibilities." [23] Two weeks later, Scherer was unanimously elected president.[24]

An essential part of the plan for the "new Throop" was its emphasis on a high-class scientific and technical school with a carefully selected student body. This required the dropping of a large number of students, and the abandonment of the high school, the normal school, and the nurses' courses. (The lower grades, discarded by Throop, would be formed into a polytechnic school.) One of Scherer's first tasks was the recruiting of a student body that would make possible the kind of institute Hale envisioned. The prospects were scattered in high schools all over the country. Therefore the new president, "the peripatetic vendor of the 'Throop idea,'" had to travel far and wide, his valise filled with lantern slides, extolling the still unknown institute. Somehow he had to persuade prospective students that the "greater Throop" would do for the Pacific coast what M.I.T. was doing for the Atlantic, perhaps even more. It was not an easy task. It was not surprising, therefore, that when he looked back on this difficult pioneer period he should compare his function to "a shaky one-horsepower generator; how shaky no one ever guessed." [25]

Still, despite all difficulties, the students, attracted by this pied piper, came from every part of the country, even from the far corners of the earth. The wealthy members of the community, inspired by its goals, donated funds for its buildings. In time, some of the best scientists in the world, attracted by Hale's belief in the future of the embryonic institute, would join the faculty. Even by January 7, 1909, a few weeks after Scherer's arrival, $70,000 had been raised, and, as Hale told Goodwin, "much more is in sight." On April 26th, ground was broken for the first building on the new campus. The foundations for the "greater Throop" were laid.

On February 5, 1910, a year and a half after Scherer's induction, the archway sculptured by Alexander Calder over the triple entrance to Pasadena Hall (as Throop Hall was first called) was unveiled. Four months later the building was dedicated. That day a telegram came from Andrew Carnegie, "May Institute fulfill highest hopes of warmest friends." [26] The following fall, Throop opened its doors to a student body reduced from the original five hundred to a small but select group of thirty-one. (It was said afterward that this "contraction of numbers in the face of a great expansion of plan was probably the boldest step ever taken by an American educational institution.") [27] The students rattled around in the great new building, and when the rains came they plowed their way to classes through mud and slime. But they shared Hale's and Scherer's enthusiasm for the new Throop, which would eventually gain worldwide fame as the California Institute of Technology.

CHAPTER TWELVE

Birth of the 100-Inch
Telescope
✳ *1906-1910*

"IT IS IMPOSSIBLE to predict the dimensions that reflecting telescopes will ultimately attain. Atmospheric disturbances, rather than mechanical or optical difficulties, seem most likely to stand in the way. But perhaps even these, by some process unknown, may at last be swept aside. If so, the astronomer will secure results far surpassing his expectations." [1]

Thus Hale wrote prophetically in 1907, a year before the 60-inch reflector was mounted or there was any guarantee of its success. Already he was thinking of an even larger telescope—a mammoth 100-inch. In building the 40-inch and the 60-inch, he had encountered one unexpected difficulty after another. It had taken five years from Yerkes's promise to the dedication of the telescope. Fourteen years were to pass between William Hale's gift of the 60-inch disk and the actual mounting of the telescope. A patient man would have found the waiting hard; the impatient Hale found it devastating. Still, nothing could quench his enthusiasm. He was now ready to launch on an even greater voyage. It would be risky, how risky he had no idea. But even if he could have foreseen the agonies involved, he would probably have gone ahead anyway.

One day in the summer of 1906, when Mrs. Hale was still in the sanitarium, he had gone out to spend the weekend at the Hookers'. All through dinner John Hooker sat silent, while Hale carried on an animated conversation with Mrs. Hooker and Ellie Mosgrove.

After dinner, as the talk turned to astronomy, Hale mentioned the possibility of a 100-inch telescope. Immediately Hooker brightened and started asking questions, and Hale was soon enthusiastically explaining the value of such a gigantic instrument.[2] It would, he said, give two and a half times as much light as the 60-inch, seven times as much as any other telescope then in use for stellar spectroscopy. With it, revolutionary results in the photography of nebulae might be expected. "Tens of thousands of these small objects, on a sufficiently large scale for the most refined measurement," could be caught, and any change in form, any sign of development could be detected. With the addition of a convex magnifying mirror and a yellow color screen, unprecedented photographs of globular clusters, planetary nebulae, the moon and planets, could be obtained. Finally, a 100-inch telescope "would enormously surpass all existing instruments in the photography of spectra of stars and nebulae, thus giving new information on their chemical composition and the temperature and pressure in their atmospheres." [3]

Hooker was enthralled by Hale's glowing account. He asked how much the mirror for a seven-foot telescope would cost. Hale estimated that $25,000 would probably cover the cost of the casting of the disk alone. Before the evening was over, Hooker said he might consider giving such a mirror to Mount Wilson. Some days later he called to say that he had decided to give $45,000 for the construction of a 100-inch mirror. Like Yerkes, however, he insisted it be the largest in the world.

Hale was elated; yet he was cautious. Perhaps he was somewhat appalled by the hazards of the undertaking. Subsequently he explained the problems to Hooker in detail, emphasizing the risk of failure. The huge block of glass for the mirror would weigh four and a half tons—four and a half times as much as the 60-inch. Could such a disk be successfully cast? No one could know. Even then, could a mounting be built to carry such a weight with the necessary precision? Then, too, there was the problem of atmospheric conditions. Would they be good enough to permit so large an aperture on Mount Wilson? Or would some other site have to be found? Finally, money must be raised to mount the telescope in a building with a revolving dome. This would cost at least half a million dollars. The Carnegie Institution, Hale pointed out, might never be willing to provide such a sum.

Moreover—and this should have been enough to give pause to anyone, even to Hale—the great 60-inch reflecting telescope would not be mounted and tested for two more years. Hale hoped, of course, but he could not be sure that it would prove successful. Even the 60-inch was a tremendous leap beyond any reflecting telescope then in existence, and a 100-inch was of course a far greater leap still. Certainly Will Hale had been right when he called his brother the greatest gambler in the world. Now, even while some people talked about his courage and his daring, George Hale himself must have recognized that many others would consider this latest venture foolhardy.

At the moment, however, Hooker refused to take any of these obstacles seriously. He advised Hale to go ahead with plans for a 100-inch mirror, on the chance that someone, somewhere, would pay for its mounting. From this time, he declared, work on the mirror would be more interesting to him than his own business.[4]

As soon as Hooker's gift was assured, Hale wrote to Robert Woodward, the president of the Carnegie Institution, telling him that the gift had been made unconditionally. "We are greatly pleased, in view of the immense possibilities of a 100-inch reflector." [5] Woodward in return expressed appreciation of Hooker's enlightened interest in astronomy. The Carnegie trustees, with the understanding that there was no commitment to mount it, accepted the gift "with unanimous enthusiasm." [6]

On September 19, 1906, the order for the glass disk went out to Saint-Gobain in France.[7] On August 28, 1907, while the 60-inch mirror was making its perilous journey up the mountain, Hale received news of the successful casting of a 100-inch disk. Less than three months later, the great disk, the size of a small room, arrived in Hoboken aboard the steamer *St. Andrew*. The New York press called it the most valuable single piece of merchandise ever to cross the Atlantic. From Hoboken it was shipped to New Orleans. From there it made its final stage overland to Pasadena. The perilous journey was completed on December 7, 1908,[8] the day the 60-inch mirror was set in place on Mount Wilson. The following day the disk was unpacked in the observatory office on Santa Barbara Street. Ritchey, who was to grind it, took one look and declared it useless. The glass seemed to be in three layers "not very perfectly fused together." Hale, too, confessed disappointment. "In fact," he wrote to Goodwin, "it is surprising that they sent it. . . . Of course we don't pay for this!" [9] Hooker was equally disgusted. At his urging, Ritchey was dispatched to France to discuss and arrange for another casting.[10]

In the meantime the 100-inch lay neglected in the Mount Wilson optical shop. The success of the 60-inch reflector and Hale's magnificent discoveries on the sun with the tower telescopes could not compensate entirely for the delay. If anything, the success with the 60-

inch made him more eager than ever for the great light-gathering power of a 100-inch and for the results it would bring.

Meanwhile, too, his relations with Hooker were deteriorating as the latter became increasingly antagonistic, even trying to turn some of Hale's associates against him. In at least one case—that of Ritchey—he succeeded, partly because of the optician's firm belief that the disk could never be used and his idea that a mirror of his own design would be infinitely superior.

It soon became apparent, however, that Hooker's antagonism was due only in part to the apparent failure of the disk. To his surprise Hale found that another cause was Hooker's completely unreasonable jealousy over Hale's friendship with Mrs. Hooker. For some time Hooker's desire to be the complete center of attention had been evident. On Sundays when George and Evelina Hale would go out to call, both Mr. and Mrs. Hooker would start to the door. If he succeeded in getting there first, he would carry his guests off to the library. If, on the other hand, Mrs. Hooker arrived first, she would take them to the living room. In time Hooker's jealousy became so uncontrollable that he refused to let any man enter the house when he was not there. Inevitably, too, his feelings affected his thinking on the 100-inch. For months, even years, Hooker's attitude would make life difficult, at times unbearable, for George Hale. Eventually, with other factors, they would be a large cause of the breakdown Hale suffered in 1910.[11]

These years since his arrival in California had been extraordinarily active ones. He had founded the greatest observatory in the world on a terrific gamble. He had made the greatest discovery of his career in the observation of magnetic fields in sunspots. He had been absorbed in all his other schemes for the revival of the National Academy of Sciences, the transformation of Throop Institute into a first-class technological school, and the organization of astronomy on an international scale through the Solar Union. He had been working at top speed without relaxing. It was not surprising if the strain of all these projects, calling for action, on top of the tumultuous years of earlier activity, was proving too great. As in a spring wound tight, it was surprising only that something had not snapped sooner. When to all these, intense personal difficulties were added, breakdown seemed inevitable, especially for one with Hale's high-strung, overanxious makeup.

When Harold Babcock arrived on Mount Wilson, in the summer

of 1909, he soon saw that all was not well with his new chief.[12] Attracted by Babcock's understanding nature and gentle spirit, Hale would often speak of his "terribly hard dreams." "Sometimes he would get up in the night and in his tormented half sleep would try to climb the picture frames on the wall." Soon others became aware of his increasing nervousness. Dr. McBride urged him to take time off. Finally, realizing that he was reaching the limit of his endurance, he promised to take the doctor's advice. "Last summer," he confided to Huggins, "I had a great deal of trouble from nervousness, and my physician told me I ought to give up work altogether for a considerable period. Instead of doing this, I have been working not quite so vigorously of late and have planned to take a trip abroad to complete the cure!

"I find in actual practice that I gain much more than I lose through these trips, partly because of the rest, and also because of the new ideas which come from the stimulus of discussion with men in various departments of research!" [13]

With this trip in view, he started planning for discussions abroad of the magnetic fields in sunspots and the 60-inch research program. If this was not Dr. McBride's idea of a rest, it at least meant a change from the routine and stress of administrative life, which, in his highly nervous state, Hale found increasingly irksome.

In London, soon after his arrival, he received one of the highest honors that can be bestowed on a scientist—election to membership in the Royal Society of London. He was forty-one years old. "My dear Lord Duke," Arthur Schuster wrote facetiously, ". . . in virtue of the power invested in me, I confer upon you the decoration of the diamond all set out in rubies. This decoration carries the title of duke with it." [14]

There was a continuous round of entertainment. Yet he felt surprisingly well. "Not even a cold, and practically no indigestion." He was excited by discussions of his research with old friends; he was enjoying meeting new ones. The experience was repeated in Paris. There was a reception in his honor at the observatory, a magnificent affair at Prince Bonaparte's, a superb "stag" dinner given by the Baron Rothschild. In all this there was only one discordant note—a renewed clash with his old opponent Henri Deslandres. At a meeting of the International Association of Academies, the Frenchman intimated that his study of the hydrogen flocculi had been more thorough than Hale's. Angrily Hale replied in his "halting" French.

He made his point. Afterward the French scientist De Gramont exclaimed, "Poor Deslandres! In spite of all his good intentions, he is always six months behind you!" [15]

The whirl continued. In Holland, in Germany, in Italy, and again in London, Hale was feted by the greatest scientific and social figures of the day. In Leiden he spent an evening with the famous Dutch physicist H. A. Lorentz, whom he invited to come to California to lecture at Throop (an invitation that Lorentz accepted some time later). "This one discussion was worth all the trouble of making this trip from America." [16] There was scarcely a scientific society that did not do him honor. There were lectures to give all the way from the Royal Institution in London to the Italian Physical Society in Rome. There were meetings not only of the International Association of Academies but also of the Executive Committee of the Solar Union and of the Italian Spectroscopic Society. Just before he was to sail for home, Oxford conferred on him an honorary degree of Doctor of Science in the honorable company of Oliver Wendell Holmes of the United States Supreme Court and Earl Grey, Governor General of Canada. He felt better than for months past. One day he even took a twenty-mile bicycle ride with Turner from Banbury to Stratford. The country was "glorious" and, blissfully happy, he forgot he had a care in the world.[17]

He was to sail from Cherbourg. On his way to Paris, he stopped at Saint-Gobain to see the new furnace and annealing oven for the second 100-inch disk. Everyone was hopeful of success. He sailed for home feeling like a new man.

Back in Pasadena, he continued to feel fine—for a while. He even felt able to do some night observing on the mountain. In the fall of 1909 the planet Mars was making a close approach to the earth. Percival Lowell's imaginative writings on the "canals" had roused intense popular excitement, and Hale was bombarded with questions from those who wanted to know if the canals were built by human beings. He doubted if there was anything in it, but he wanted to see for himself what Lowell might be calling "canals." One clear night he turned the 60-inch on the red planet. He saw some definite markings—"one of those so-called canals"—but they were totally unlike Lowell's drawings. These observations were made visually.[18] Later he took photographs. He examined the famous marking known as Dawes' Forked Bay. With a low power, it looked like Lowell's drawing "except that the 'canals' extending from the two extremities were vague and diffuse and by no means narrow and sharp." With high power, he could see the two forks as rounded and irregular in outline, "made up of interlacing and curved filaments." These were so fine that they could not possibly be seen with small aperture or low powers. He was convinced that Lowell's "canals" were largely a figment of his imagination. He reported his findings to the noted Italian expert on Mars, Giovanni Schiaparelli.[19] Nevertheless, the majority of the public remained unconvinced. It preferred to believe in a Mars inhabited by fantastic beings.

These were Hale's last observations with the 60-inch for a long time to come. Soon his headaches increased, his indigestion returned. He had trouble sleeping, and when he finally slept, it was only to be tormented by frightful nightmares. Unable to concentrate for any length of time, he became increasingly nervous.

One of his chief problems, from both a technical and personal point of view, continued to be the 100-inch. The two seemed inextricable. In February, 1910, a cablegram arrived from Saint-Gobain; the second disk had been cast. It had been poured from three huge pots of molten glass. Where the layers met, great clouds of air bubbles had been trapped. The men at Saint-Gobain had buried the huge slab in a manure pile for annealing.[20]

Meanwhile, as they waited anxiously for news from France, work on the grinding machine was completed, and Andrew Carnegie arrived on his long-planned visit to the mountain. He was in high good humor, looking forward with boyish anticipation to the trip. With him were his wife and daughter Margaret, who was just about Margaret Hale's age.

It was a wonderful day; the first half of the way they jogged along in full sunshine. But the last half was marred by a chilly fog that drifted over the mountain and soon turned to driving rain. "Mr. Carnegie," a reporter noted, "was unable to discover any new world to conquer through the big telescope." [21] The next morning they awoke to a snow-covered landscape.

Still, Carnegie was highly enthusiastic over his tour of the observatory. When Hale told him about the plans for the 100-inch, he asked him if he had put the matter before the trustees. Hale said he did not feel he could ask for more; he had already been treated so liberally. Afterward Carnegie declared expansively: "We do not know what we may discover here. Franklin had little idea what would be the result of flying his kite. But we do know that this will mean the increase of our knowledge in regard to this great system of which we are a part." Afterward he had his picture taken, standing proudly in the 60-inch dome.

This visit occurred on March 17th, and, as it happened, Henry Fairfield Osborn, the paleontologist, was there too. An account of this visit was preserved by the noted astronomer Henry Norris Russell, a research associate at Mount Wilson, who had it from Osborn: [22]

"After hours of enthusiastic showing of the work and equipment," Osborn recalled a night in the 60-inch dome after Carnegie had gone. "Hale suggested that we might make a night of it and watch the observing. So we kept it up, until, at perhaps two o'clock, I took advantage of a pause to say, 'This is grand, but I am worried about one thing.'

" 'What's that?'

" 'Why, the most precious instrument here, and the one most difficult to replace.'

" 'What do you mean?'

" 'I mean George Ellery Hale.'

" 'Pooh-pooh-pooh . . . I know I'm tired; but in a few weeks everything will be finished here. I'll go for a fishing trip and come back in good shape for the (Solar Union) meeting.'

" 'Hale, I happen to know something about things of this sort, and I don't like the looks of you. Leave this work to your associates now (they are fully competent to finish it) and get away to Oregon as soon as you can.'

"But he could not be persuaded, and kept on over my protest. A month or two later I had a post-card—'Dear Osborn. You were right. G.E.H.'"

The breakdown Osborn had predicted soon forced Hale to give up all work on the mountain. Other astronomers who knew him well and had watched him were not surprised. Schlesinger at Yale wrote, "I suspect it to have been a recurrence of your propensity to take the hills on the high." [23]

In June, Scherer told Fleming, "He has rather a bad case of brain congestion and exhaustion." Nevertheless Hale stubbornly refused to go on vacation alone. Finally Scherer agreed to go with him. On July 10th they started for Klamath Lake. There they fished, loafed, did exactly as they pleased. On their return, while still unable to work, Hale thought he saw some improvement. But Scherer was concerned. He wrote to Fleming: "The whole thing is a tragedy of which I can speak to you more freely when you get home. . . . He is badly broken." [24]

Nevertheless, Hale felt unable to leave Pasadena for the year of idleness prescribed by Dr. McBride. In January, 1910, the meeting of the International Union for Cooperation in Solar Research, which he had been planning for months, was to be held on Mount Wilson. Astronomers were coming from every part of the globe. He had even persuaded Carnegie to pay the traveling expenses of many of the foreign delegates. This was to be the fourth meeting of the Solar Union. The first had been held at St. Louis. At the second meeting in Oxford, in 1905, the Solar Union had been formally organized. The third had taken place in Paris (at Meudon) in 1907. The life of the union had been short, yet a wide interest in solar research had been stimulated. Knowledge had been increased through the establishing of a new system of wavelengths, a chain of spectroheliographs around the world, and an international study of solar rotation and radiation. As this, the fourth meeting approached, Hale could see enormous possibilities not only in solar research but also in the broadening of the international program to include all astronomical research.

The time came. Thirty-seven delegates arrived from foreign countries, including all parts of Europe, and from Japan and Canada; forty-seven came from the United States to make this the largest astronomical meeting ever held up to that time. There were astronomers "both orthodox and heterodox"; there were physicists and even a few mathematicians. From Pulkova Observatory in Russia came Backlund and Belopolsky; from Tortosa, Spain, Father Cirera, a Jesuit; from Edinburgh, Frank Dyson; from Marseilles, Charles Fabry; and from Paris, the Comte de la Baume Pluvinel. From Potsdam in Germany came the noted director of the Astrophysical Observatory, Karl Schwarzschild; from Groningen, J. C. Kapteyn, and from Berlin, H. Struve. The long and notable list included still others whom Hale had known since the beginning of his astronomical career—astronomers like H. F. Newall, who arrived with his wife from Cambridge, and from Catania, Sicily, A. Ricco, with whom he had climbed Etna sixteen years earlier.

From the beginning Hale had hoped, but never quite believed, that so many would be able to make the long trip. Said Harold Babcock, "Though he himself would have rejected such an idea, others believe that no one but Hale could have initiated and carried out such an undertaking." [25] Local newsmen, overwhelmed by the influx of foreign visitors, wrote ecstatically of the arrival of one hundred and seven of the "world's greatest stargazers" and of the hazardous trip up the mountain, made in wagons, on muleback, and on foot. Still, despite the rigors of the trip, the astronomers, parched by thirst, famished with hunger, revived quickly, once they reached the top—to the astonishment of the reporters. Seeing the great tower telescope "piercing the air over 100 feet above the tallest trees," they "made a grand rush for this marvel of the age." [26]

At the meetings various programs were discussed, but the most vital was the decision to expand the aims of the Solar Union to include the study not only of the sun but also of other stars in a program that would cover the entire field of astronomy. This, as Hale stated, was in harmony "with a procedure which I believe to be nearly always desirable, i.e. to advance into wider fields only after it has been shown that real success can be attained in a single subject." [27]

Everything about the 1910 meeting was a huge success except for one thing. Hale, to his intense sorrow, was unable to share in most of the proceedings. He attended only the garden party, the dinner he himself gave, and at the last moment, to the surprise of everyone, one meeting on the mountain. Afterward he was exhausted. That night, just as the astronomers were preparing to look through the 60-inch, he was forced to go off to bed. He was profoundly disappointed.

Meanwhile, during annealing, the second 100-inch mirror had been broken, and work had to begin all over again.[28] Nevertheless Hale remained optimistic: "I have no doubt of ultimate success." Carnegie, likewise optimistic, wrote to Sir George Reid in a flamboyant way that would have amused the subject: "Professor Hale is a genius and is making great strides. He tested a new plan recently and discovered a cluster of 16,000 new stars in Pegasus. A second trial disclosed 50,000 new worlds the eye of man had never seen. In short, the whole world is soon to listen to the wonderful discoveries of Mt. Wilson Observatory. . . ." [29]

In contrast, Hooker's gloom continued to darken the optical shop. Ritchey, pessimistic as always, continued to urge the merits of the composite mirror he had designed. Always temperamental, he now became increasingly difficult. Hale had suggested that the first disk might be usable after all, but both Hooker and Ritchey opposed the idea violently.[30]

In the midst of the controversy, Hale agreed to leave for Europe —but insisted that Evelina and the children go along. Absolute rest was the object. In London he refused all invitations, and spent long quiet hours in that "sanctum sanctorum" of British academic, scientific, and literary life—the Athenaeum. He spent other happy hours in old bookstores, or wandering through the familiar halls of the National Gallery. The family went sightseeing. At the suggestion of Dr. William Welch of Johns Hopkins, Hale also went to Oxford to see Dr. William Osler, who examined him carefully, and gave it as his opinion that with needed rest he would be completely recovered in six months.[31]

But even here Hale could not escape the haunting problem of the 100-inch. A letter arrived from Woodward in Washington, supporting a test of the first disk.[32] But at the same time Hooker wrote that nothing further was to be done "until after you have seen the Frenchmen. . . . I cannot bind myself to further responsibility at this time." [33] Hale, highly upset by this ultimatum, collapsed. The pains in the back of his head became excruciating; depression engulfed him.

Dr. McBride had forbidden him to stop at Saint-Gobain. Yet, driven by his concern over the disk, he now insisted on spending several weeks in Paris. It was an utterly cold and miserable interlude. He avoided the academy, the observatory, and all meetings with scientific men. Despairing of recovery, he wrote to Goodwin, "I can't see any real improvement in my condition."

Ironically, during these gloomy days when his passion for science was throttled, honors poured in on him— "far more than my share"—from the Royal Institution, the Physical Society of London, the Amsterdam and Vienna academies, as well as degrees from Cambridge and Berlin. Out of his pit of loneliness, he commented wistfully, "In every case it has meant a friendly feeling on someone's part, and I never knew a time when friendly regard counted for more." [34]

One day, unable to stay away from Saint-Gobain, he talked with M. Delloye, a member of the firm, and found there was a good prospect that a disk cast on December 5th would turn out well. Afterward he wrote hopefully to Hooker, who had become increasingly impatient "partly because of the lapse of time and partly for other reasons." [35]

By this time, Hooker's original promise of $45,000 had been expended in the endless experimental work at Saint-Gobain. He had also agreed to put up an additional $15,000 in payment of Ritchey's trip to Paris, and other expenses. He now refused to pay this additional amount. Moreover, he had not yet paid the final $10,000 of the original $45,000. In Hale's absence, he called on Adams, acting director of the observatory. He insisted that if he paid this $10,000, Adams must sign this release: "Received of J. D. Hooker in full satisfaction of the balance remaining unpaid and covers all demands and obligations of the said J. D. Hooker to the Carnegie Institution of every kind and character." [36]

Adams, angered by this ultimatum, cabled to Hale, "Hooker completes $45,000. Wants release all claims. . . . Shall I sign? Answer." Hale, beside himself with worry, replied, "If there is any way to get from him a sufficient sum to prevent a deficit this year . . . and stave off the question of release, do so by all means." [37]

As a result of this latest bombshell, Hale again came close to collapse. Evelina Hale was furious. She wrote to Adams, asking him to send no more news. "And I wish you would tell Mr. Hooker you had this from me." "I wish," she exclaimed, "that glass was in the bottom of the ocean." [38]

Meanwhile Adams had been making further mechanical tests of the first 100-inch disk. He found that it behaved like a homogeneous cylinder and concluded that this justified parabolizing and perfecting its figure. Ritchey disagreed. But as soon as Hale learned of Adams's conclusions, he cabled him to go ahead.

In the light of Ritchey's attitude, however, the situation remained tense. "I really think," Adams wrote, "we shall have to consider the question of Ritchey's connection with the observatory in the near future with due regard to the condition of his mind. I honestly do not believe that he is fully responsible for his actions at the present time." [39]

Meanwhile a curious correspondence that would have amazed Hale was taking place, as Hooker wrote a long, fawning letter to his "partner" Andrew Carnegie. In it he claimed credit for the decision to put the rejected disk on the finishing table. He added that he had paid all the funds he had promised. "I can say that I urged this matter and think all are greatly encouraged that a suitable surface for a fine reflector can be had out of it. . . . The whole disc rings when hit, indicating it to be flawless." He concluded, "I trust Mr. Hale will come back to us sound and well, that he may take up this highly interesting work and carry it to a glorious prominence. We hope to see it the most notable of anything in its line in the great round world. I have the utmost faith in its success." [40]

Carnegie, receiving this extraordinary letter, could have had no idea of the true state of affairs. He wrote back, "My dear Pard . . . Our chief solicitude now is for the genius Hale. If we only get him back fully restored to health, all will be well. Delited to hear of your success with the discarded disc. It will be a great triumf if you make the stone that was rejected the chief cornerstone." [41]

December passed. The long days in Paris were brightened by the arrival of D. H. Burnham, the architect from Chicago. In his company Hale's spirits rose; soon he was visiting churches, cathedrals, parks, and museums without trouble to his head.

Nevertheless his wife felt that a stay in Egypt was the only escape. She booked passage on a boat sailing from Genoa in January. They went by way of Mentone on the Riviera. It was a lovely place, and here, for a time, with the sunshine, the oranges and lemons, the flowers, Hale felt at home. The "freedom, for a time, of correspondence regarding the Hooker question helped greatly." The children, too, were flourishing. "It is marvelous to see William, eating everything, and thriving without a trace of his old intestinal trouble." [42]

An odd thing, however, happened during this visit. It was something he himself could never explain entirely. One night, when he was sitting in his room, out of nowhere a little man suddenly appeared, and soon was advising him on the conduct of his life. Sometimes Hale had had a ringing in his ears. Now the visitation of this little elf seemed to be connected in some way with that ringing. After this, his first visit, he came often, in many widely scattered places, until he became almost a mascot. [43] Hale rarely spoke of these visitations. But years later he described them to Dr. Leland Hunnicutt in Pasadena, hoping perhaps for some realistic explanation of the phenomenon.

Soon the time came to leave Mentone for Genoa. On the way the train stopped in Ventimiglia; at the station Hale bought a paper, the *Petit Niçois*, published in Nice. As he leafed through it, a notice caught his eye. It stated that Andrew Carnegie had just given ten million dollars to the Carnegie Institution. Now, as on that day nine years before when he had seen the notice of the founding of the Carnegie Institution, he found it hard to believe his eyes. He jumped back on the train and, as soon as they reached Genoa, rushed up to the first newsboy and bought a copy of the *Corriere della Sera*. There, to his "intense surprise and delight," as he wrote to Carnegie, he found confirmation of the news. He saw, too, that Carnegie had emphasized his interest in Mount Wilson and the 100-inch telescope. In an almost dangerous state of elation, Hale dashed around and bought up all the papers he could find. In all of them he found the same wonderful news.[44]

Later a copy of Carnegie's letter of gift arrived from Woodward. "I hope," he said, "the work at Mt. Wilson will be vigorously pusht, because I am so anxious to hear the expected results from it. I should like to be satisfied before I depart that we are going to repay to the old land some part of the det we owe them by revealing more clearly than ever to them the new heavens." [45]

Yet this magnificent news, Hale soon discovered, was nearly as disturbing as the troubles with Hooker. Again his head mutinied. On January 22nd he wrote in his diary: "Bad night, awoke several hours thinking of 100-inch. Drove about city in morning. Walked with Holder in afternoon. . . . Peculiar tingling in foot, much more marked than last June." [46] Despite every effort, the news had set him off "on a wild whirl of scheming." Mrs. Hale watched anxiously. Carnegie's gift, she noted, sent him "all to pieces in a day's time. . . . He immediately began to make plans for years ahead and so worked himself up to almost the breaking point. I wished Carnegie could keep his millions to himself." [47]

News of the gift spread rapidly. When Richard Maclaurin, the president of M.I.T., heard the news he sent Robert Woodward his congratulations and said that he noticed with particular pleasure the references to Hale (who had been made a member of the M.I.T. corporation. Then he added, "I wish he [Carnegie] could be induced to found something in honor of Hale at this Institute." [48] At this point Hale was only forty-two—rather an early age for memorials!

CHAPTER THIRTEEN

From Ancient Egypt
to a New Universe

✳ *1911-1917*

IT WAS THE END of January, 1911, when the Hale family sailed for Egypt. At last Hale was reaching this fabulous land of which he had dreamed all his life. Perhaps the original source of inspiration was old "Doc" Williams, a student of ancient history, who had transmitted his love to George Hale. Certainly the flame had been fed by William Hale's visit to Egypt in 1897. Yet, without doubt, Hale's passion, not only for Egypt but also for archaeology, had been crystallized in his friendship with James Breasted, the Orientalist. They had met first when both men became members of the original University of Chicago faculty. In those early years their meetings were rare. Yet each time Hale found himself increasingly attracted by Breasted's vivid accounts of his investigations of the hieroglyphs along the Nile and by the ardent spirit that had led the archaeologist down cataracts and through sandstorms in frightful heat to achieve his goal. Also, among the books Hale had bought in 1897 was Flinders Petrie's absorbing volume *Ten Years Digging*.

For a period after Hale moved to California, when Breasted was in Egypt, the two friends did not meet. Some time around 1911, however, probably when Hale was planning this Egyptian trip, the friendship was renewed. They came together in a period when both, for different reasons, were profoundly discouraged not only with their work but also with the course of their personal lives. Now, in their meetings, in their letters, the enthusiasm of the one served to buoy up the other.

Breasted, in particular, just at this time, had the feeling that his dream of recording the history of the great Nile Valley through hieroglyphs might never be realized. As the only professor of Egyptology in an American university, he had an intense feeling of isolation. There seemed to be no one with whom he could talk—until he again met Hale on a day he never forgot. Years later he wrote: "I don't think I ever would have had the courage to go on but for you. It is now nearly twenty years since you came out here to see me, and took me down to the University Club for lunch. What you said that day gave me a new lease on life." [1]

From the beginning, the two men shared not only similar enthusiasms but also a similar outlook. Both were scientists, but both were humanists too. "It was hardly strange," Charles Breasted writes in his vivid biography of his father, *Pioneer to the Past*, "that my father should be so strongly drawn to this ingenious, charming, altogether modest, apparently invincible young man. Their paths had crossed infrequently. Yet it was typical of Hale that, whenever they had met, he had never failed to minimize his own activities and to astonish my father with questions revealing a remarkably intimate acquaintance with the nature of his work." [2]

Now in Egypt for the first time, Hale must have longed for Breasted as a guide. Yet, much as he wanted to see the wonders of Egypt, he found he was up to very little. They went to Luxor, and there again he saw Dr. William Osler, who "overhauled him and found no new symptoms." Afterward he wrote to Walter Adams, "My hypothesis is that in spite of frequent set-backs, I am actually much better and that some fine day I will wake up to find myself all right." [3]

In Egypt, too, he was delighted to see his son Bill's fascination with ancient artifacts. He watched him picking up bits of glass and pottery, pieces of mummy cloth that the eleven-year-old boy saved carefully and labeled for his "collection." "And this," he informed Goodwin, "without in the least interfering with a perpetual flow of boyish spirits and never-ending desire to ride for countless miles the fastest donkey of the bunch!" [4] He began to wonder whether Bill might not become an archaeologist, and three years later, when Bill was fourteen, he would write in his diary, "Decided with William that he should go into archaeology, with broad training covering all the early civilizations of the Near East and a background of general science (evolution.)" In Bill's case, as in everything else in his life, George Hale could not resist the impulse to plan!

From Egypt the family journeyed on to Rome. Here Hale tried to settle down to a life of idleness in exile, but he found it impossible to keep interested in anything—except those things that were forbidden. Until recently he had read with pleasure "a great variety of books." Now the little elf that had first appeared to him in Mentone stood at his side, demanding attention and suggesting that the book he was reading was unimportant. "How to escape this new form of torture, which is incessant, I do not know," he exclaimed in a letter to Goodwin.

Evelina Hale, worried over her husband's condition, continued her censorship of letters, in order to protect him from worry, though she expected "to get him into all kinds of scrapes." [5] As a result, weeks went by without any apparent news from California. Occasionally she would show him a letter she considered innocuous enough, and he was touchingly grateful. One such piece of heart-warming news came from Throop. In March, 1911, Theodore Roosevelt, former President of the United States, arrived in Pasadena to lecture. He spoke of Hale with the greatest warmth. "I want," he exclaimed, "to see institutions like Throop turn out perhaps ninety-nine out of every hundred students as men who are able to do given pieces of industrial work better than anyone else can do them. The one-hundredth I want to see with the kind of cultural scientific training that will make him and his fellows the matrix out of which you can occasionally develop a man like your great astronomer, George Ellery Hale." [6] Hale was amazed that Roosevelt had even heard of him! Afterward Charles Holder wrote, "It made the hearts of the 'Club' swell to busting before the finest audience ever seen in Pasadena to hear him give you a fine and deserving send off." [7] (The "Club" was composed of Holder, McBride, Scherer, and Hale.)

In this letter, too, Holder hinted at the "new plan" for Throop. Hale, puzzled, asked his wife what she knew. Finally she showed him some of Scherer's letters. He learned then that, on a proposal from Scherer, a bill had been introduced into the state legislature to make Throop part of the state's higher educational system, under the name The California Institute of Technology. On February 7th, Scherer, in his element in this kind of fracas, had written excitedly from Sacramento, "The bill went in this morning, and will probably become law before you get this. At any rate Throop stock has gone up 1000%. Advertisement? Every paper in the state full of Throop, and favorably. My first taste of politics.

"I've thought of you every waking hour since the lightning struck," he exclaimed. The next day, still jubilant, he had written again, and now Hale read: "For every whack I gave for Throop, I added one joyously for you and your father." Three days later he reported a "perfectly delicious" clash and reconciliation with the powerful Benjamin Ide Wheeler, president of the University of California. "Bless your old soul," he exploded, "I wish you were here to help me paint this town scarlet." This outburst was signed "Kid McCoy." [8]

Later, however, Wheeler changed his attitude. If such a state institution were established in the south, he feared a duplication of effort with the state's engineering colleges. He doubted Scherer's claim that there would be no extension of activity beyond that set down in the bill. He suggested that parallelism in two sections of the state would bring deplorable rivalry.[9] Therefore, according to Scherer, Wheeler had used his power with the legislature, and the bill was defeated. Nevertheless Scherer, as he reported to Hale, decided to fight the thing through, "largely on account of the superb advertising" they were getting.[10]

Throughout the fight, Scherer was strongly supported by the Southern California newspapers, who praised him for his "valiant fight." Four days after the bill's defeat, the members of the senate expressed regret at the vicious attacks, and moved to reconsider the question.[11] Meanwhile, however, Scherer had telegraphed to Gates, the senator from Los Angeles, asking him to withdraw the Throop proposition "absolutely." Evidently he had concluded that to become a state institution, too many political sacrifices would be required. Thus he wrote exuberantly to Hale, "We retain our independence, are known favorably in every corner of the state, as the institution of which the University is mortally afraid, and the loyalty of our friends has been intensified to the tune of $250,000 from four men—Fleming $150,000; Culbertson $50,000 and General Sherman $25,000; the remaining $25,000 being the Laird's gift secured by three others. . . ." [12]

Thus Throop failed to become a state institution, and over ten years were to pass before the name—The California Institute of Technology—proposed then, would be adopted. Still, Scherer's optimism was well founded. The resultant publicity, combined with Roosevelt's visit, did indeed give the embryonic institute a tremendous boost. Hale longed to return to Pasadena to share in the fight. "Wheeler be damned," he exclaimed.[13] But his wife felt the time was not yet. Instead they went to London.

They arrived there in the midst of a dense fog. In the long days that followed, despair again engulfed him. Then one day in May his old friend the astronomer H. F. Newall arrived to carry him away from the "despairing gloom" to his house at Madingley Rise. There, in the peace of the country, with the great elms waving overhead and the dome of the Cambridge Observatory rising in the distance, he slowly improved. Twenty years later he would look back gratefully on this luminous interlude. "The demons (of unknown fears) slipped into hiding, never again to emerge in their original fury." [14]

Finally, in Mid-June, the Hales sailed for home. He was feeling much better, and was sure he would soon be entirely well. For some weeks he stayed with his brother in Chicago. There he began receiving letters from Dr. McBride that were filled with Cassandra-like warnings.[15] Hale's case, McBride declared, was one of brain exhaustion caused by twenty-five years of a "pretty steady strain in one direction. No brain thus exhausted could recover its normal vigor in a few months or even a year. . . . I think you will probably never be able to work as you have worked, and your breakdown will prove a happy event if you learn from it that in the future you will need do, not what you like to do, but what you have to do in order 'to be saved.'"

With every sentence, Hale's spirits must have sunk to a lower depth. McBride's letters were terrifying in their implications. "Knowing as I do that you are worth saving, I am writing this because I know what you cannot possibly know, of the uncertainties of your future. As simple as it may seem to you, as well as you seem to be, I can tell you that your improvement is only on the surface, and that you are now, and will be for some time, living on the edge of a precipice."

McBride then insisted that Hale go to a sanitarium in Bethel, Maine, run by an acquaintance, a Dr. Gehring. Under this unremitting pressure, and beginning to consider his situation hopeless, Hale finally agreed. He arrived at Bethel late at night in a pouring rain. He drove to the Prospect Hotel through the "descending torrents," only to discover that most of the hotel had been burned the previous Saturday—a desolate sight and a depressing introduction. Nor was the part of the hotel that remained conducive to cheer.

The following day, after a dreary night, he started for the sanitarium. He described his experiences to Harry Goodwin. The "castle" with its "dungeons" was guarded by two great stone posts.

Inside the dungeon gates he was met by the ruling "wizard," a man with a piercing eye that "transfixed his very soul," and by a "wizardess of terrifying aspect." The following day the "wizard" called on him in his room. Soon, after being "set upon and mauled," Hale learned that the root of all his trouble lay in his "guts." He learned other things too; he was to take "vile liquids, many in number and boundless in quantity." He was to repair to the "fiendish wood pile, and saw wood for his Satanic holiness." He was to lie for hours on the floor, with a billet placed under his spine. "And I am to suffer other tortures and trials without limit." [16] The picture, oddly, as he drew it for Harry, had a certain enchantment.

Gradually he settled down to the routine of sawing wood and gardening. He received daily massage and took medicine nine times a day. Occasionally he was allowed to go fishing. This routine occupied every moment. Soon, too, another form of treatment was begun, based on the doctor's belief in the principle of the subconscious self and the value of suggestion. First he would give a short lecture on the subconscious. After that he would solemnly close the transom and intone: "Close your eyes and lie entirely limp. Your body is quite limp. Your body is quite relaxed and inert. If I lift your hand it falls back dead and motionless. Your legs begin to feel as heavy as lead." (But they don't, said the patient to himself.) "Your extremities become somewhat numb and seem to be asleep. You involuntarily take long, deep breaths." (Not I, said the patient.) And so on and on the voice droned, while Hale lay, convulsed with hilarity, unable to take the whole business seriously. Later he delighted in describing these periods of "floating on the bosom of the great sub-conscious," and on his return home he reenacted many of the scenes for Ellie Mosgrove's benefit.

Nevertheless, after several sessions, he had to admit to some success, as the doctor, directing many suggestions to the "arch enemy" in the back of his head, gave his subconscious self a lot to ponder on. He even sent his wife a book on the subject, suggesting that such treatment might help to remove her headaches and periods of depression that so often followed his own. As a result he also began to appreciate more fully how important a role the subconscious had played in his own life, and how much he drew on it when developing new ideas. Such ideas nearly always came in the morning, or when he was in a dreamy state, often even while he was shaving. "I remember what a lot of good scheming I used to do on the train between Chicago and Williams Bay, when I was about half asleep." [17]

Slowly he improved, and the terrible haunted feeling that had plagued him in Rome disappeared. One day he wrote to Margaret, who was about to celebrate her fifteenth birthday; he confided that he hoped to be allowed to go home by October. "Ask Laddie [the dog] to polish up his smiling countenance, as I want him to use it when I come into the yard. . . ." [18] Then, thinking of the bees buzzing in their Pasadena garden, he composed a little poem for her birthday:

IN MY GARDEN

What argosy of fairyland
With whirling wings and sides of gold
Comes sailing down from mountains cold
To joys found here on every hand?
Look, in that rose she comes to land
Safe harbored from aerial seas.
But not for her is idle ease
Though lotus eaters line the strand.
With honeyed freight stored in the hold
She mounts again on agile wing,
A rover bent on voyaging
With none to check adventure bold
To ports and treasures all untold.

As the end of his stay at Bethel neared, signs of improvement appeared. Suddenly he found himself humming tunes as he had not done for over a year. In contrast to Dr. McBride's macabre advice, Dr. Gehring told him he could take up regular work without fear. Indeed, he said that a restricted program was harmful; the great thing to avoid was fear that work would hurt. He concluded that there was nothing organically wrong; the trouble was merely "functional."

So George Hale returned home. In Pasadena, however, Dr. McBride insisted he must never put in more than three hours a day at the office. So once more the familiar pattern of overconcern that had dominated his life returned. Trying to heed McBride's dire warning that a second breakdown would be extremely serious, he gave up all work at home. On days when his head began to ache, he "knocked off altogether." Gradually he adjusted to his new regime—or at least acquiesced in it. By the end of December, 1911, he could write to Harry Goodwin that he was enjoying a schedule that ran from nine to twelve thirty in the morning.

In the dark years of 1910 and 1911, he had published no papers and written few letters. Now he was eager to return to his research. But his staff, delighted to have him back, kept him busy with constant demands—with designs for instruments and buildings on Mount Wilson and new plans for research. In the fall of 1912 new observatory offices were completed at the Santa Barbara Street location. Whenever he was there, he became the center around which the observatory life revolved. Elizabeth Connor, the librarian, who would join the staff in 1916, said of the effect of his presence in the observatory: "He always came in the front door, and his 'Good morning,' addressed to whomsoever might be in the hall, seemed to send a current throughout the whole building so that everyone sat up straighter and worked more briskly." [19]

Other members of the staff noted that he had the rare ability of making each person feel that he or she was the most important individual in the world and that the thing he most wanted to do was to talk to that person to find out how he or she was getting along. The question in hand was all that mattered.

Paul Merrill, who joined the staff in 1919, had this to say: "Dr. Hale was at the same time one of the most enthusiastic and one of the most conscientious persons that I ever knew. His enthusiasm was led by a very vivid imagination. However, he always managed to stick close to the facts. He never soared aloft in a great flight that couldn't be sustained by actual experience later. That is why people came to have a great deal of confidence in his ideas, and why his recommendations carried very heavy weight. His enthusiasm was the very objective kind. He was just as enthusiastic for schemes of other people, if he believed they were sound, as he was for his own. His own feelings were kept in the background, and his own desires. What he seemed to be expressing was his eagerness for the progress of astronomy. That was so genuine, that no one who came in contact with him could doubt his real feelings on the subject."

Only once in all his experience at the observatory did Merrill find Hale lacking in enthusiasm over a research project that Merrill had proposed. Still, despite his doubts, Hale told him to go ahead anyway. When the scheme worked out, Hale was quick to acknowledge his error. "Merrill," he said, "I was entirely mistaken." [20]

Harold Babcock, speaking of Hale's genius for personal relations, commented: "He had the gift of inspiration in an extraordinary way. You could talk to him for ten minutes, and leave him, feeling as though you were walking on air, able to do anything. In any research project his mind leaped ahead to see the possibilities. He was far in advance of anyone else. His excited interest in everything was conveyed to all with whom he came in contact."

Like his father in his childhood, they soon learned, however, that Hale continued to want things "yesterday." This trait had changed little since his boyhood. "The men in the shop used to laugh about him," Babcock said with a smile, "they loved him so—they couldn't do anything but admire his tremendous drive and energy. They used to say if Mr. Hale comes in here and wants something, he wants it yesterday! Not tomorrow."

Not only those who knew him in the observatory but also those who met him for the first time outside were impressed by his impatient energy as well as by his magnetic spirit. "His was the most vivid personality I have ever known," said one, while others spoke of the gentleness of his spirit, his unfailing consideration for others. Elmer Grey, the architect, said, "He was the most interesting man that ever came into my office." [21]

Margaret Harwood, the astronomer, referring to his extraordinary memory, spoke of their first meeting. She was introduced to him in a crowd by E. C. Pickering. Years later Hale saw her again at another meeting. He knew her immediately, who she was, where she came from, and what she was doing. When she was appointed to the Maria Mitchell Observatory in Nantucket, he wrote to congratulate her.[22]

As the astronomer Frederick Seares, one of his closest associates at the observatory, put it in his article "The Scientist Afield," when he spoke of Hale's outlook: "Hale's approach to something new was always logical. With him, however, the logical element somehow lost its stiffness. . . . The logical formulation was shaped by the imaginative powers of a temperament essentially artistic. Problems thus presented became something engaging, to be undertaken as adventures of the spirit; and to Hale they were indeed such adventures, entered upon with a kind of joyous gaiety." [23]

It was this spirit of "infective enthusiasm" that his colleagues quickly caught. Few of them knew of the difficulties in his life. Many could not understand his illness. They were aware only of his buoyant spirit, his generous nature, his sensitive consideration for others, his eagerness to be "on the way."

Over and over in his letters, his concern for others, his compassion appear. From his own experience he knew the problems of illness and the difficulties it brought in its wake. When a member of the staff had a breakdown in 1911, he consoled him, "I know from my own experience that anything which tends to raise the old sensations of fear must be avoided." And to W. J. Hussey he wrote in 1915, after Hussey's wife's death: "I venture to warn you against falling with me into the neurasthenic quagmire that engulfed me five years ago. Even now I have not escaped it entirely, and the loss of time has been most serious and disappointing. Beware of a similar fate while you still have an opportunity to avoid it." [24]

Yet, in these months after his breakdown, much as he enjoyed the renewed contact with his staff, much as he appreciated sharing their problems, he found it exceedingly taxing. After his mornings at the observatory he felt unequal to creative thinking or, indeed, to activity of any sort. In the afternoon he was forced to lie down and read his favorite detective stories. "It seems absurd that only three or four hours a day should so nearly use me up." Gradually, however, despite Dr. McBride's strictures, he began to do a limited amount of writing. Gradually, too, he returned to his first love, the sun.

Even as the 60-foot tower telescope for studying the sun was completed in 1908, Hale had decided characteristically that it was just not big enough. It was the old story. He had written to Cleveland Dodge, trustee of the Carnegie Institution, describing his plans, explaining his hope of investigating the question of the relationship of sunspots to magnetic storms on earth.[25] For the first time, he had pointed out, it would be possible to make a magnetic survey of the sun, recording the daily changes in the areas, intensities, and polarities in sunspot fields. By comparing these records with simultaneous records of terrestrial magnetism, he hoped to detect a relationship— "if it exists." To make such a survey, however, he needed a much larger solar image. He must have a tower telescope 150 feet high; such an instrument would give an image 16 inches in diameter, and would cost between $30,000 and $40,000. But solar activity was decreasing, and there was need for haste. This time the trustees had evidently appreciated what Woodward called "the results of Hale's indefatigable labors." To his surprise and joy they had voted the appropriation for the new tower telescope immediately.[26]

Because in this telescope the mirrors would be mounted 160 feet above the ground, an outstanding problem was the need to ensure protection from the wind, and to make the huge tower steady. Hale puzzled over the solution for some time. Finally the answer came (again, apparently, "out of the blue") when he was walking down the trail with Abbot. Suddenly he stopped and exclaimed, "I know! I'll put breeches on it!" And so in the finished structure an inner skeleton tower that carried the instruments was encased in an outer skeleton tower that carried the dome—with enough space between the two to prevent contact. The solution proved highly effective.[27]

This telescope, completed in 1912, soon proved its superiority over the smaller, 60-foot, tower. During the warmer hours of the day the solar image was much sharper than with the smaller tower, and observation could be carried on throughout the day. In it the sunlight is reflected from a coelostat mirror (which turns to follow the sun as the earth rotates) at the top of the tower to a secondary flat mirror. From there it is carried vertically down the inside of the tower to the laboratory at the tower's base. Here the sun's image is about 16½ inches in diameter (compared to the 5-inch image obtainable with the 60-foot tower). Here, as Hale noted, "any part of this large image, such as a small sunspot can be held indefinitely on the slit of a powerful spectrograph 75 feet in length." This spectrograph is mounted in a concrete-lined well, 10 feet in diameter, excavated 80 feet down in the rock beneath the tower. At the base of the well a grating ruled with 14,000 lines to the inch is mounted. When the sunlight reflected down from the top of the tower reaches this grating it is broken up and reflected back through a lens mounted near the bottom of the well. The resulting spectrum is received near the slit in the laboratory at the tower's base. As a result of the great focal length, the dispersion is so great that the light that descends through a slit only three thousandths of an inch wide is returned as a spectrum about 40 feet long, from red to violet. The distance between the D lines of sodium in the third-order solar spectrum is over an inch.[28] Moreover, as compared with the old Kenwood spectrograph of which it was the "direct successor," the new 75-foot spectrograph gave results fully twenty times as accurate.

As soon as it was finished, Hale set to work to try to solve a question that would vex him always. His discovery of the magnetic fields in sunspots had led him logically to the question "Is the sun as a whole a magnet?" He knew that finding the answer to that question would be extremely difficult, if not impossible.

As early as December, 1908, he had written to Robert Woodward of preliminary results that seemed to favor an affirmative answer. The results—the displacement of spectroscopic lines in light from the region of the sun's poles—were so small, however, that there was question whether they might be due to instrumental error. Now, in 1912, back at work with his new tower telescope, he returned to the study of the problem. The magnetic effect, he soon found, always seemed to be present in both hemispheres of the sun. Therefore, he thought, it must represent a general phenomenon, as distinguished from the local magnetic effect he had discovered in sunspots. Still, the whole business was intensely puzzling. Sometimes different observers obtained different results. On occasion he thought he had definite proof, and would consider making a preliminary statement. Then some new snag would appear and he would decide to wait. Nevertheless he enjoyed the challenge thoroughly. He told Evelina Hale, "There are some delicious complications, which always prove more interesting than the anticipated phenomena, provided they are worked out in the end." [29] Moreover, if he could prove the presence of the field, he saw "lots of fun ahead in making a complete survey of that field, in determining the sign of the charge, in estimating the sun's period of rotation as given by the motion of the field, and so on."

Related questions also occurred to him. Could other stars have magnetic fields? If so, would it be possible to detect them in a planetary or in a spiral nebula? At first thought, this might seem hopeless because of the faintness of the nebulae and the high dispersion required. Nevertheless he decided to make the attempt, and soon he was designing apparatus for its possible detection with the 60-inch telescope. Soon Babcock was making the test, but the equipment then available was not sensitive enough to show anything on his photographic plate. Many years were to pass before the magnetic fields in stars that Hale had suspected would be detected as a result of the development of modern electronic techniques.

While Hale was trying to detect a magnetic field in the sun, "a wonderfully bright and efficient Hollander" arrived at the observatory. His name was Adrian van Maanen, Hale set him to work on the puzzle. At first van Maanen failed to find any evidence of a field. But as he gained experience, he began to show results that agreed with those of Jennie Lasby, who had made many of the earlier measurements—except that his measurement of the magnitude of displacement was systematically lower. The evidence seemed to point to a weak magnetic field. If these results could be relied on, this would be, as Hale told Woodward, "the most important result hitherto obtained by the observatory, on account of its bearing on solar theory and support of the fundamental physical hypothesis of Schuster that every rotating body is a magnet." [30] (This hypothesis is no longer considered valid.)

These observations were made at sunspot minimum when the sun's surface was largely free of spots. The results, therefore, Hale felt, could not be affected by the immensely powerful spot fields. Some of his colleagues, however, remained skeptical; he himself wanted more evidence. Until the end of 1915, the observations continued without any certain result. By that time, Hale, as well as everyone else at the observatory, was forced to turn attention from the little star by which we live to the exploding events on one of the small planets that moves around it. It would be some years before he could return to his fascinating puzzle.

Meanwhile, another problem related to magnetic fields had been absorbing Hale's interest. With the 60-foot tower, small spots had been beyond the range of observation, and this had been one of his primary reasons for wanting a more powerful instrument. From 1908 to 1913 solar activity had been at a minimum (in the eleven-year sunspot cycle.) By 1912, however, just as the new 150-foot tower telescope was completed, the first small spots began to break out in high latitudes. As the cycle developed, Hale soon noticed a remarkable phenomenon—so remarkable that for some time he found it hard to believe.

On July 11, 1914, he wrote to Evelina Hale in Madison, Connecticut, "I got a very curious new result yesterday, which has given me much to think about. Since the sun-spot minimum the new sun-spot vortices are whirling in the opposite direction—that is, in the southern hemisphere the vortices before the minimum were whirling in the direction of the hands of a clock. Now they are turning the other way. There are a few exceptions, and in the northern hemisphere the change is not so certain. But on the whole, there seems to be no doubt of a genuine change of some kind, which is bound to prove interesting when investigated."

It had long been known that spots usually occur in pairs and that at the beginning of a spot cycle new spots first appear at relatively high solar latitudes, and gradually approach the equator as the cycle develops. At sunspot maximum nearly all spots appear in two zones of latitude a few degrees north and south of the equator. As the spot activity wanes, these zones persist. After the minimum is passed, spots appear in two different areas: the spots of the new cycle begin to appear in high latitudes, but a few spots from the old cycle persist in low latitudes.

From 1908 to 1913 sunspots had been on the wane. During this period only twenty-six spot groups had been observed magnetically. "But these sufficed," as Hale was to write, "to reveal the polarities then characteristic of northern and southern hemispheres. With but two exceptions, all of these groups showed that preceding spots in the northern hemisphere were of south polarity (with south-seeking poles), while their following spots were of north polarity. In the southern hemisphere the order was reversed—preceding spots were of north, following spots of south polarity."

In 1912 this rule persisted. Then, to the surprise of everyone, the phenomenon that Hale had described in his letter to his wife appeared. As the first small spots began to break out in high latitudes, they showed a reversal in polarity from that of the preceding cycle. In the vast majority of cases, preceding spots in the northern hemisphere showed north polarity, while those in the southern hemisphere showed south polarity. "Some extraordinary change had occurred in the sun." [31]

This was an amazing result, which was confirmed repeatedly but for which no explanation was offered until forty years later.

Meanwhile, on earth, Hale was faced by problems that at times seemed even more baffling and disturbing than those he found in the sun. Bill came down with an attack of bronchitis, and his father "could not shake off the fear of tuberculosis." His wife, suffering from back trouble, nervousness, and depression had gone to Boston for treatment. Margaret, away at school in the East, was unhappy. Perplexed by these difficulties, pathologically upset by his own illness, still worried to distraction over the 100-inch, he suffered a relapse that, in March, 1913, sent him back to Dr. Gehring's sanitarium. There, in peace and quiet, the strain was relieved. Soon the driving pain in the back of his head was alleviated. He was able to finish a paper on the sun's magnetic field and to prepare his speech for the fiftieth-anniversary meeting of the National Academy of Sciences in April. At the end of two weeks he wrote to his brother, "I am certainly in fine form, and have done some work since I came, other than chopping in the woods and floating on the great subconscious." [32]

Before long he was on his way to Washington for the academy meeting. Yet, back in Pasadena once more, the tension increased and the emotional stress returned. In the summer of 1913, Dr. McBride decided his patient must give up work entirely and take another trip abroad. The International Association of Academies was to meet in St. Petersburg, and with that objective he finally agreed to go, and McBride went with him. After the meeting he traveled south to the ancient cities of Troy, Mitylene, and Athens. It was a journey into a glorious past he had dreamed of for years, and it was to inspire him with new dreams for his adopted city of Pasadena.[33] He returned home ready to conquer the world.

It was not long, however, before he was experimenting in another direction. A new 13-foot spectroheliograph had been installed in the 60-foot tower telescope, which had been rebuilt. With it he was soon (October 24, 1915) composing stereoscopic views. The results were superb, and he was elated. He wrote to Evelina, "You can look right down into the swirling craters above sunspots, and see the great walls of prominences rising overhead like palisades. It certainly is a pleasure to see all these phenomena in actual relief, after imagining them so long at different levels." This, as he told his brother, was "something we hardly anticipated in the Kenwood days."

"The next logical move," he added, as his mind leaped ahead to another far-reaching scheme, "is to apply moving picture methods, and I shall begin soon with a simple case—sunrise on a lunar mountain. There should be no difficulty in obtaining a series of pictures on any good night showing the sunlight as it first strikes the tip of a lunar peak and then moves down its walls, while the long shadows creep across the plains below. In fact, it may even be possible to arrange *stereoscopic* pictures of this kind, and also of solar vortices. The difficulty with the sun is merely that the good seeing does not last long enough to get a full set of pictures."

In 1914 Hale was only forty-five. Yet even then he was considered one of the greatest scientists in the world. A poll had been taken by *Technical World* to determine the twelve foremost living American men of science, and George Hale was elected one of the twelve. The magazine planned a series of articles, and Forest Ray Moulton, his former colleague in Chicago, was selected to do Hale's profile. He now wrote asking for material.

"I don't like to appear in the guise of a celebrity," Hale replied, "as I am no genius, and what I may have accomplished is merely the result of a very deep interest in science and a lot of hard work. The popular notion which I have encountered abroad that ample funds have always been directed my way, would not impress anyone familiar with the details of the early history of the Yerkes and Mt. Wilson Observatories." He spoke of his interest in departments of science other than astrophysics. He described his library with its emphasis on biography, art and architecture, poetry, the history of thought and civilization, European politics, Greek art and life, and Egyptology. He told of his work as president of a "local art association," and his plans for the establishment of an art school. He referred to the organization of Throop. Finally he spoke of his enjoyment of golf, tennis, and especially of fly fishing—to show, as he said, "that science does not bound my entire horizon." It was, he acknowledged "a motley throng of interests." [34] Moulton's article was published in 1914. [35]

The next summer, in August, 1915, Hale, urged on by Harold Babcock, agreed to take time off for a camping trip to the High Sierras. The astronomer St. John, and Bill Hale, now fifteen, went along, also Ernest Babcock, geneticist from the University of California.

On the return, from Camp Curry, Yosemite, Hale wrote to Margaret in a letter full of the joy of being in the mountains: "We had a great time, with fine trout-fishing in Lyell and Dana Creeks and in Elizabeth and 'Dog' Lakes. Everywhere we could see the effects of glaciers, some of the mountain surfaces, high up on the peaks, being so highly polished that they shone like mirrors in the afternoon sun." They followed a trail from Tuolomne Meadows up along the side of Cathedral Peak. All along the way he reveled in the beauty of the scenery, the wonder of the mountains. "Later we came into another splendid forest, with magnificent pines and firs and spruces, and frequent carpets of meadow grass and wild-flowers. The trail wound on for miles and finally we saw the enormous rock mass of the Half

Dome above us on the right and knew we were approaching the valley. . . ."

"William," he exclaimed, "enjoyed the whole trip hugely, especially the expedition to Mt. Dana (13,000 feet) which he and three of the others climbed. The two old men—St. John and I—stayed in camp, as befitted their grey hairs." Years later, William recalled that trip as one of the delights of his life, and one of the rare occasions when he was able to join his father on an expedition without any ostensible aim. Together they fished for the trout that "found their way into the pan within a few minutes of their fatal glance at the floating fly." And they shared an understanding they had rarely known.[36]

Back in Pasadena after this interlude, Hale returned to the plaguing problems of the 100-inch disk. In May, 1911, John Hooker had died, leaving his promises on the disk unpaid. Yet, as a result of Carnegie's munificent gift, the Carnegie Institution, had now been able to make funds available for carrying on the work. Despite Ritchey's "unrelenting gloom," the grinding of the first disk went on. It was a tremendous job. The disk had an area of nearly 8,000 square inches; each one of these inches had to be polished true to within a couple of millionths of an inch. It was a nerve-racking job, and the longer Ritchey worked at it, the more difficult he became.

As Walter Adams told W. H. Wright at Lick some years later: "I might say, for your information that the mirror would never have been completed but for Hale's insistence. Ritchey had a notion that the glass had a weak diameter and buckled under its own weight in any position. He wanted to throw it overboard and build a mirror in thin sheets. The 'weak' diameter proved to be astigmatism due to the stratification of the air in the testing hall. Hale and I carried out most of the tests of the mirror at various stages photographically by the Hartmann method, for we really could not trust Ritchey to keep an unprejudiced mind during the progress of the work."

At the beginning, Ritchey had been entrusted with the design of the 100-inch telescope mounting. As his hostility and his unwillingness to accept suggestion grew, Hale was forced to relieve him of the responsibility—with a resulting improvement in the design. Finally he also had to take him from the task of polishing the disk.[37] He had long known that Ritchey suffered occasional strange attacks. With the strain of working alone in the optical shop for months on end, these attacks increased in intensity, and Hale now learned that they were epileptic. He concluded, as he told Turner, that "anyone familiar with Ritchey's conduct would feel that the most charitable view is to assume that his illness is at the bottom of everything." [38] In the end, therefore, a large share of the final work of giving the mirror its final paraboloidal curve was done by the able optician W. L. Kinney.

By 1913 the piers for the new telescope had been completed. It had been a tremendous job, and Adams ranked them "close to the pyramids as a monument of human endeavor—human endeavor in this case being represented by George Jones" (the very able and powerful superintendent of construction). The mounting, designed first by Ritchey, then extensively modified by Hale, Pease, and F. L. Drew, had been built at the Fore River shipping plant in Quincy, Massachusetts, "where machinery sufficiently large to build better battleships was available." It had been shipped by way of the Panama Canal to San Pedro Harbor. The more refined parts of the instrument had been made by the skilled machinists in the observatory shops. In February, 1916, the sections of the great telescope tube had arrived safely after a long voyage from the East Coast, in a time of war when shipping was already becoming dangerous. [39]

The job of getting all this massive equipment to the top of Mount Wilson had been hazardous. A special road had to be built, and even that special road was precarious. On the way up the mountain, Walter Adams, riding in the truck built to transport the heavy mounting, almost met a fatal end. The truck was carrying a load of cement when, at a precipitous point known as Buzzard's Roost, it toppled over the edge and hurtled down into the canyon some three hundred feet below, carrying the driver, Tom Nelson, with it. Adams and the assistant driver jumped to safety. Miraculously, the driver was also spared. [40]

All the parts, too, for the great dome, designed by D. H. Burnham of Chicago, had to be transported up the treacherous trail, but by this time Hale was far away in Washington, conceiving another project of a very different sort.

Meanwhile, however, he had been working in his spare moments on his small book *Ten Years' Work of a Mountain Observatory*. It was to be published in 1915.

As he pondered the design for the book, he wrote first a rough draft for an introduction that would express something of his own philosophy as a scientist, and his firm belief that a scientist's life is one of the best. "His path, it is true, is steep and beset with difficulties, but it leads to heights which continually unfold new prospects of ever increasing charm. Nature has hidden her secrets in an almost impregnable stronghold, and guarded them by every artifice derived from infinite experience and endless ingenuity. . . .

"For ages her defenses have been building, and these must all be scaled. The task of the investigator is like that of a band of explorers, forced to penetrate a wild and rugged country, bristling with obstacles, and defended by occupants whose secret isolation may afford their only hope of existence. Peak and cañon, mountain torrent and tangled forest, dispute the entrance and impede the advance. To outlying scouts must be entrusted the task of discerning means of scaling cliffs, forcing the passage of swamps and thickets, and bridging streams."

Nor was the explorer's role so different from that of the scientific investigator. "The parallel is a closer one than it may seem. The difficulties of research are increasingly great as we ascend the scale of Nature's products. . . ." [41]

In the end, apparently, Hale decided against an introduction for his book. Instead he began directly with a comparison of the old methods and the new that had inspired the building of Mount Wilson, "planned for the exploration of unfamiliar fields" where "its mode of attack and its means of progress must grow with its work and develop with the disclosure of new and unexpected possibilities." He emphasized once more that its chief object was to "contribute, in the highest degree possible to the solution of the problem of stellar evolution" and the origin of the earth. [42]

National and International Outlook

✳ *1915-1919*

ON THE 8TH OF MAY, 1915, shortly after the annual meeting of the National Academy of Sciences, Hale picked up the New York *Times* and read the flaming headlines:

LUSITANIA SUNK BY A SUBMARINE, PROBABLY 1260 DEAD: TWICE TORPEDOED OFF IRISH COAST: SINKS IN 15 MINUTES; CAPT. TURNER SAVED, FROHMAN AND VANDERBILT MISSING; WASHINGTON BELIEVES THAT A CRISIS IS AT HAND

For months past he had fumed at Wilson's "wavering, inconsistent policy." At the academy meeting a few days before the *Lusitania* disaster, he had talked with Dr. William H. Welch, the president, and other academy members of the need for scientific preparedness. Yet few of them shared his sense of urgency.

Now, as the war clouds spread rapidly across the Atlantic, he thought of the numerous signs that had pointed to the Kaiser's intention to dominate the world by military force. He thought of the days in Berlin in 1893 when he had watched the Kaiser riding up and down Unter den Linden, and had first sensed the violent German militaristic spirit. He recalled the day, twenty years later, when in 1913, on his way to Europe, he had talked with a German officer and had listened to his indiscreet references to "our war with England" months before the actual declaration of war. These and other incidents had, he felt, prepared him for the tragedy of 1914, and had convinced him that the United States must ultimately share in that tragedy.[1]

In July, 1915, he wrote to Dr. Welch, who was about to sail for the Orient with Dr. Simon Flexner of the Rockefeller Institute, "The

Academy is under strong obligation to offer services to the President in event of war with Mexico or Germany." [2] He urged him to learn the opinion of the Academy Council before he sailed. This letter reached Welch in San Francisco. He wired back, "Hardly possible to obtain opinion of Council before sailing Saturday morning for Honolulu." [3] Obviously he felt that action could wait until his return in December. He considered war unlikely while Wilson was President. Then, as Hale wrote, "he vanished into the misty depths of the Pacific."

On the day Hale wrote to Welch, it was announced that Secretary Josephus Daniels had selected Thomas Alva Edison to head a Naval Consulting Board to "aid in developing war devices to assist in perfecting the Navy as a fighting machine." [4] It was suggested that Henry Ford, Charles Steinmetz, Hudson Maxim, Orville Wright, Simon Lake, and Alexander Graham Bell be asked to join the Edison committee. These men were inventors all—there was not a pure scientist in the group—and Hale was convinced that their limitations could prove disastrous. He felt the urgent need for an academy committee with the widest possible scientific representation. Yet with Welch on his way to China, there was nothing he could do but wait—and watch while the situation in Europe became increasingly desperate for the Allies, and share his brother's conviction, "It becomes clearer every day that the U.S. ought to have declared war—*to save our own hides*." [5]

In February, 1916, he watched the drive on Verdun with growing anxiety. "The Allies," he wrote to Harry Goodwin, "*must* win or we will be in for it—hot and heavy in our turn. . . . As for the President, I had utterly given him up, and vowed myself ready to vote for Roosevelt or anyone else with any backbone." [6] Eight months had passed since he had telegraphed to Welch. With the possibility of a diplomatic break with Germany increasingly imminent, he felt that other scientists must now share his sense of the impelling need for action. In March he wrote to Robert Woodward. [7] The plain duty of the academy, he urged, was to organize American men of science in the nation's service. In case of war, they should be in a position to act at once. To Woodward and to the more dispassionate Welch, it is likely that Hale appeared in the guise of a somewhat fiery monster, crying for blood. But, as usual, he found halfway measures impossible.

On March 24th the *Essex* was torpedoed. In April the *Sussex* was sunk. "How Wilson can crawl out of a break with Germany is more than I can see," Hale exclaimed.[8] Yet, as tension increased to the breaking point, he was glad to find that more and more people shared his view.

On April 24th, shortly after the sinking of the *Sussex*, the Army and Navy Ball was to be held in the sail loft of the Navy Yard. Martha persuaded her brother, then in Washington, to go along. It was a grand affair, attended by admirals, ensigns, dowagers, and debutantes. War and all thought of war seemed far away. The President and his bride, Edith Galt Wilson, were there. Colonel W. W. Harts, Martha's dashing husband, who was Wilson's aide, was at the President's side.[9]

The ball was held just at the time of the meeting of the National Academy of Sciences. At that meeting Hale was renominated as foreign secretary (a position to which he had first been elected in 1910). Before accepting the nomination, he rose to state that unless he could be allowed full liberty to express himself fully on any subject, he could not accept the position. No one objected, and he was unanimously elected. He then presented his resolution offering the academy's services to President Wilson in case of a break in diplomatic relations with Germany. The resolution was adopted, and a committee was appointed to call on the President.[10]

The appointment was made for April 26th. At the appointed hour of quarter to one, Hale, joined by Walcott, Woodward, Welch, and Edwin G. Conklin of Princeton, appeared at the White House. Wilson welcomed them. He listened as Welch described the academy's place as adviser in scientific matters to the government and as Hale told of its potentialities and the importance of scientific research in national defense. When they had finished and had presented the academy resolution, Wilson agreed that it would be advisable to collect information on investigators and laboratories willing to contribute to the war effort. He suggested that a committee be formed "to undertake such work as the Academy might propose." Yet, evidently fearing that this move might be construed as a step toward war, he asked that on "no account" should his oral approval of the plan be made public.[11]

Hale automatically became chairman of the new committee. In his little red diary, he scribbled in pencil: "National Service Research Foundation—Object: The promotion of scientific research in the broadest and most liberal manner, for the increase of knowledge and the advancement of the National Security and welfare." Soon he was writing to Scherer: "I really believe this is the greatest chance we ever had to advance research in America." [12]

By June, 1916, Hale and the members of the organizing committee, after some weeks of intensive work, had completed a plan to promote research. The plan was approved by the Academy Council. But before the National Research Council (as it was soon called) [13] could become effective, the President's public approval was needed. At this point, however, Wilson was unwilling to do anything that would jeopardize his neutral position. He was running for a second term on the platform that he had kept the United States out of war, and Hale realized it was useless to approach him directly. However, an oblique approach through his "silent partner," the redoubtable Colonel House, might work. "It is a pity," Hale confided to his brother, "to have to enter the back door of the White House, but all agree it is one way to get results." [14]

On June 22nd, therefore, he set out for Lake Sunapee, New Hampshire, to see the powerful colonel.[15] He found a small man with a slight, almost undistinguished figure, white haired, white moustached, with profoundly penetrating eyes. He found him in a good mood, willing to listen. He emphasized that the council had gained the promise of cooperation from leading scientific societies and of support from heads of universities, schools of technology and medicine, research foundations, industrial-research laboratories. "We have received the promise of service from every man of science whose aid we have sought," he said. At the end of their meeting, House agreed to speak to the President.

Hale still felt that the President must be pressed even more strongly. He decided to attack from another direction. In the coming national election, Charles Evans Hughes was running against Wilson. If Hughes *should* win, his support would be needed for the embryonic council. Working with all the instincts of a trained politician, Hale told James Garfield, Hughes's manager, of their plans and hopes. Garfield, in turn, stated that Hughes recognized fully the fundamental role of science in any plan of preparedness, and promised his candidate's support. This, of course, was exactly what Hale wanted, not only for its own sake but also for its effect on President Wilson. Immediately after his interview with Garfield, he wrote to House, telling him what had happened and saying that he was preparing a statement for the press that would be published before

Hughes's letter appeared. He asked for authorization to announce that the President had approved the council's plans; he also requested a formal letter of approval from Wilson to Welch, the academy president. He emphasized that this was not a question of endorsing the exclusive action of a single society but the action of a group that represented all American scientists. He set a deadline of July 27th for mailing the announcement to the Associated Press.[16]

This technique worked like magic. Colonel House replied by telegram and a special-delivery letter in which he promised to consult the President "immediately." [17] On July 24th, three days before the deadline, the President wrote to Welch, "I want to tell you with what gratification I have received the preliminary report of the National Research Council, which was formed by my request under the National Academy of Sciences. The outline of work there set forth, and the evidences of remarkable progress towards the accomplishment of the object of the Council, are indeed gratifying." [18]

At the same time Wilson telegraphed to Hale, who set to work immediately to implement the council's plans. One of his first moves was to telegraph to Gano Dunn, president of the J. G. White Engineering Corporation, the United Engineering Societies, and the Engineering Foundation, arranging an appointment at the Biltmore Hotel that evening.[19] With Dunn came the noted inventor Michael Pupin, who some years later, at Hale's urging, would tell the fascinating story of his life in his book *From Immigrant to Inventor*. The three men worked out a cooperative plan between the Engineering Foundation and the National Research Council. (The Engineering Foundation had been founded by Hale's old friend, the builder of telescopes, Ambrose Swasey, with the object of stimulating, directing, and supporting scientific and engineering research. Pupin was a vice-chairman of the foundation.)

"It did not cost me much effort," Pupin writes, "to persuade Dunn that one of the biggest tasks which the Engineering Foundation could take up was to grubstake the Council during its formative period." [20] As a result, the Engineering Foundation Board endorsed the council for a year, and (from September, 1916) the administrative organization as well as the total income of the foundation was turned over to the embryonic council.

Dunn and Hale had first met some years before, and a strong friendship had quickly developed between them. When asked about that friendship long afterward, Dunn, after a long pause, said slowly and deliberately, emphasizing each word: "The most apt description I can give of the personality of George Hale is borrowed from Elihu Root, who called him the 'lambent flame.' . . . He had a genius for working with other men, who instinctively followed his leadership because of the creative instructiveness of his ideas, and his capacity as a great teacher. In all his work his sense of humor was irresistible and his word was his bond." [21] Hale and Dunn had indeed much in common. Both were by nature and preference experimentalists. Both were intensely interested in microscopy. Both felt that in pure science and in engineering there is poetry, and emphasized always the importance of a broad cultural background for scientists and engineers alike. Together they would work to carry out their ideas to the lasting benefit of the growth of science and engineering in America.

Shortly after the meeting at the Biltmore, Hale sailed for Europe with the rotund and jovial Dr. Welch, who proved to be a wonderful companion. As he puffed on his cigar, he seemed able to talk exhaustingly on any subject. They were anxious to learn the "general character of the war services rendered by the scientific men of these countries." Welch, who had founded the first pathological laboratory in America and who was also a founder of the Johns Hopkins Medical school, was particularly concerned with the military hospitals and with medical problems created by the war. Hale was concerned with the nature of the physical and chemical problems of war that the Allies faced. For many years Germany had led the world in the utilization of scientific advances. Now, with imports suddenly cut off, the Allies were forced to try to fill the appalling gap. By 1916 they had been at war for two years. Already they had made advances in airplane design, in the development of methods of aerial photography, in the paramount problem of finding sound-ranging devices to combat the German submarine menace. (In one month the Germans had sunk as much as 900,000 tons of Allied shipping.) But much remained to be done. The more he learned, as they traveled through England and France, the more eager Hale became to see the United States enter the war on the side of the Allies. Meanwhile, he vowed to do all he could to have his country better prepared scientifically if and when that day should come. In fact, he could hardly wait to get home. When their departure from France was delayed by a submarine scare, he became highly upset. He feared he would not reach New York in time for the dinner that had been planned for leading members of the Engineering Foundation and a few National Academy members, representing the Research Council.[22]

"He is very tired," Welch wrote significantly in his diary, "and seems almost on the verge of a breakdown, due I think in part to the depression from the discouragement of our plans." [23] As soon, however, as he discovered that they would make it after all, his dark mood changed, and once again he was his usual gay self. "He is much relieved by the decision," Welch noted, "and recovers from his depression."

The dinner, held on September 20, 1916, was a huge success. At last the National Research Council was launched. Hale became its chairman; C. D. Walcott and Gano Dunn were made first and second vice-chairmen, respectively. At last it could begin a career that would do more than any institution founded up to that time to organize and develop science on a national scale.

It was an important step. Previously, organizations like the American Association for the Advancement of Science, the American Philosophical Society, and of course the National Academy, had held meetings once or twice a year. But there had been little concerted planning for the general advancement of science and the national welfare. Now, under Hale's leadership, American scientists would have the chance to develop cooperative research on an unparalleled scale, first for war and later, as he had hoped and planned from the beginning, for peace. The movement was to have great and lasting effects. From this time science was to become an increasingly powerful force in American life.

As usual, however, the initial course of the new organization was destined to be a rough one. Returning to Washington in the fall of 1916, Hale found himself embroiled with the Advisory Commission of the National Defense Council (which had been set up in August, 1916.) [24] The commission was directed by Hollis Godfrey, an engineer who was president of the Drexel Institute in Philadelphia, and, according to Hale, "a hopeless visionary." Hale feared he intended to set up a separate scientific organization that might supersede the National Research Council and that could only cause endless confusion. Hale wrote anxiously to the President.[25] But Wilson, absorbed in the national election, failed to answer. Again Hale turned to

House. He urged him to induce Wilson to announce that the National Research Council, "acting in cooperation with the Council of National Defense," would be of great service to the country.[26] In this way he hoped all doubt of its position as the appropriate body to deal with research problems connected with national defense and industrial development would be dispelled. But House also failed to answer.

Hale was frantic. Yet he realized that, at the moment, he could do nothing. At the end of November he returned to Pasadena. Soon after his arrival, a letter came from Welch: "You should know that Godfrey has done everything to create the impression that you are personal enemies and [are] trying to destroy each other." [27] Although to Hale's inflamed imagination the situation now took on the proportions of a major war, it was actually not crucial. "Godfrey," Welch added, "has posed as a representative of science and fallen down, as might have been foreseen."

Nevertheless Hale was perturbed. He was sure the council and all his work would be overthrown. He suggested a talk with House, but Welch, insisting there was nothing to fear, considered this inexpedient. In the heat of the fracas, Hale decided impulsively that he ought to resign as chairman of the council, and urged Welch to take his place. He argued that the thing to do was to settle down quietly to his astronomical work. He also pleaded ill health. "Various causes (certainly not astronomical work, of which I have done so little) have conspired to develop some of my old symptoms recently, and the impossibility of doing justice to the work of the Council [has] aggravated them." [28]

Even as he wrote, however, he must have realized that he had no intention of adopting his own suggestion. On February 1, 1917, Germany had proclaimed unlimited submarine warfare; on February 3rd diplomatic relations with the United States were broken. Hale had immediately telegraphed to Wilson, placing the Research Council at his service.[29]

Just before he again left for Washington, word came that the Council of National Defense had adopted a resolution requesting the cooperation of the Research Council.[30] Hale, immensely relieved, went east with a light heart.

Washington was a maelstrom, and he was soon plunged into its center. He had rented offices for the council in the Munsey Building at 1329 E Street N.W., and the days, crowded with interviews and meetings, were never long enough. He was swamped by an endless stream of physicists, engineers, botanists, psychologists, and chemists, eager to solve the wide variety of problems that would be raised in the event of the entry of the United States into war. There were also many would-be scientists and inventors eager to submit weird contrivances "guaranteed to sweep submarines from the seas"—death rays and the like.[31]

The problems that had to be solved were manifold. They ranged from the treatment of disease to the manufacture of optical glass for gunsights, rangefinders, and periscopes; from chemicals needed for high explosives to the "scores of other products developed in Germany after long years of investigation," and now suddenly rendered inaccessible. The list of those who arrived in Washington to help in the solution of these problems would read like a *Who's Who* in science in America. They came from universities and technical schools, from government and from industry.

One of the first to arrive in the fall of 1916 was Hale's old friend Arthur Noyes from M.I.T. He came to help in the preliminary organization of the council and to direct all work in chemistry, particularly the impelling problem of nitrogen fixation. Like Hale, he soon found that if these days were intensely busy, they were also intensely exciting.

Years later, in 1933, Hale was to write to Isaiah Bowman, who had just been made chairman of the Research Council. As he looked back on the days at the beginning of his venture, he recalled an "exhilaration which I have never felt in equal degree except in the course of scientific research.

"When I first took the job I had no funds for the Council, no office rooms, no friends (except Stratton) in Government Departments—little, in fact, but the pleasant difficulty of overcoming the prejudices of the chiefs of military and naval bureaus against 'the damned professors.' It was a bully game, and I wish I could try it again."

On April 6, 1917, war was declared. On May 29th a Franco-British commission landed in New York. It included the great British physicist Ernest Rutherford; a commander in the royal navy named Bridge; Charles Fabry from Marseilles, and Henri Abraham from the University of Paris. One night Hale gave a dinner in honor of the commission. It was a gala occasion, but with the most serious undercurrent. Ellie Mosgrove, who had come on from California with Mrs. Hooker, was there, and she never forgot the speech that Rutherford gave that night. His dramatic words remained graved in the memories of all who heard them: "Our backs are to the wall. We must have help." No one could escape the frightening urgency of those words. Ellie said afterward that all through the dinner she shivered. Unless the submarine menace could be overcome, Britain would collapse of starvation.[32] Hale, profoundly stirred, became even more determined to push the solution of the difficult problem of a submarine-detecting device. As a result, the astronomers at Mount Wilson would soon be turning their eyes from the stars to this pressing danger under the seas.

By the time the Franco-British commission arrived, the physicist Robert Millikan had come to Washington from the University of Chicago to take his post as the chief officer in charge of the council's research program. Hale bestowed on him the imposing title of "Third Vice-Chairman, Director of Research and Executive Officer of the National Research Council." He had known Millikan since his early days at the University of Chicago, and had always been attracted by his charm, his enthusiasm, and his boundless energy.

Millikan, the son of a Congregational minister, was born in Morrison, Illinois. After his graduation from Oberlin in 1891, he stayed on as a tutor in physics. From there he had gone on to Columbia, where he was a student of the Serbian physicist Michael Pupin; then (after a year in Germany) in 1896 to Chicago, where he became a member of the Physics Department headed by Albert Michelson. In 1906, with Henry Gale, he wrote A First Course in Physics.

In his years at Oberlin and Chicago, Millikan had shown his skill as a teacher as well as the breadth of view that was evident in his writings. He had shown, too, his skill in his famous oil-drop experiment by which he determined the mass of the electron. At the anniversary celebration of the National Academy of Sciences in 1913 (which was also the year of his academy election) he had received the Comstock prize for his "rare acumen and experimental skill" in providing the most direct and the most convincing proof of the existence of "electrical atoms or elements." Ten years later he was to receive the Nobel prize in physics.

Now, as he joined the council in Washington, Millikan was to show his executive ability as he assumed a large share of the load of managing the organization. Not long after his arrival new difficulties arose with a Mr. Saunders, who had been appointed by Secretary

Daniels to the acting chairmanship of the Naval Consulting Board. Saunders characterized the whole enterprise of the Research Council as the extreme of presumption, abused its organizers and even threatened them if they did not desist. But this kind of attack, as Millikan was to point out, "only aroused Hale's fighting spirit."

Meanwhile new activities were being organized, old ones strengthened, and new offices rented. "I am still in the shuttle and it will not stop," [33] was Hale's constant refrain. He found a more or less permanent place to live, in the Grafton Hotel on Connecticut Avenue near L Street. Here, in the late fall of 1917, Evelina Hale joined him, and they settled down for the "duration." From England, H. H. Turner wrote to Walter Adams, who had again taken over the acting directorship of the observatory: "Yes! We must let Hale be at Washington as much as we can. I expect he will end up as President of the United States; he will make a good one if he does." [34]

The dawn of 1918 brought with it one of the severest winters in Washington's history. As the work of the council continued to grow, Hale could see its possibilities in countless directions. One of his goals was the formation of a Research Information Committee that would survey the progress of research in countries the world over. A fundamental role in this program was to be played by scientific attachés appointed to embassies in London, Paris, Rome, and Washington. They were to work in collaboration with the Army, Navy and State departments to learn what was happening along scientific lines abroad and to relay it home. As chairman of the committee, Hale decided on a young psychologist, Robert M. Yerkes.

The two men had met some years before at the American Academy of Arts and Sciences in Boston. Hale had been attracted not only by the keenness of Yerkes's face but also by his name. Yerkes had admitted a distant relationship to the traction magnate, but, as Hale soon discovered, he was of quite a different cast. He was at the time an assistant professor at Harvard, and with some diffidence he had asked Hale if he could spare the time to discuss a new research project. Hale agreed to meet for dinner at the Harvard Faculty Club, and there Yerkes described his dream of establishing a laboratory for the biological study of primates, "including our seemingly nearest relatives, the great apes." The experimental approach in the field of psychology was still suspect, and the majority of Yerkes's colleagues had ridiculed his "notions." But Hale was entranced, and offered to do anything he could to promote the scheme.

Up to this time, Yerkes had been torn between two careers, one in teaching, the other in research. "Hale's expression of faith, optimism and spirit of high adventure" set him on the course of research in comparative psychology that was to make his "colony of primates" internationally famous.[35]

When war broke out, Hale recalled their meeting, and immediately proposed that Yerkes help in the organization of the council's psychology division. As a comparative psychologist, he represented a general field of scientific effort that few scientists, and almost no military men, could think of as potentially useful in connection with war. But soon, with a small group of psychologists, he was proving the practical value of psychology in time of war. To this task that of the Research Information Committee was soon added.

Meanwhile Hale was trying to reach an even more fundamental goal—the establishment of the National Research Council on a permanent basis. If it was of value in war, its role in the life of the nation should be even greater in peacetime. The bearing of its activities, as Hale was to write, "reaches down to the very foundations of national welfare. The problems of peace are inextricably entangled with those of war, and if scientific methods and the aid of scientific research were needed in overcoming the menace of the enemy they will be no less urgently needed during the turmoil of reconstruction and the future competitions of peace. Thus the very agencies of war will become powerful factors in the competitions of peace, and the research methods from which they sprang will play a far larger part in the world than ever before."

As he looked forward to an age in which science would play an increasingly dominant role in the life of the nation he concluded prophetically: "Our place in the industrial world, the advance of our commerce, the health of our people, the output of our farms, the conditions under which the great majority of our population must labor, and the security of the nation will thus depend, in large and increasing measure, on the attention we devote to the promotion of scientific and industrial research."

On March 26, 1918, he wrote to President Wilson, enclosing a proposed draft of an "Executive Order" that defined and authorized the specific duties of a permanent council.[36] He emphasized that it involved no extension of privileges or powers. After three weeks the answer came.[37] Hale's first reaction was that Wilson did not understand what was wanted. But soon he learned from Walter Gifford that the order had gone to the National Defense Council and had been turned down there. "Some member," Millikan notes, "had assumed an attitude of definite hostility and opposition, and no one of the others had felt disposed to bestir himself to the extent of making a fight for it." Hale had never expected the request to go anywhere except to the President. He discovered, however, that House had been ill, and had therefore been unable to push the order through.

As soon as he learned this discouraging news, he hurried to New York, where House was staying. He was accompanied by John J. Carty, director of research of the American Telephone and Telegraph Company, a witty and wise man to whom Hale had been attracted as soon as they had met. House insisted he had no intention of letting the matter drop just because of "Cabinet" opposition that came from David Houston, William Redfield, and Josephus Daniels. "All right," he exclaimed, "that simply means that we shall have to put it through over their dead bodies." [38] And this was about what happened. House telephoned the White House, and the following day President Wilson, after listening to his powerful adviser, made a few changes in Hale's phrasing and sent the order back to the council. Late on the afternoon of May 10, 1918, Hale had an appointment with the President. The next day Wilson signed the order. In it he asked that the National Academy of Sciences perpetuate the National Research Council. "The work accomplished by the Council in organizing research and in securing cooperation of military and civilian agencies in the solution of military problems has demonstrated its capacity for larger service." [39]

Then followed a list of the council's functions, with the first, and most important, the stimulation of "research in the mathematical, physical and biological sciences, and in the application of these sciences to engineering, agriculture, medicine and other useful arts, with the object of increasing knowledge, of strengthening the national defense, and of contributing in other ways to the public welfare." [40] So the National Research Council was born to a permanent and a long life. Simon Flexner declared, "It is a most important triumph." Thereafter the order was printed with the academy's charter, and the council took its place as a dominating force in the development of science in America.

Throughout this strenuous period, Hale had more to do each day than he ever had time to do. Yet, in general, he was well and happy. Margaret had joined them, and during her mother's absence in California was staying with him at the Grafton Hotel. Each evening she would pick him up at his office and they would go for a ride in "Oscar," her Dodge coupe. Sometimes they drove through Rock Creek Park; in the spring of 1918, with the flowering dogwood and redbud in bloom, it was particularly beautiful. In the evenings they often read aloud. Occasionally they went to the theater.[41] One night they saw a "screamingly funny" performance of *Penrod*. For a while they forgot the "Boche on the very banks of the Marne." [42]

In June came a welcome interlude—the chance to return home by way of Green River, Wyoming. The Mount Wilson astronomers had set up camp there in preparation for the coming total solar eclipse. Hale, feeling like a boy again, left Washington with Margaret.

On June 8th he recorded in his diary: "Observed total eclipse at Green River, Wyoming." [43] At the last moment thin clouds spread over the sun's face. Nothing was seen of the corona. Yet this eclipse inspired him to return once more to that haunting problem—the detection of the corona without an eclipse. He designed a new form of apparatus with a photoelectric cell. A month after his return to Pasadena he scribbled: "Believe corona device will work." [44] But again, as in his earlier attempts, his experiments failed. He began to think of other ways of trapping the corona. One of these was prophetic. In the annual report for 1918 he wrote: "It is perhaps barely possible to send meteorological balloons, carrying automatic photographic apparatus, directed toward the sun by a gyrostat and finally controlled automatically by the heat of the sun's image, to a height sufficiently great to allow the corona to be photographed without an eclipse. But the technical difficulties would be considerable." [45]

Forty years later, Hale's prophecy on the use of balloons for solar research would become a reality when Martin Schwarzschild sent his balloons up to record magnificent photographs of the solar granulation and of sunspots. Yet only twelve years were to pass between that eclipse of 1918 and 1930 when the French astronomer Bernard Lyot finally succeeded in obtaining the corona without an eclipse, the prize that Hale had so long sought. He made his observation in the clear atmosphere of the Pic du Midi in the Pyrenees at an altitude of 9,400 feet. He succeeded on his first attempt with a special kind of telescope equipped with a simple lens of very fine optical quality, which deflected all the photospheric light from the image. As Otto Struve pointed out in 1947, "The method that is now being employed by Lyot and his successors is not very different, in principle, from the one used by Hale. Only greater precautions against scattered light by dust on the surfaces, and by small bubbles or scratches and other defects in the optical parts, were required to insure complete success." To which he adds, "Hale's failure to photograph the corona in full sunlight was undoubtedly the greatest disappointment of his scientific life." [46]

In his diary, on June 11, 1918, three days after his coronal notes, Hale made a brief but pregnant entry: "Goddard here. Began arrangements for his work." Fascinated by Robert Goddard's experiments on rockets, Hale offered him a place to carry on his experiments in the Mount Wilson laboratory on Santa Barbara Street. The Signal Corps had made an allotment of $20,000 and advanced it to the Smithsonian Institution, where Dr. Walcott had been the first to support Goddard's rocket ideas when many people scorned them as fantasy. Now his task was to explore the military possibilities of rockets. By August he was ready to make the first demonstration of his rocket in a canyon below the observatory. It soared upward, and landed 250 feet beyond the edge of the canyon. This was Goddard's record of the event: "They all jumped at the first shot. Then Hale shook hands and said he would send a good, strong report to Washington. Hale's assistant said: 'Prettiest thing I ever saw, to see that come flashing up over the canyon.'" [47] Later, Goddard made other demonstrations (first in Pasadena, then in Aberdeen) of his device that promised to revolutionize warfare. Yet, fortunately for the world, if unfortunately for the development of the rocket, the war in Europe was ending. Two years after he began his experiments at Mount Wilson, a paper entitled "A Method of Reaching Extreme Altitudes" was published in the Smithsonian Annual Report. [48] In that paper he made the startling statement that some day a rocket might be shot to the moon. The full importance of this "visionary" work would not appear for over thirty years, after both Hale and Goddard were dead.

Soon after his return to California in that summer of 1918, Hale drove down to Coronado Beach for a vacation. He lay on the beach but he could not rest. His head teemed with plans. When the war should end he looked forward to a resumption of old friendships and renewed cooperation with foreign scientists. He dreamed of establishing an international organization that would first include the Allies and might later be extended to include all the nations of the world. In April, 1918, he had written of his dream to the British physicist Arthur Schuster, whose belief in international science was so great that, as a friend put it, "If it were possible to hold an Interplanetary Conference, say on the planet Jupiter, it would be very possible that Dr. Schuster would be asked to act as the earth's representative." Hale was sure of his sympathy and enthusiastic support.[49]

Over the years they had met at meetings of the International Association of Academies—in Vienna in 1907, in Paris in 1910, at Petrograd in 1913. At these meetings Hale had become increasingly aware that the association bore a marked similarity to the National Academy in the era before his attempts to revive it. Its members were largely devoted to mutual admiration, and its contribution to the advancement of research had therefore been slight. It did not, moreover, represent in any way the many international organizations devoted to research in special fields. Before the war Hale had proposed that one way to make it a more lively, functioning organization would be to establish a central scientific headquarters at The Hague. He had even asked Carnegie for a building fund. Nothing had come of the proposal. Now, therefore, he decided, was the time to consider a plan for an entirely new international organization— devoted to promoting the larger interests of science and research "potentially strengthened through a greatly enhanced appreciation of their national importance." [50]

Taking the National Research Council as a prototype, he suggested to Schuster that the proposed "Inter-Allied Research Council" devote itself to "large undertakings requiring the concerted efforts of scientists of different nations working in diverse scientific fields." He suggested, too, that it be led by the great national academies, the Royal Society of London, the Paris Academy of Sciences, the Academia dei Lincei in Italy, and the National Academy in the United States.[51]

Under the parent organization, Hale proposed the formation of international unions representing the various branches of science and technology. He felt that the feasibility of such international undertakings had been clearly shown in the Solar Union, which had first met in St. Louis in 1904 and had been formally organized at

Oxford in 1905. Gradually the union had grown to include all phases of astrophysics. During the war its activities had been suspended. But with the war's end he looked forward to its revival and to its future influence in a world where international cooperation would become increasingly vital. Already he had suggested the possibility of combining with the Solar Union other cooperative astronomical programs—the Carte du Ciel, Kapteyn's Selected Areas, time standards, astronomical ephemerides, the distribution of astronomical programs, the orbits of minor planets. In June, 1918, he wrote accordingly to Schuster on the formation of "a single international Astronomical Association" that would come under the aegis of his proposed international council.

"Advances in research," he noted, "are not the result of lucky chance, but are due to the suggestions and joint efforts of many able men. In the search for ideas we must cast our nets as widely as possible, because the initial conceptions underlying new devices are often found far from home; sometimes in very unexpected places."

In 1918, with the war still raging, the National Academy had assembled for its April meeting in Washington, and Hale had presented his plan for an "Inter-Allied Research Council." It was unanimously approved. Yet, when Hale discussed it afterward with Millikan, he found him unimpressed. Millikan apparently considered this latest move one of overorganization, overplanning. He obviously felt that at times Hale was "carried away by his role of initiator and promoter." [52] Still, he conceded that the many schemes Hale promoted succeeded. Seldom had he known anyone who examined a situation more thoroughly or anyone who sought the advice of others more continually.

On September 17th, Hale boarded the *Lapland* as an official academy delegate to the proposed council that was to meet in London. His companions, also academy members, were Simon Flexner and W. F. Durand, the engineer, who had held the office of scientific attaché in Paris. They crossed with a convoy transporting twenty thousand troops. In contrast to the potential dangers in the sea, Hale, unconcerned, immune to physical fear, reveled in the "calm, provincial joys of Anthony Trollope." Nearing England in a wild storm, they were escorted into port through the submarine-infested waters by a fleet of destroyers. [53]

The following day they found themselves in war-ravaged London. The meeting called by the Royal Society, and held at Burlington

House, was presided over by the noted physicist J. J. Thomson. There were delegates from Belgium, France, Great Britain, Italy, Japan, Serbia, and of course the United States. It was an impressive gathering.

At the first meeting Hale outlined the plan for what he was now calling the Interallied Research Council. It was warmly endorsed by all the delegates, and a plan was drawn up that would include research for national defense, as well as for industry and pure science. "In fact," he commented, "all the essential elements of the National Research Council are included. . . . While she may sail under another name, she is just the sort of vessel we wanted to build." [54]

But as usual in any international proposal, there were difficulties. When it was suggested that the Germans should ultimately be asked to join, the French and Belgians objected vociferously; they demanded that Germany be excluded from all international organizations by a clause in any peace treaty. The war, of course, was not yet over, and feeling was bitter. In the end, however, a compromise was reached and a meeting of the provisional Allied committee was set for Paris in November. [55]

In the interlude Hale enjoyed himself thoroughly. He spent happy, peaceful days with Newall in Cambridge and hours in stimulating discussion with J. J. Thomson at the Master's Lodge and with Lord Rayleigh in London. He discussed ideas for international cooperation with Arthur Schuster. [56] He went off for a weekend with the great astronomer James Jeans, at Bexhill, in surroundings that conjured up visions of Shelley and of Keats, who had written "Ode to a Nightingale" in this beautiful place. [57] He even found time to work on a paper entitled "The Nature of Sunspots," which he was to give before the Royal Society. This proved to be as successful as any lecture he had ever given. "J.J.," who presided, "went so far as to say" that the day would be a memorable one in the history of astronomy. [58]

Suddenly came the news of the Armistice. With everyone else, Hale found it hard to believe that the war was over. He watched the dramatic effect on the war-weary Londoners. He listened to them celebrating as he sat writing in the Athenaeum. There were hundreds of German guns in St. James's Park, and some of these had been dragged up and set on fire by the celebrators. Later he watched cabs, crowded with men and girls, rushing through

Piccadilly—once again lighted—"amidst a chorus of yells." Horns, drums, cymbals, even banjoes, joined in the babel.[59]

In Paris a few days later, he found a scene even more ecstatic as the liberation of Alsace and Lorraine was celebrated. "I was more than pleased myself to be in Paris on such a day," he wrote patriotically, "and to feel that we had helped to make it possible." [60]

On November 26th, the long-planned organizational meeting of the International Research Council (as it was to be called) was held in Paris. At last the new organization was born to what promised and indeed was to be a long and brilliant life. Charles-Emile Picard, French mathematician, was elected president, and the physicist Arthur Schuster was chosen as general secretary, in charge of the central office, which was to be in London. Again Hale stressed that the council's primary purpose should be "to promote international cooperation in scientific research rather than to hold congresses for the reading of papers, though it may combine both functions." To promote such cooperation, it was agreed that a group of international unions, dealing with various branches of science, should be formed. This decision was to be acted on at the meeting the following year in Brussels, when, too, the International Astronomical Union would be formally organized. "The I.R.C.," Hale told Harry Goodwin, "is under way." To which he added exuberantly, "The experience was the most interesting I have ever been through." [61]

"It is very satisfactory," he told Elihu Root, as he looked into the future, "to see a group of scientific men gathered from all parts of the world, working together in the most cordial manner and with the most complete disregard for national boundaries. Such men, in the course of time, are bound to have wider influence in national affairs, because the value of their knowledge and experience is being increasingly recognized by the state. I therefore believe the encouragement and extension of international cooperation in research to be a favorable field of action for those engaged in the promotion of peace."

The International Council would become the chief representative of its various unions, including Astronomy, Geodesy and Geophysics, Chemistry, Mathematics, Physics, Scientific Radio, Geography and the Biological Sciences. As a result, its name would eventually be changed to the International Council of Scientific Unions. Under its aegis, joint commissions, such as that on Solar and Terrestrial Relationships, would be set up and the High Altitude Research Station on the Jungfraujoch would be established. Under it, too, such worldwide cooperative efforts as the International Geophysical Year would eventually be carried out. So, despite Millikan's doubts, another of Hale's dreams had come true.

In December, Hale returned to Washington to find Margaret radiant over her engagement to Paul Scherer, James Scherer's son, and deep in plans for their wedding on Christmas Day. When the news had reached her father in London, he had written to tell of his joy in her happiness and of his love: "No one can rejoice more heartily than I do in your great and well deserved happiness, and no one can feel more certain that Paul will justify your best expectations." [62] To Evelina he had written proudly, "No one can blame him for falling in love with Margaret who is pure gold."

Margaret had written: "My dear, dear Daddy, what an angel you are to appreciate Paul, and my happiness! I hadn't realized how much your cable would mean to me until it came today. You have helped Paul to make me the happiest mortal in all this wide world." [63]

Characteristically, Hale now began planning for their future. He had visions of a little bungalow in Pasadena that he might give to his daughter as a wedding present. He hoped, as he told Jim Scherer, that Paul, who had left Throop to join the navy, would return to finish college before they were married. "If he still thinks of going into science, he may have a great chance in Throop's laboratories."

That Christmas was a gay one. There were the plans for the wedding. Then Bill arrived from Princeton. One night there was a gala dinner, and on another a theater party. Then, on Christmas Day, Margaret and Paul were married.

Meanwhile life and work in Washington went on. With the war's end, most of those who had rushed to Washington were now equally anxious to rush home to peacetime jobs. One of these was Robert Millikan. As a result, Hale found himself swamped with work. Only the loyal Noyes remained at his side.

In a way, the organization of research for war had been accomplished so quickly, and the war had lasted such a relatively short time after the entry of the United States into it, that many of the

FOURTH CONFERENCE

INTERNATIONAL UNION FOR COOPERATION IN SOLAR RESEARCH

August 30 - September 3, 1910

Mount Wilson

1. ELLERMAN	12. HARTMANN	23. CORTIE	34. LAMPLAND	45. KAPTEYN	56. HILLS	67. CHANT
2. H. C. WILSON	13. KÜSTNER	24. TURNER	35. HALE	46. MRS. FLEMING	57. LARMOR	68. EVERSHEIM
3. ST. JOHN	14. SLOCUM	25. RUSSELL	36. BELOPOLSKY	47. WATSON	58. COTTON	69. ROTCH
4. LARKIN	15. SAMY	26. KAYSER	37. DESLANDRES	48. SCHLESINGER	59. DYSON	70. W. MITCHELL
5. TOWNLEY	16. KNIGHT	27. ADAMS	38. SCHUSTER	49. HUMPHREYS	60. BARNARD	71. STRATTON
6. V. M. SLIPHER	17. WOLFER	28. MILLER	39. CAMPBELL	50. MADRILL	61. KING	72. H. D. BABCOCK
7. FOWLE	18. FATH	29. AMES	40. RICCO	51. J. F. SANFORD	62. NEWALL	73. RITCHEY
8. COBLENTZ	19. RYDBERG	30. BACKLUND	41. MRS. KAPTEYN	52. CHRÉTIEN	63. PRINGSHEIM	74. BRACKETT
9. FROST	20. HEPPERGER	31. KÖNEN	42. KOSLER	53. DE LA BAUME PLUVINEL	64. LEUSCHNER	
10. IDRAC	21. FOX	32. PICKERING	43. K. SCHWARZSCHILD	54. FABRY	65. J. S. PLASKETT	
11. HAUSSMANN	22. FOWLER	33. FOWLER	44. MC AIDE	55. ABBOT	66. GALE	

Delegates to the Fourth Conference of the International Union for Cooperation in Solar Research, 1910. (*Courtesy of Mount Wilson and Palomar Observatories*)

Aerial view of Mount Wilson Observatory. (*Courtesy of Mount Wilson and Palomar Observatories*)

Dome of the 100-inch Hooker
Telescope.

The 100-inch Hooker Telescope. *(Photos courtesy of Mount Wilson and Palomar Observatories)*

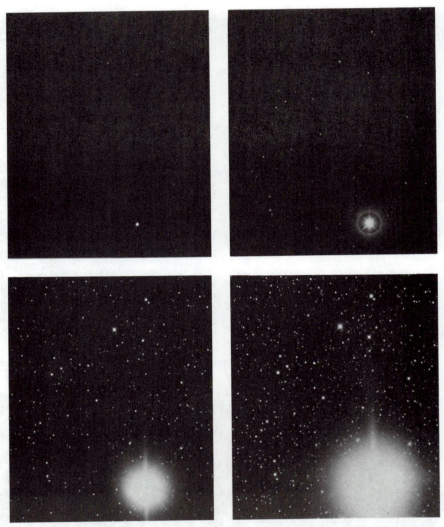

A series of photographs of the star Eta Aurigae taken with the same telescope but with increased exposure times. The photographs also indicate the advantage of increased light-gathering power in a telescope. *(Courtesy of Mount Wilson and Palomar Observatories)*

Constellation of Orion. The Orion Nebula is in the sword handle just below the three bright stars in the belt.

The Orion Nebula photographed with the 100-inch Hooker Telescope. *(Photos courtesy of Mount Wilson and Palomar Observatories)*

Early view of Throop Institute campus.

Modern view of the California Institute of Technology campus. *(Photos courtesy of California Institute of Technology)*

George E. Hale, Arthur Noyes and Robert Millikan at the Norman Bridge Physics Laboratory. *(Courtesy of California Institute of Technology)*

Principal facade of the Huntington Library building. *(Courtesy of Henry E. Huntington Library and Art Gallery)*

National Academy of Sciences building, Washington, D.C. *(Courtesy of National Academy of Sciences)*

Members of the Committee for Intellectual Cooperation of the League of Nations, at Geneva in 1922. *Left to right:* A. De Castro (Brazil), George E. Hale (U.S.A.), K. Bonnevie (Norway), Gilbert Murray (Great Britain), Mme Marie Curie (France), L. De Torrès-Quevedo (Spain), Henri Bergson (France), J. Destrée (Belgium), F. Ruffini (Italy), G. de Reynold (Switzerland), D. N. Bannerjea (India). *(From a collection of photographs made by George E. Hale)*

council's goals had not been achieved. During the war the spirit of cooperation between scientists of every cast had been wonderful. They had come from universities, foundations, industry, and government to pool their efforts in a common goal. Thus the council had become a focus of scientific activity unlike anything before known. Science itself had gained stature in American economic life. Now Hale hoped to carry this spirit over into peacetime activities. Throughout the war he had worked to promote industrial research. But overemphasis in this direction, he knew, was dangerous. Now, with peace, he felt that the council must try to obtain more funds for pure science and to encourage every form of investigation, whether it had a practical bearing or not.[64]

Another scheme on which he had long been working, even before American entry into the war, was a plan for national research fellowships.[65] In comparison with most other professions, research scientists were still very poorly paid. (Among the most poorly paid of all were the astronomers.) As a result many of them gave up and turned to more lucrative lines. The fellowships Hale had proposed would enable those with Ph.D.'s who had proved their ability to continue in research "until thoroughly committed to it, beyond the danger of giving up under the pressure of teaching or other duties." He had first proposed the idea for such advanced fellowships at Throop in 1916.[66] In January, 1917, he had suggested it to Julius Stieglitz of Sigma XI, honorary scientific society.[67] When nothing had come of this proposal, he had turned, with the support of Arthur Noyes and J. J. Carty, to the Rockefeller Foundation. George Vincent, the president, was sympathetic, but at the moment could do nothing to implement his sympathy. Undaunted, Hale had turned in May, 1918, to the Carnegie Corporation, with the support of Carty and his old friend Charles Walcott. But Millikan, who accompanied them to New York, was lukewarm. He told Hale, "I am myself very doubtful of having the National Research Council award these fellowships, and still more about the wisdom of having the N.R.C. designate the permanent research appointments at the various institutions." [68] (Later, Millikan would favor the proposal wholeheartedly.)

Now, in 1919, with the war over, Hale decided to tackle the Rockefeller Foundation again. To his surprise, the foundation approved the fellowship plan almost immediately.[69] Thus one of the most important schemes in higher education in science in America was

CHAPTER FIFTEEN

A Temple of Science

AT THE WAR'S END, Hale had been attracted by many tasks. An outstanding one was the fulfillment of his dream for a permanent home for the National Academy of Sciences. Soon after his election to the academy in 1902, he had attended his first Washington meeting in the old red-brick Smithsonian Institution building.[1] From the beginning he had thought how wonderful it would be if the academy could have a home of its own in Washington. He had first proposed the idea in 1906. The older members, he soon observed, were not enthusiastic. Still, he continued to talk about it at every opportunity. In such a building, he pointed out, weekly or monthly meetings could be held and lectures by leading scientists could be given. There, public exhibits, tracing the history of science and illustrating current research, could be held. There, even, a popular illustrated journal of science might be sponsored. "I feel very strongly," he told Walcott in 1908, "that the Academy must accomplish much more than it does at present, if it is ever to occupy its proper position in the scientific world." It should, he urged, "acquire in time a commanding influence of a broad and liberal character, favorable alike to the development of research and the public appreciation of science."[2]

Again and again before the war, and before the establishment of the Research Council, he had tried to obtain funds for an Academy building.

One of those with whom he had discussed his dream was D. H. Burnham, the architect, his father's friend, who was then chairman of the National Art Commission. Burnham agreed to mark out a space for an academy building in his Washington plans.[3] Another, and, as it turned out, the most influential, was the statesman Elihu Root, trustee of the Carnegie Corporation, whom Hale always considered one of his most valued and beloved advisers. Root was one of Carnegie's closest friends. Hale had hoped that at the time of the academy's fiftieth-anniversary celebration in 1913, Carnegie might be persuaded to make a gift to the nation of an academy building "of such a character as to unite the interests of the various scientific institutions" he had founded. He had hoped, too, that Root might help to do the persuading.[4] But nothing had come of his attempts to interest Carnegie in his project.

In 1913 he had gone from Dr. Gehring's sanitarium to Washington for the fiftieth-anniversary celebration. "In my opinion," he had written, "the chief advantage of this celebration will not be accomplished unless it marks the beginning of a new epoch in the history of the Academy." He was still convinced that it was not accomplishing more than a very small fraction of what it ought to do for science in the United States.

In comparison with many of the European academies of which he had made a special study, it had accomplished little. In England, for example, the Royal Society had assumed "such a commanding position in British science" that it was constantly consulted by the government on questions of importance, and put in control of such enterprises as the National Physical Laboratory. Hale looked forward to the day when the National Academy could assume a similar vital role. He felt its prestige "could be enormously enhanced if the public could be led to regard it as the source of information regarding scientific advancement in any field." To accomplish this end he had made countless proposals. One of these was for a series of lectures on evolution to be supported by the William E. Hale Fund given in his father's memory.[5]

In 1914, this course had been inaugurated by the great English physicist Ernest Rutherford, who spoke on "The Constitution of the Elements." On Rutherford's return to England, Hale wrote gratefully: "I never before saw so much interest or enthusiasm at any scientific lecture. . . . I never was so stirred in my life. . . . The rush of your advance is overpowering, and I do not wonder that Nature has retreated from trench to trench, and from height to height, until she is now capitulating in her inmost citadel." [6]

Rutherford's lecture was followed by others on stellar evolution, on geologic transformation, on the development of living forms. In

1919 two lectures would be given by James Breasted on the development of civilization and its transition to Europe.

Meanwhile, however, Hale had published a series of articles in *Science* on "National Academies and the Progress of Research." In them he traced the organization of the academies of the world, and considered their relationship to national welfare.[7] In 1914, he had sent the first of these articles to Carnegie, in the hope of awakening his interest in an academy building. Soon afterward he called on Carnegie himself in New York. "Inactivity," he commented in a classic understatement, "does not appeal to me."

On the appointed evening he had arrived at the door of the huge Carnegie mansion at Fifth Avenue and Seventy-second Street. He had never before asked Carnegie directly for money, and he felt some misgiving. He was ushered into the huge living room in which Carnegie, looking like an elf, was waiting. He was seventy-nine, and sleepy, obviously not much interested in what his visitor had to say. But gradually, as Hale ranged from "stars to Rutherford's electrons," he showed more animation. At last Hale rose to leave. Thus far he had said nothing of the main purpose of his visit, but now, standing at the foot of the stairs, he had exploded into an account of his hoped-for building. It might, he said, cost $750,000. Carnegie listened quizzically, and at mention of the academy, he murmured, "Oh, that's just one of those fancy societies." Briefly and rapidly Hale tried to change his opinion. But it was late, and he saw that Carnegie was not listening. He did, however, make one concession. He agreed to consult Root and Henry Pritchett, trustee of the Carnegie Institution, "apparently quite oblivious of the fact that they had asked for the same building the previous year." [8]

Hale, encouraged, had gone back to the University Club to bed. But he soon found his optimism to be premature. The following day, a Sunday, he had picked up Pritchett and returned to the Carnegie mansion. They arrived at noon. Hale quickly sensed a change in the atmosphere. They found Carnegie with his formidable secretary Mr. Bertram, who immediately began to discuss Carnegie's plans for a peace palace at The Hague. Hale, sensing the chill in the air, saw that conditions were unfavorable for him and for science. Finally, with some hesitation, he brought up the academy and his building. To his surprise, Carnegie attacked him "savagely." "Besides," he declared, "it is Sunday, and no time for such a matter!"—"a rather amusing view for him to take." From Bertram's occasional remarks it was obvious that "he was mainly responsible for the remarkable change in Carnegie's attitude." [9]

By this time, however, Hale's own rare anger was roused, and when Carnegie intimated that Hale had taken advantage of his hospitality Hale replied without "mincing words." The argument went on, with Carnegie attacking and Hale defending the academy. It might have gone on this way indefinitely, but at a critical point Hale pulled from his pocket a letter from Root urging Carnegie support for the building. The philanthropist's expression changed. He admired Root's judgment tremendously. He hesitated, then slowly smiled. Gradually the tension lessened, and soon the Laird was his old friendly self again. He even assured Hale that the money was available if the corporation could be convinced of the project's importance. As they left, Pritchett, who had offered Hale only feeble support throughout the stormy interview, promised more help later.

Hale boarded the night train for Washington, worn out by the conflict. The next morning he called on Root in the Senate. Root promised to see Carnegie the following week. On his way back to California, Hale wrote out a brief statement on the academy building and sent it to Carnegie. In Pasadena he amplified it. He wrote: "The history of the Academy which I have sent you proves that it has already done much in this direction, but the possibilities of the future are enormously greater. In many departments, American science is now on a par with that of the first countries of Europe. The Academy can stimulate the development of the weaker branches, support and elevate the stronger ones, encourage local scientific societies in every state which now lack appreciation and guidance, preserve memorials of men of science and illustrate in an historical exhibit the contributions made by this country, inform the public of scientific progress in every department, and act, more worthily than is now possible, as accepted representative of America in the great field of international research." [10]

To accomplish all this and more, an academy building was vital. "As the clearing house of American science, and its official center in both a national and an international sense this building would be a contribution of the first importance to our intellectual and material progress."

It was late at night when he finished his letter, but, unable to wait, he drove to the post office and mailed it. In a week there came an equivocal letter from Pritchett, to whom he had also written for support. "It is a discouraging business, and I do not think anything can be done under the present situation." [11]

Soon afterward a letter from Carnegie confirmed Pritchett's dark outlook and killed all immediate hope. "The project of a building," he wrote, "to cost in the neighborhood of $900,000 is a very large one, and would require a large endowment to make it productive." [12] In view of growing demands, he said he could make no pledge for the future. To Hale, it again seemed obvious that Pritchett, whom Breasted characterized as "a roll top desk sort of man" had failed to back him up sufficiently.[13]

A month later, unexpectedly, a letter came from James Scherer, who was visiting the Carnegies at Skibo. He wrote, "The Laird is *worried* about you and that building, and Mrs. C. is decidedly on your side." He had had "three large confabs" with Carnegie. Carnegie, defending his position, had argued that Hale was putting the cart before the horse by asking for a building before its function was sufficiently developed. To this, Scherer retorted that Hale was merely imitating Carnegie, who was trying to make the peace movement permanent by building the palace at The Hague. And Mrs. Carnegie agreed. She argued that he ought to have his building, that it would be a great thing for the academy, and that with the Carnegie name and influence, the thing would go far. Scherer recounted that, as they came out from breakfast, Carnegie had turned and winked, then nudged him and said, "She's got me!"

At the letter's end, Scherer counseled: "Let the thing simmer for a while is my advice; and the pot may boil yet, but I don't expect Pritchett to supply much fuel!" [14]

For many months Hale had been forced to "lie low." Yet he tried to be philosophical. At forty-six he had learned that it often takes many years to reach a goal—that "a big undertaking doesn't succeed without an indefinite amount of hard work." [15]

In the interim he had turned to other projects. One of these was the founding of the *Proceedings of the National Academy of Sciences*—a journal that he hoped would provide for prompt publication of the latest results in American research and that might also help to break down the barriers between different sciences.[16] Arthur Noyes had agreed to accept the chairmanship of the Board of Editors. This was in 1914.[17]

By that time, too, Hale had been forced to turn his thoughts to the more immediate role the academy should play in time of war, and to the founding of a research council.

In 1918 that war had ended. Five years had passed since his interview with Carnegie. Now he was seeking a home not only for the academy but also for its newborn child, the National Research Council. The need for a building was more imperative than ever. Early in February, 1919, therefore, he returned to the Carnegie Corporation. This time he was asking for $5,000,000. (The ante had gone 'way up since his original application.) For two months he heard nothing. Then, on March 28th, the news he had awaited for thirteen years came. The Carnegie Corporation had approved his request—on condition that a suitable site first be found and purchased.[18] Hale, after walking the streets of Washington, had decided that a site on Meridian Hill above Sixteenth Street would be magnificent. He had even made some rough sketches of the kind of building he would like to see there. The cost of the land, he found, would be about $170,000.

Now he set about raising the money with a simplicity and directness that amounted to genius. He sat down at his desk and dashed off a series of telegrams to benefactors all over the country, asking for sums that ranged from $10,000 to $100,000. The story was told afterward that the money was subscribed overnight; and while this account was somewhat exaggerated, the money did come pouring in "like pennies from heaven." In a casual note to Frost, Hale described what actually happened. Three subscriptions of $10,000 and one for $5,000 came by return telegram.[19] Then Millikan went campaigning. This produced three more subscriptions of $10,000 each. Thus, almost as soon as they had begun, they were well on their way. With the help of Gano Dunn, J. C. Merriam, and James Angell, the balance was subscribed before the next corporation meeting on May 29th. The Carnegie group was astonished, and Hale, as he told Margaret, was feeling "pretty jolly!"

Meanwhile, too, he had gone to New York to discuss the design of his new building with Bertram Goodhue, the architect to whom he had turned for advice on Throop's architectural plans. "His transparent honesty, his dislike of the traditions that kill originality, his boyish enthusiasms—who could escape the charm of that versatile personality? It was a delight to enter his office, and very hard to leave it." [20]

In New York that day Hale told Goodhue of his vision for the academy in which the historical development of science in America, as well as the results of latest research, could be presented in a way that would appeal to the general public, as well as to the scientist. Goodhue, quickly catching his enthusiasm and his vision, promised to draw up a design with a site on Meridian Hill in mind. [21]

Weeks later, however, this site was abandoned, and a four-acre tract on B Street (now Constitution Avenue) facing the Mall was chosen instead. Goodhue was acutely disappointed. The place was a muddy waste, and a building in this location, he felt, could not avoid being dull and formal. For a time he wanted to abandon the project entirely. It took all Hale's powers of persuasion to induce him to change his mind. Finally, however, he agreed to go on with the plans for a building of "extreme simplicity and refinement."

As the building progressed on the drawing board, Hale shared eagerly in every phase of the design. He discussed with Gano Dunn, chairman of the Building Committee, his ideas for the bronze panels on the front of the building that would portray the founders of science. They debated at length the selection of eight episodes in science to be shown on the bronze doors. Many details, such as the marble pediment above the front door that symbolizes the evolution of man "under the invigorating rays of the sun," and Herter's painting of Prometheus stealing the divine fire from the sun, reflected his own interests. [22]

Ground for the new building was broken in 1922. For months Hale had poured his heart and soul into its planning. Yet, when the cornerstone was laid, he was in Egypt, banished once more by Dr. McBride's edict. On his return to Washington in February, 1923, he hurried down to see the new marble structure, already one story high. "It is bully to see it rise," he exclaimed in a letter to his brother. [23] But progress was delayed by the depression that engulfed the country in 1923. It was not until April, 1924, that the dedication could be planned. On March 1st Hale left for the East. On the way, he scribbled ideas for the exhibits in the new building and for the dissemination of scientific knowledge, in which he hoped the academy would play a leading role.

He reached Washington on March 5th, just in time for the

Academy Council meeting that was to be held in the new building. From the first he was entranced by its beauty. He felt it surpassed any building in Washington, with the possible exception of the Lincoln Memorial. Under Goodhue's skillful pencil, it had retained the singing lines of Greek architecture, yet it had a unique quality. It was simple and symmetrical, long and low. Windows that ran the full height of the first two floors imparted a feeling of height. The only decoration was the simple Greek lettering on the main façade of the building. The quotation from Aristotle was of Hale's choosing: "The search for truth is in one way hard and in another easy. For it is evident that no one can master it fully nor miss it wholly. But each adds a little to our knowledge of Nature, and from all the facts assembled there arises a certain grandeur." [24]

For nearly twenty-two years Hale had dreamed of the moment when this building would be completed. At times it had seemed an impossible goal. Now at last it was a reality. All the time he was in Washington he found it impossible to stay away. One moonlit night he knocked on the door of his old friend Arthur Noyes, who had also come on to Washington for the celebration. When Noyes answered, Hale suggested that they go for a walk.

Soon they found themselves at the foot of the marble steps, gazing up at the beautiful building gleaming white in the moonlight. They climbed the steps slowly, wonderingly. At the top they looked across at the Lincoln Memorial, also shining in the moonlight, and they must have thought of the President who had helped to found the academy during the dark days of the Civil War. That building was a magnificent memorial. Yet this, the National Academy, was a living, working center for the present and the future—for the world. From this time it would play an increasingly important role in the scientific life in the country.

But Hale, in these days after his arrival, was doing more than admire his building, as he worked night and day on details that had to be completed before the dedication. He helped Francis Pease, who had come on from Pasadena, to mount the coelostat for the spectroheliograph in the dome. He arranged a grating in the basement so that visitors could see the shining spectrum projected in the central hall and watch the daily change in sunspots. Here, too, the Foucault pendulum swung to show the earth's rotation. Thinking of his boyhood, he mounted an exhibit of living infusoria, swimming in a drop of water. These were only a few of the exhibits that showed the advances in industrial, technological, and pure research and that would broaden the horizons of thousands of visitors for years to come—until they were abandoned during World War II.

In these exhibits, in the academy itself, Hale hoped to express that philosophy of life and faith in pure science that he summed up at the conclusion of his article on the academy: "Science . . . spreads before the imagination a picture which no artist could hope to rival. Science does not seek a formula with which to reproduce the sculpture of Praxiteles or Rodin, the paintings of Rembrandt or Turner, the poetry of Homer or Keats. It recognizes here, as it does in religion free from dogma, a domain beyond its own. But its appeal is to the imagination as well as to the reason. The painter, with common pigments and bristle brushes, creates on canvas a great portrait or landscape. The architect with blocks of primeval mud hardened into rock beneath an extinct ocean, builds a great cathedral which stirs us by its majesty. Science, revealing with its instruments of metal and glass the widest sweep of nature, inspires the imagination by vistas of the stellar universe, the exquisite life of the microscopic world, the successive stages in the evolution of the earth and of man. . . . And the sweeping picture that science spreads before us is unmatched in its appeal to the imagination and its stimulus to progress." [25]

Shortly before the dedication, Bertram Goodhue arrived in Washington. He was ecstatic over this, his latest and, as it proved, his last creation. That night he returned to New York. Two days later, three days before the dedication, Hale was shocked to receive a telegram that told of Goodhue's sudden death the night before.

Then, too, there were other difficulties. As Hale became increasingly upset by the "row" (the nature of which is not now clear), Dunn cautioned, "The whole building isn't worth possessing if it deprives us of your advice and leadership." [26] Yet somehow everything was ready on time. On the 28th of April, 1924, the most illustrious body of scientists in the country gathered in the great central hall. In the audience were many of Hale's old friends, as well as members of his family. In the same year, in answer to his long campaign to include a broader range of scientists in the academy membership, James Breasted, an archaeologist, had been elected. He, of course, was there. There too were J. J. Carty and Gano Dunn, engineers both, in a category previously barred from membership. Gano Dunn, as chairman of the Building Committee, was the main speaker.

Hale had refused to take any active part in the celebration. Yet, in the course of his address, Dunn declared, "If there should be removed from our fair project, from its scientific ideals and beneficent usefulness to the American people, the part it owes to the gifted vision and tireless devotion of George Ellery Hale, our temple could not stand.

"It is a temple indeed, a temple—as written above us," he went on as he looked up at the dome, where the words Hale had chosen were graved, to "Science, Pilot of Industry, Conqueror of Disease, Multiplier of the Harvest, Explorer of the Universe, Revealer of Nature's Laws, Eternal Guide to Truth."

The dedication was a complete success, but the celebration was darkened by tragedy. The following day, Hale's old friend of Yerkes days, Ernest Fox Nichols, was demonstrating his recent work on radio waves when suddenly he stopped and sank down on the bench. His wife rushed to his side and gave him a hypodermic. But he was beyond help. Michelson, the academy president, rose to announce death from a heart attack and the adjournment of the academy session. Slowly the academicians walked in procession through the great bronze doors, following the body of their colleague.[27]

Three days later the academy adopted a unanimous resolution, presented by Dr. William Welch and seconded by Gano Dunn. It read: "That the National Academy of Sciences, earnestly desiring to possess a portrait in oils of its fellow member, George Ellery Hale, as a permanent memorial and an adornment of the walls of the fine building which it owes in large measure to his unselfish and untiring efforts in furthering the material and intellectual interests of the Academy these many years, and whose preeminence in science, universally recognized throughout the scientific world, has added distinguished honor to the Academy, requests Doctor Hale to sit for a portrait to subscribe for which will be counted a pleasure and privilege by members of the Academy and by friends and admirers." [28]

Hale was profoundly touched by this honor. The following year when the portrait painted by Seymour Thomas was finished, it was unveiled in the Reading Room just west of the library. It hangs there today, taking its place on the academy walls with the magnificent panel painted by Albert Herter that shows the founders of the academy with Abraham Lincoln.

CHAPTER SIXTEEN

Into "the Depths of the Universe"

✳ *1917-1923*

ON JULY 1, 1917, while war still raged in Europe and Hale was deep in his work for the National Research Council, the 100-inch mirror was brought safely up Mount Wilson. By the first of November everything was ready for the crucial test. Hale, back from Washington, was on the mountain. With him were Walter Adams and Alfred Noyes, the British poet, who was spending a summer as a visiting lecturer at Throop.

The inside of the dome was dark. The walls, the telescope, the stairs were painted black. The "feel" of the place was mysterious. "The whole scene," Noyes wrote afterward, "was worthy of the brush of a modern Rembrandt—the giant muzzle thrusting out through that dim arched roof and throbbing as it moved against the rotation of the earth; and the little throng of men, American, English, French, Italian, below it. This 'little throng' included all those—astronomers and opticians, carpenters and electricians, machinists and iron workers, who had helped in the long years of building." [1]

Hale, with Adams, climbed the long flight of narrow black iron steps to the observing platform. On the floor below, where a dim red light glowed, the night assistant pushed the control buttons. Simultaneously three different sets of motion were started. The observing platform rose and turned; the dome, its slit open wide to the star-filled sky, revolved in the opposite direction; the telescope itself turned until it pointed to the brilliant planet Jupiter.

As soon as the telescope was set on Jupiter, Hale crouched down to look through the eyepiece, desperately eager to know if all the years of effort had been successful. He looked and said nothing; only the expression on his face told of the horror he felt. Adams followed. His expression was a mirror of Hale's. They were appalled by what they had seen. Instead of a single image, six or seven overlapping images filled the eyepiece. "It appeared," said Adams afterward, "as if the surface of the mirror had been distorted into a number of facets, each of which was contributing its own image." [2]

At the moment it seemed that Ritchey's dire predictions were right and that the numerous air bubbles in the glass had ruined the final disk. Was this really the case, or could there be some other cause? Someone mentioned that the workmen had left the dome open during the day, and the sun might have shone on the mirror, distorting it with its heat. But was this sufficient to cause the frightful image of Jupiter they had just observed? They could not know; they could only wait and hope. For the first two or three hours they sat on the dome floor, looking into the eyepiece occasionally. The image improved slightly, but not enough to offer much hope. Large mirrors cool slowly, and the night had just begun. There was nothing to do but continue to wait.

While they waited they talked. The talk ranged from war to poetry. Suddenly Hale looked up to remark to Noyes, "You poets write of many things, but you never write of the development of science in the modern world. You write of wars. Why don't you write instead of the fight for knowledge?" Out of that chance remark came Noyes's epic poem of the growth of astronomy over the centuries. He called it *Watchers of the Sky*. Long afterward he said that it changed the course of a large part of his life, and cost him more than ten years of work.[3]

In the prefatory note to that book, Noyes, inspired by Hale's vision, wrote: "The story of scientific discovery has its own epic unity—a unity of purpose and endeavor—the single torch passing from hand to hand through the centuries; and the great moments of science when, after long labour, the pioneers saw their accumulated facts falling into a significant order—sometimes in the form of a law that revolutionized the whole world of thought—have an intense human interest, and belong essentially to the creative imagination of poetry." [4]

In the prologue he gives a picture of that memorable night on Mount Wilson:

> Up there, I knew
> The explorers of the sky, the pioneers
> Of science, now made ready to attack
> That darkness once again, and win new worlds . . .

For more than twenty years,
They had thought and planned and worked.[5]

He described the ten years that it had taken to make the mirror,
with no sure guarantee of its final success:

Where was the gambler that would stake so much—
Time, patience, treasure on a single throw?

On and on the lines ran until he came to the account of that night
he had shared—the crowning moment of Hale's years of "hope and
fear." He described the "chief's" observation of the first "seeming insignificant stars." Yet while he speaks of drama, Noyes fails to record
the real drama of that night in the dome. He does not describe the
dark hours after that first appalling observation when only the passage of endless hours could tell of the success or failure of the huge
telescope. It was that drama that neither Adams nor Hale would
ever forget.[6]

After a while they walked outside and gazed down at the lights
twinkling in the valley below. Then slowly they walked down the
hill to the Monastery. Agreeing to meet three hours later, they went
in to bed. Hale lay down without undressing. But he could not
sleep. An hour later he got up and tried to read a detective story.
But this too failed. At 2:30 A.M. he returned to the 100-inch dome.
Before long, Adams arrived and confessed that he too had found
sleep impossible. Once again they climbed the long flight of steps to
the dome floor, then the narrow flight to the observing platform.

By this time Jupiter was out of reach in the west. They swung the
great telescope over to the brilliant blue star Vega. Almost afraid to
look, Hale again crouched down and looked into the eyepiece. He
let out a yell. The yell told Adams all he wanted to know. "By Jove!"
Hale exclaimed. Adams took his turn at the eyepiece, and echoed
Hale's exclamation. There shone Vega, a gleaming point of light, the
most beautiful thing they had ever seen. There was no longer any
doubt about the success of the telescope. At last Hale knew—and
soon all the world would know—that all the years of effort, all the
expense, all the agony had not been wasted.

In his little red diary, Hale made the cryptic note:

Friday, November 2, 1917—With Alfred Noyes to Mountain.
First observations with 100″—Jupiter, Moon, Saturn.
November 3, 1917—To Pasadena.

On that same day he reported to James Jeans in England. "We got a first test of the 100″ last night, and its performance was very promising. I think there is good reason to hope that it will be of great service in extending the work done with the 60″ reflector." [7] Yet, if he was hopeful, he was cautious still.

After this, as everyone at the observatory became involved in the war, work with the 100-inch was delayed. But at the war's end, as soon as he could get away from Washington, Hale returned exuberantly to the mountain. Before long the 100-inch was hard at work, proving its unqualified success and justifying his long gamble. Soon he was able to compare the performance of this new instrument with the older 60-inch that rose nearby—under identical atmospheric conditions. (No one had been sure that, even if the optical and mechanical difficulties could be overcome in building this huge telescope, sharply defined images could be obtained with so large an aperture. The question had been always, "Would the atmosphere prove sufficiently tranquil?")

Now clear-cut tests could be made. The 100-inch mirror has an area 2.8 times that of the 60-inch; therefore it receives nearly three times as much light from a star or other heavenly body. "Under atmospheric conditions perfect enough to allow all this light to be concentrated in a point, it should be capable of recording on a photographic plate, with a given exposure, stars about one magnitude fainter than the faintest stars within reach of the 60-inch. The increased focal length, permitting such objects as the moon to be photographed on a larger scale, should also reveal smaller details and render possible higher accuracy of measurement. Finally, the greater theoretical resolving power of the larger aperture, providing it can be utilized, should permit the separation of close double stars beyond the range of the smaller instrument."

In all these directions the tests indicated that the predictions Hale had made to Hooker thirteen years earlier would be realized. As a result of the increased light-gathering power, millions of stars would be added to those already known; spectroscopically, stars and nebulae inaccessible to the 60-inch could now be studied.

By September, 1919, Francis Pease had already used the 100-inch on a night of extraordinary "seeing" to take the best photographs of the moon ever made with that telescope. A few days later all possibility of observing was destroyed by smoke from the forest fires that raged nearby. On September 22nd the Mount Wilson Log read, "Shapley observing. 'Pittsburgh seeing.' Closed up on account of ashes falling on mirror." [8]

As the work progressed and the promise grew, Hale exclaimed, "I cannot feel envious of anyone under present conditions." Every night was an adventure. Once, turning the 100″ on the moon and observing the marvelous detail in its craters and mountains, he suddenly felt transported back to his childhood and to Jules Verne's fantastic *From the Earth to the Moon*. He wrote to his brother: "The effect in the big dome was striking enough—the huge tube half visible in the moonlight, pointing up through the shutter opening, with two of us riding on the tube, looking down into the craters. It was like the giant instrument on the Rockies 'in Missouri,' when the sight was so startling that the 'unfortunate man disappeared.'" [9]

After this he examined with the greatest delight the first photographs of nebulae taken with the new telescope. Immediately he could see their superiority over the 60-inch plates. In such objects as the Ring Nebula in Lyra and the Great Nebula in Orion he could see a vast amount of spectacular detail heretofore invisible. He saw, too, that in the sharpness of the images the results were equally magnificent.

"It is evident," he exclaimed after studying these and other promising results, "that the new telescope will afford boundless possibilities for the study of the stellar universe.

"The structure and extent of the galactic system, and the motions of the stars comprising it; the distributions, distances, and dimensions of the spiral nebulae, their motions, rotation, and mode of development; the origin of the stars and the successive stages in their life history; these are some of the great questions which the new telescope must help to answer."

He concluded: "In such an embarrassment of riches the chief difficulty is to withstand the temptation toward scattering of effort, and to form an observing programme directed toward the solution of crucial problems rather than the accumulation of vast stores of miscellaneous data. This programme will be supplemented by an extensive study of the sun, the only star near enough the earth to be examined in detail, and by a series of laboratory investigations involving the experimental investigation of solar and stellar conditions, thus aiding in the interpretation of celestial phenomena." [10]

When Hale returned to Mount Wilson from Washington, one of the most crucial and most pressing astronomical problems was the question of the size and extent of the universe. In the next few years the view of that "universe" would be radically altered. In that revolution the 100-inch telescope would play a leading role. Hale had planned this great telescope before the 60-inch was mounted, even before there was any certainty of its success. Now the 60-inch had raised questions it could not answer, and Hale's great gamble was to pay off. On that gamble the future reputation of numerous astronomers would be built.

Over a period of years from 1908 to 1928, Hale told the story of the advances that created the revolution. The story is amplified in his correspondence.

In 1908, in *The Study of Stellar Evolution,* he described the spiral nebula in Andromeda: "Persistent attempts to measure the distance of this nebula from the earth, made with the most powerful of modern instruments, have totally failed. We may therefore say that the distance is almost inconceivably great, and that therefore the dimensions of the nebula are so enormous as to be quite beyond comparison with those of the solar system." [11] The only thing that astronomers could say at this point was that this nebula must be "exceedingly remote." But how far from the earth no one had the faintest idea.

Six years later, in 1914, when war broke out in Europe, Kapteyn, who had been devoting much of his time to a study of the structure of the universe, returned home. The year before he had received the Bruce Medal from the Astronomical Society of the Pacific, and Heber D. Curtis of the Lick Observatory had given the main address. In it he spoke of the current state of knowledge on the perplexing problem of the magnitude of "the universe," by which he meant our own galaxy: "When we attempt to visualize the dimensions of this universe of stars, we must confess that our knowledge is most fragmentary and imperfect; at present we know the distance of only those stars which form the inner fringe of the stellar universe in which our sun is but a very inconspicuous unit." [12] By direct methods, he noted, the distances of stars out to about 150 light-years could be measured. Beyond this, however, there could be little confidence in direct results. Beyond the confines of a sphere about 1,000 light-years in diameter, he exclaimed, "we really know nothing today." Beyond this anything was pure guesswork. Moreover, any hope of ever penetrating any further than 2000 light-years, he felt, was slim. (A light-year is equal to the distance traveled by light in a year and is approximately equal to 6,000,000,000,000 miles.)

In 1913, the prospect of sounding the distant reaches of space appeared dim to most astronomers. Yet, already, a means of penetrating those reaches had been found. In 1912, Henrietta S. Leavitt, at the Harvard Observatory, was investigating the variable stars known as Cepheids in the Magellanic Clouds. She found in them an exact correlation between their period and their average brightness. She was able to show that the period varies in a regular way with the luminosity. From these observations it became possible to determine the distances of remote objects like star clusters and nebulae. Cepheid variables became known as "the yardsticks of the universe."

In 1914, Hale had a meeting with a young and exceedingly energetic astronomer named Harlow Shapley. Shapley came to the meeting primed to discuss all the latest astronomical results, including, no doubt, the question of these variable stars, in which he was deeply interested. Instead, as he recalled long afterward, Hale discussed music, opera, anything but astronomy. He was surprised, therefore, when soon afterward an invitation came to join the Mount Wilson staff.[13] Shortly after his arrival he became fascinated by the globular clusters, many of which had been revealed by the 60-inch for the first time. These relatively stable and compact spherical groups of stars, held together by their gravitational attraction, usually contain several hundred thousand stars. He was particularly interested in their distribution. He observed that they appear to lie mainly in the direction of Sagittarius in the southern hemisphere of the heavens. In the northern hemisphere they were relatively sparse. He suggested therefore that our galactic system must be in the form of a flattened disk or watch-shaped aggregation of stars, with the sun lying at a considerable distance from its center. He also observed that the globular clusters are distributed symmetrically with respect to the plane of the Milky Way. With the help of the Cepheid variables, he was able to derive a distance for the huge globular cluster that lies in the constellation of Hercules. He concluded that it lies at the fantastic distance of 36,000 light-years.[14]

"If this measure is correct," Hale was to write, "and there is much independent evidence to support it, we take a tremendous leap into space and time when we reach out to this cluster. Light travelling at the rate of 186,000 miles per second requires 1.2 seconds to reach us from the moon, 8 minutes to come from the sun and 4⅓ years to cross the space between us and the nearest star. Our views of such objects are thus contemporaneous, or nearly so; we see them as they are now or as they were within a few years. But the Hercules cluster is in another class. The light that left it 36,000 years ago, travelling at the rate of nearly six million million miles per year, has only just reached us. Thus we cannot say how the cluster appears today, or whether it has existed at all since the dawn of our civilization. There is every reason to believe, however, that if we could see the present cluster—as astronomers will see it 36,000 years hence—it would appear essentially as it does in our photographs of its remote past. For 36,000 years is as a day in the cycles of the universe, where millions of years bring little change." [15]

Shapley's observations were made with the 60-inch in 1916 and 1917, just as the United States was entering World War I, at a time when much of the regular work of the observatory would be disrupted by that war. They were also made just before the new 100-inch went into action. The far-reaching effect of these and other results would therefore not become apparent until after the war. In 1917 Hale wrote in the *Annual Report*, "The classic problem of 'island universes' is again under vigorous discussion and much attention will be paid here to its further investigation." That discussion would reach its climax soon after the war's end.

Meanwhile, in March, 1918, out of the hurly-burly of wartime Washington, Hale took time to write to Shapley to congratulate him on his paper in the *Publications of the Astronomical Society of the Pacific*. "The outline," he noted, "seems to show that you have struck a trail of great promise. The distribution of the globular clusters with reference to the galactic plane seems to indicate their organic connection with our own system, though the hypotheses regarding their dispersion in the 'region of avoidance' will need a lot of proof." [16]

However, he added, "I think you are right in making daring hypotheses, and in pushing the work ahead as you have done, as long as you stick to J. J. Thomson's definition, and are prepared to substitute new hypotheses for old as rapidly as the evidence may demand.

"I confess," he continued, "that I still entertain many doubts about the nature of the spiral nebulae and their relationship to our own system. I also feel a bit skeptical regarding your hypothesis to account for the novae in spirals. We must evidently give more and more attention to all phenomena relating to spirals and accord them a large place in our revised observational program. . . ."

Two years after this, Hale published an article on "The New Heavens" in *Scribner's*. In it he surveyed the current state of astronomical knowledge: "Hundreds of millions of stars have been photographed, and the boundaries of the stellar universe have been pushed far into space, but have not been attained. Globular star clusters, containing tens of thousands of stars, are on so great a scale (according to Shapley) that light, travelling at the rate of 186,000 miles per second, may take 500 years to cross one of them, while the most distant of these objects may be more than 200,000 light-years from the earth." [17]

Then he added, "The spiral nebulae, more than a million in number, are vast whirling masses in process of development, but we are not yet certain whether they should be regarded as 'island universes' or as subordinate to the stellar systems which include our group of minute sun and planets, the great star clouds of the Milky Way, and the distant globular star clusters."

Just about the time this article was published, Shapley's name was proposed for the directorship of the Harvard Observatory. Hale, aware that many of his astronomical colleagues considered Shapley's ideas too radical and the young astronomer too daring, wrote to President A. Lawrence Lowell, recommending Shapley for the position: "The true value of Shapley's recent researches, which call for radical and sweeping changes in the older views of the extent and structure of the universe, will be determined within a few years. While opposed by certain astronomers, he is strongly supported by some of the ablest investigators, whose independent confirmation of his results seems very significant. In my opinion, most of Shapley's work will prove to be a very important contribution to astronomy, even if it should not be substantiated in every detail. I believe him to be sufficiently open-minded to pursue his investigations for the sole purpose of learning the truth rather than with the object of establishing a favorite hypothesis. He certainly possesses the knowledge, ability and industry needed for the directorship of Harvard Observatory. His daring, now criticized by some, but encouraged by me in his work here, may prove to be one of his strongest qualities. No great progress is ever made without constructive imagination, and Shapley has this quality in high degree." [18]

In 1921, Shapley left Mount Wilson to become director of the Harvard Observatory. That same year his famous debate with Heber D. Curtis took place at a meeting of the National Academy of Sciences in Washington.[19] The subject was "The Scale of the Universe." The debate was actually in two parts. Having found a distance of 36,000 light-years for one of the nearest of the globular clusters, Shapley suggested that the disk-like galactic system might have a diameter of perhaps 300,000 light-years. "These great dimensions," Hale noted after the debate was over, "have been denied by Curtis, who argues in favor of a galactic system about one tenth as large. But more and more evidence is accumulating in favor of the larger conception of Shapley, which has already found wide acceptance among astronomers.

"The question at issue," he emphasized, "is the size of the galactic system of stars to which the sun belongs." On this question, Shapley's estimate proved to be more nearly correct than that of Curtis, although his values were later modified when interstellar absorption was accounted for.

But the question of "island universes" was another matter. In 1922, in an article on "The Depths of the Universe," Hale wrote, "The question has not yet been settled whether these [the spiral nebulae] are not farther from us than the more distant stars or whether they should be regarded as 'island universes,' isolated in the depths of space and comparable in size with the galactic system. Curtis, who holds the latter view, estimates their distance to range from 500,000 to 10,000,000 light years, while Shapley, van Maanen and others believe them to be much nearer.

"Interesting arguments," he concluded, "have been advanced on both sides." [20] Despite those arguments, no final conclusions were reached. Afterward the advocates on each side returned to work.

One of these was an aristocratic, pipe-smoking astronomer named Edwin Hubble, who had been at the Yerkes Observatory and had joined the Mount Wilson staff in 1919. He was particularly interested in the nature of the various types of nebulae, and especially of the spirals. At the end of 1923, he made his first breakthrough at Mount Wilson when he identified a Cepheid in a spiral nebula. The key was at last provided for determining its distance. The crucial question, "Could the arms of these spirals be resolved into stars, or were they nebulous objects?" as Shapley had suggested, was about to be answered definitively. In 1925, Hale wrote another article for *Scribner's,* in which he surveyed the latest results. He called it "Beyond the Milky Way." In it he wrote: "The beautiful photographs published by the Lick Observatory in their superb volumes on the nebulae and the exquisite plates taken on Mount Wilson with the 60-inch reflector by Ritchey, Pease, van Maanen, Hubble and Humason show countless minute knots along the spiral arms. But how shall we recognize these as true stellar images, or distinguish them with certainty from the many faint stars that belong to the galaxy?" [21]

By 1925, this question had been answered. It had been answered because the great 100-inch telescope was there, ready and waiting to push back the frontiers of the universe. "Hubble," Hale noted, "has recently answered this question in a conclusive way. Concentrating his attention upon the outlying parts of such great spirals as Messier 33 and the Andromeda nebula, rather than upon the amorphous nebulosity of the central regions, he has taken a series of photographs with the 100-inch Hooker telescope for the express purpose of determining their resolvability. These are covered with stellar images, which on a plate of Messier 33 are the smallest ever secured —only five or six tenths of a second of arc in diameter." From his study of these images on plates of Messier 33 and of the Andromeda nebula, Hubble observed that some of them fluctuated with the "undeniable" light changes of Cepheids. (Temporary or "new stars" had been observed in the Andromeda nebula. But no Cepheid had ever been detected.) This, as Hale called it, was Hubble's "capital discovery." From this discovery it became possible for the first time to measure the distance of these remote objects. From those measurements Hubble was able to derive a distance for the Andromeda nebula of 850,000 light-years *—a fantastic distance compared to any that the majority of astronomers had deemed possible. It was, Hale pointed out, "more than four times as far as the most remote globular cluster and well beyond the boundary of the galactic system."

This discovery was announced at a dramatic meeting of the American Astronomical Society held from December 30, 1924, to January 1, 1925. Hubble himself was not there, but when his paper was read

---

* A revised distance scale shows that the Andromeda nebula may be at a distance of over 2,000,000 light-years.

all the astronomers at that meeting knew, as Allan Sandage was to point out in his introduction to the Hubble Atlas of Galaxies, that the controversy on "island universes" was settled once for all. "It proved beyond question that nebulae were external galaxies of dimensions comparable to our own. It opened the last frontier of astronomy, and gave, for the first time, the correct conceptual view of the universe. Galaxies are the units of matter that define the granular structure of the universe." [22]

So, the answer to the puzzling question "What are galaxies?" was at last given. As Sandage writes, "No one knew before 1900. Very few people knew in 1920. All astronomers knew after 1924." The answer was this: "Galaxies are the largest single aggregates of stars in the universe. They are to astronomy what atoms are to physics. Each galaxy is a stellar system somewhat like our Milky Way, and isolated from its neighbors by nearly empty space. In popular terms, each galaxy is a separate island universe unto itself."

As the 100-inch telescope went into action, one of those who would contribute most to its reputation was a young man who had long been connected with the mountain. His name was Milton Humason, and he had begun his life on Mount Wilson as a bellboy at the hotel. There he had done everything—from washing dishes and working in the corral to shingling cottages. Anything, he apparently felt, was better than going to school. In time he became a driver of the pack train, working with the burros and mules, packing equipment of all sorts up the mountain. Thus he came into close contact with many of the observatory people, especially those who usually rode rather than walked up the trail. One of these was Hale, whom he compared long afterward with some of the other astronomers. "He was always the really nice person we liked," he said. "He was always gentle, always kind, always had a smile for everyone."

From the beginning Humason loved the life on the mountain. He was delighted, therefore, when some years later he was asked to become the observatory janitor, and when this, in turn, led to a position as night assistant on the mountain. Work with the 6-inch gradually led to the 10-inch, then to the 60-inch and finally to the 100-inch telescope as the astronomers—notably Paul Merrill, Harlow Shapley and Walter Adams—recognized his skill as an observer and his patience and perseverance on spectroscopic observations that often required extremely long exposure times. At first, when Adams had proposed that Humason be asked to join the staff, Hale, noting his lack of formal education, had demurred. But soon, as his ability became increasingly evident in his observation in nebulae of the large velocity or "red shifts" (as they came to be called), Hale was quick to acknowledge that, as Humason recalled, "he had been all wrong when he opposed my appointment." [23] These observations which showed the recession of the nebulae were to play an essential role in the theory of an expanding universe. Years later the value of Milton Humason's work would be recognized when he received an honorary doctor's degree from the university at Lund in Sweden.

Meanwhile, in this revolutionary period of the 1920's, other radically new results were being obtained at Mount Wilson, again proving the immense value of the new 100-inch telescope. In 1920, Albert Michelson arrived to carry out his earth-tide and velocity-of-light experiments and to test his interferometer method on the measurement of star diameters. Even with the full aperture of the 100-inch, the interference fringes were beautifully defined. Using this method, Francis Pease and J. A. Anderson measured the diameter of the giant red star Betelgeuse—an extremely difficult feat. Its angular diameter was 0″.045. (A foot rule viewed at a distance of about 800 miles would subtend this angle.) At the great distance of Betelgeuse from the earth, this diameter represented an astounding value of 300,000,000 miles. "When you remember," said Hale, "that our sun is less than one million miles in diameter, this looks like a very much enlarged edition."

Such spectacular results raised a familiar problem. Michelson's 20-foot interferometer could measure only a limited number of stars. A larger interferometer would require a larger telescope. In addition, other startling results were being obtained with the 100-inch, and these too could be investigated further only with greater light-gathering power. Soon, therefore, Hale was laying plans for a 50-foot interferometer. He began to dream, too, of a larger telescope.

Oddly, just at this time, when the 100-inch was proving its magnificence, Hale was forced to do something about the man who had never believed in its powers, so that even after its success he continued to cause trouble and generate discord. This, of course, was George Ritchey. Sometimes he was sanctimonious, at times even sadistic. One night, for example, he was helping Henri Chrétien, the French astronomer, who was using the 60-inch for the first time. He put a photographic plate in the telescope's plate holder. For several hours Chrétien guided the telescope with infinite care in order to hold the star image on the plate. When he was about to develop the plate, Ritchey informed him that it was a blank. He had not wanted to risk a good photographic plate on a novice. When Hale heard this tale from Adams, he was incredulous. Yet he was still unwilling to dismiss a man with whom he had worked for thirty years.[24]

Adams, however, had long since reached the limit of his patience. He was particularly disturbed by Ritchey's attitude during the war when he had signed himself "Commanding Officer, Mt. Wilson Observatory." Moreover, he had "assiduously spread the report throughout the Optical Shop and among visiting Ordnance officers that the 100″ telescope is a failure, and his attitude toward the instrument is such that we cannot count on his doing anything of value with it. . . . Altogether the situation is an almost impossible one for the future."

In the end, therefore, Hale gave in. He concluded that for the good of the observatory, as well as for his own peace of mind, he must ask for Ritchey's resignation. On October 13, 1919, Ritchey's connection with the observatory was terminated. Hale found it one of the hardest experiences of his life. Yet, once it was done, he found "the whole observatory rejoicing."

About this time, however, he was faced with a decision of quite another sort. In December, 1919, Robert Woodward was to retire as president of the Carnegie Institution. Elihu Root, chairman of the board, asked Hale to accept the position. In declining, Hale recommended instead the paleontologist John C. Merriam of the University of California, who had been acting chairman of the Research Council.[25] "No other man of my acquaintance," Hale wrote, "combines the diverse qualities required. His devotion to research and his constructive imagination enable him to conceive great research projects. His power of organization, and his ability to enlist the enthusiastic support of others, permit him to carry his projects into effect. His catholicity of interest, and his intimate knowledge of paleontology, anthropology, archaeology, botany, zoology, and other branches of science, give him an outlook of extraordinary range. . . ."

The Carnegie trustees, however, were apparently determined to have Hale as president. For a short time, he was again tempted. But once more his desire to pursue his own research won out.

Yet the problem of retaining the Mount Wilson directorship remained a puzzling one, especially when it was complicated by ill health. He was finding so little time for his own research that he was often in despair "over the condition of things." Just when everything would appear to be going well, he would be faced by collapse. One exciting night he had taken Root, Pritchett, Merriam, Myron Herrick, and Ulysses Grant up the mountain to see the 100-inch. He did everything possible to make the visit successful. When the observing was over, they walked down to the Monastery, and there, in front of a huge log fire, Root and Herrick discussed the League of Nations while the others listened intently. It was late when they turned in, and Hale was worn out. The following day he could hardly get down the mountain. On arriving home, he had to go to bed.

Nevertheless, if he was often tempted to relinquish the directorship of the observatory, he realized that in this position he could carry out large-scale experiments with powerful apparatus that might otherwise be impossible. One day he wrote to his brother Will about a new scheme that would require large new magnets, special electric vacuum tubes, apparatus for high-voltage discharges, "and probably a larger aperture than any we now have." "So," he concluded realistically, "I had better not give up too easily the opportunity I now have to attack problems with ample means." [26]

Nor could he limit himself to the observatory. His mind remained as extraordinarily fertile as ever, and once an idea had been born, he naturally felt committed to its development. One such scheme in this period after the war was that of Throop, which he had helped to nourish from its second infancy and which, in 1920, had been rechristened The California Institute of Technology. [27] In the years since the adoption of its "new and daring policy," it had won a recognized place among the nation's scientific and technological schools. Its student body, grown from thirty-eight in 1910 to 440 in 1920, overflowed the campus facilities. [28] In addition to the original Pasadena Hall (its name now changed to Throop Hall) and the new Gates Chemistry building, plans were under way for other buildings that had been on the original blue prints. The plans, drawn up before the war by Elmer Grey and Myron Hunt, had been elaborated by Bertram Goodhue, who had come to Pasadena in 1915 to discuss the special problems posed by a school of engineering and science. At the war's end, some of the money for the building program had been promised; now more had to be obtained. This, on his return to Pasadena, was one of Hale's chief concerns.

Yet, even more important than buildings was a first-rate faculty. Progress had been made with appointments such as that of Harry Bateman, the English mathematician, and Arthur Noyes, Hale's old friend from M.I.T., who had worked on the Research Council during the war and who had agreed to come to Pasadena on a part-time basis.

In September, 1917, a young mathematician named Warren Weaver was appointed assistant professor. Like Harold Babcock, on his arrival at Mount Wilson, Weaver felt there an underlying current of excitement. "It was entirely clear, even at this early moment, that something strange and wonderful was astir here. We had early vision of a new and special sort of institute which would be intellectually compact; which would be imaginative, flexible and superb in quality; which would be dedicated not to size or noise or the routine levels of technological training, but rather to outstanding quality and to the supreme adventure of advancing basic knowledge about Nature." [29]

Finally, in 1919, Noyes decided to make the final break with the East and come to Pasadena on a full-time basis. It was not an easy decision. At M.I.T. he had built the finest physical-chemistry laboratory in the country, and he dreaded the move. Nevertheless he was soon sharing Hale's enthusiasm for making the West Coast institute a leading center of learning. Increasingly Hale found himself relying on his old friend's quiet, wise judgment, not only for the institute but also in his plans for Pasadena and the world. More than anyone he had ever known, Noyes lived up to the lines he loved to quote:

> Those love truth best who to themselves are true
> And what they dream of dare to do.

The winning of Noyes was Hale's first great victory for the institute after the war. The conquest, lasting over a period of many years, had not been easy. Yet, compared to his next battle, it was simple. On March 3, 1920, James Scherer handed in his resignation.

"The fundamental reason," he informed Fleming, "is an exhaustion of the nervous energy requisite to carry on the responsibilities of my position." [30]

But the twelve years since his arrival as president had done their work. The foundations had been soundly laid. "The first $100,000 is the hardest," he had said.[31] Those who followed could build on those foundations.

As soon as Hale learned of Scherer's decision, he announced his sense of genuine loss, both for the institute and for himself. Still, Scherer's resignation had not been unexpected, and Hale and Noyes had often talked of a successor. Now, one day in March, Hale, thinking of Millikan's skill as an executive during the war, asked, "How about Millikan?" Noyes was enthusiastic, and shortly afterward they arrived on Millikan's Chicago doorstep. The "smiling Bob" welcomed them exuberantly and led them into his office in the Ryerson laboratory. Immediately Hale plunged into a glowing description of the opportunities in California. Noyes joined in. Millikan listened but made no promises. At the last moment, however, just as they were leaving, he said he would think it over.[32] Later they all met for lunch with Norman Bridge, chairman of the institute's board. Bridge argued as forcefully as Hale had done. "But I am afraid, Hale commented, "the chances of his acceptance are very slim." The University of Chicago was doing everything in its power to hold him, and Hale knew he would not leave Chicago for a place that offered poorer facilities.

Hale now became convinced that he was waging a battle he could not afford to lose. Back in Pasadena, he tackled the problem from a familiar angle—that of money. He wrote first to Arthur Fleming, who had offered $1,000,000 to the institute but had not yet fulfilled his promise. Millikan, he insisted, must be offered inducements he would find impossible to decline. He pointed out that, even as president, Millikan would be unwilling to give up his active participation in research. Therefore, he urged the need for endowing an outstanding department of physical research at the institute—better than that Millikan had at Chicago. To guarantee this and to assure Millikan's coming, Hale even promised to guarantee personally a sum sufficient to yield an income of $5,000. This was to be a contribution to the "Millikan Fund for Physical Research." [33] (Only in case of his death, he provided that the income should be used instead by his wife. He regretted this limitation, but "the principal sum would represent such a large part of my estate that my various obligations would not permit me to do otherwise." He also specified that "my proposed guarantee be not made public before my death.")

On hearing of this "generous offer," Millikan wrote that he was overwhelmed by his friend's sacrifice. "I should certainly be very loath to make a move which would involve unfortunate consequences for you or any of your family. . . ." [34] Gradually, Hale persuaded others to join him in his "Million Dollar Fund." But the money was slow in coming in, and Millikan still would not commit himself. One day, in January, 1921, realizing another approach was needed to inspire the reluctant physicist with the incomparable advantages of the West Coast, Hale invited him to lunch with a few members of the institute faculty and of the observatory staff to discuss possible ways of cooperation between the two institutions. [35] (Millikan was in Pasadena to give his annual course of lectures.) Whether the luncheon had the desired effect is unknown. But less than a month later Millikan sent an ultimatum to Chicago. [36] He insisted on an increase in the physics budget there—a move in which he was backed by Michelson, the department head. If this ultimatum were turned down, he promised to accept the California position.

Meanwhile, however, two other threats appeared in the East— one was the offer to Millikan of the presidency of the Carnegie Corporation, [37] the other the directorship of Simon Flexner's proposed central laboratory of physics and chemistry. The California trustees had offered Millikan the presidency of the institute and the directorship of physical research. Fleming had offered his entire fortune, if necessary, to the cause. But Hale knew he had powerful competition. He took off for New York in the greatest haste. (This was his second trip in a very brief space of time, and crossing the continent was not what it would be by air fifty years later.)

He met Millikan at the University Club on the night of May 15th. They talked until long after midnight. The next day the debate was resumed: this time Flexner, Welch, and Pupin joined in. By late afternoon Hale was exhausted. That night he had dinner with Gano Dunn, Pupin, and Millikan. Dunn and Pupin urged Millikan to accept Hale's offer. During the afternoon Flexner, unwilling to give up hope for his laboratory with Millikan at its head, said he was going to John D. Rockefeller personally. At the same time, as Hale and

Millikan knew, Harry P. Judson, president of the University of Chicago, was going to Rockefeller for funds to keep Millikan there. The whole business was becoming more and more like a scene from *Alice's Adventures in Wonderland.* "It is very difficult," said Hale, "to decide what to do especially with Noyes and Fleming out of reach. I have never wrestled with so difficult a problem, and I can do nothing else until it is settled." [38]

The following day, May 17th, was still inconclusive. One proposal was that Millikan should divide his time equally between Chicago and California. But Hale feared that unless he identified himself completely with California, the enormous psychological value of his appointment would be lost. With it, he felt, would go the possibility of developing the institute into the leading research center for physics and chemistry in the country.

Yet, tired though he was, he was undefeated. "I am doing some tall scheming and already have some new ideas that may lead somewhere." [39] These ideas seemed to bubble up from an inexhaustible well. They included a great wireless station on Mount Wilson that would help Millikan in his study of problems of atmospheric electricity, "while we (at Mount Wilson) could supply data needed to trace any relationship with solar phenomena"; C.I.T. stations for the study of terrestrial magnetism; a special laboratory for research at the lowest temperatures; a laboratory set up in cooperation with the Edison Company for researches requiring very heavy currents; a great library of physics and chemistry; new schemes of cooperation with the observatory. So on and on his dreams of the possibilities ran. To Hale they seemed irresistible, and he never doubted, as Millikan seemed to do, that the money for all such projects could be found.

Another week passed. Hale stayed in the East. For a few days he took time off to go fishing near Scranton. But on his return he called on Elihu Root. Root appeared older than the last time he had seen him, but he had lost none of his spark. He listened intently to Hale's plans and said he favored support by the Carnegie Corporation. This, of course, was what Hale was angling for. That night, back at the University Club, he outlined plans that Root could submit to the corporation at their next meeting.[40] Then he called Millikan in Chicago, asking him to return to New York to go over the latest scheme. Still Millikan appeared uncertain; and Hale feared the odds favored Chicago. "But," he exclaimed vigorously, after another assault on

Millikan in Chicago, "we haven't given up the fight by any means." [41]

Then, suddenly, unexpectedly, Millikan capitulated. At this point Hale did not know why. He could only wonder and rejoice. Perhaps, like William Hale in the early days, the physicist was just worn down. Perhaps he realized, too, that Hale would never give up the chase. After this long siege of wooing, there seemed nothing to do but "to yield to Hale's suit." [42]

That night Hale wrote to Adams: "It has been the most difficult campaign I have ever been through, involving scores of interviews and discussions, with all the cards against us." Millikan, it turned out, had finally succumbed when the terms agreed on were backed by Norman Bridge's offer of $200,000 for the physics laboratory and by Arthur Fleming's promise to turn over the balance of this fortune to the institute. This proved to be better than anything the University of Chicago could offer. "So," said Hale, "while I regret more than I can tell my long absence from the Observatory, I am sure it will profit far more by the results of this campaign than by anything I could do on the ground at the same time." [43]

In the fall of 1921, Millikan arrived to become director of the Norman Bridge Laboratory of Physics and chairman of an Executive Council, which he chose to set up instead of accepting the presidency. As he left Chicago, President H. P. Judson warned him, "Millikan, if you go way out to California you will lose your most stimulating contacts. It is the end of your scientific career." [44] These fears proved to be unfounded.

In 1921, the institute's future was no longer a gamble, and Millikan was coming in on the crest of the wave. Scherer, after his first twelve difficult years, commented, "I did the spade work." Now Millikan, with all his drive and energy, all the warmth of his buoyant personality, could carry on.

Moreover, the attitude of the country and the world at large toward science was changing radically. By 1924, Hale could note in his diary: "Within recent memory a great change has come over the U.S. Science, once smilingly regarded as a useless creature lurking in grey cloisters is now recognized as a national asset. Scientific research a great profession." In this revolution Hale had played a leading role. From it the institute would benefit enormously in these years and in the years to come after World War II when the even greater scientific explosion was to occur.

On a Saturday night at the end of January, 1922, with one of the worst storms in years brewing, the Norman Bridge Laboratory was dedicated. Hale spoke on "A Joint Investigation of the Constitution of Matter and the Nature of Radiation." [45] As he spoke, his audience saw a man of fifty-three, still youthful-looking, still modest despite a worldwide reputation. They listened closely as he looked back over the years since he began work in Chicago, to trace the growth of science and emphasize the importance of new methods and instruments of research in that development.

As he described the fundamental advances that had resulted in astronomy and also in physics and chemistry, Hale spoke, of course, without knowledge of the extraordinary atomic conquests still to be made, but with faith in a future in which understanding and knowledge of the tremendous energies and pressures in sun and stars would provide physicists and chemists a key to the unlocking of the atom. Through such programs as the joint attack he was now urging, he looked forward to Pasadena's role as an international center of research and scientific progress.

As a member of the institute's Executive Council, he would continue to take an active part in its development in the twenties and into the thirties. With his old friend Arthur Noyes he would be responsible for many important decisions and for contributing to the addition of many able men to the faculty, as it branched out into geology and seismology, biology, and even astrophysics. One of these was Thomas Hunt Morgan, the geneticist. When Hale first suggested the possibility of getting Morgan to leave Columbia, Millikan called the idea a pipedream. But soon, as he saw the chance of Morgan's coming, he was enthusiastically joining Hale in his conquest.

Still, as always, in these years after the war, Hale's greatest desire was to devote all his time to astronomy. By February, 1921, he could report to Goodwin, "I am having great sport getting back to work, dividing my time between office and house, where I still keep my measuring machine and work on spot spectra. I think this scheme, if it can be kept up, will be much easier for my head and will enable me to do something in my own field." [46]

Unfortunately, at this time, this was not to be. Again the trouble with his head returned. In April, in answer to a letter from Hale,

Kapteyn wrote, "It will be hard for a man of your temperament to be compelled by your body to curb the impetuous desires of your mind. Thank heaven that it is a 'curbing' not a 'giving up.' And then you ought to be comforted a little at least by the fact that it has been given to you, even now, to achieve ten fold what most men achieve in their whole life." [47]

Late in 1921, however, before he had to give up entirely once again, Hale obtained another interesting and significant result. This was his observation of "invisible" sunspots which, as he told Goodwin, he had had in his notebook for many moons, but had searched for systematically only recently. Now, by finding local magnetic fields, he had been able to detect spots three or four days before any spots were visible on the sun, and for two or three days after they disappeared. "I think," he noted, "they will prove very interesting as observations increase our data." And, some time later, he commented, "Thus we now have a means of studying sunspots in their embryonic and post-mortem states." [48] This, he hoped, would "assist materially in revealing the cause of spot formation."

Soon, too, the problem of the cycle of solar spottedness that had puzzled him in the years before the war recurred. On October 6, 1922, he wrote to H. F. Newall from the Hôtel du Palais in Biarritz, "Just a line to say that Ellerman has observed what appears to be the first spot of the new cycle—a little one (single) at N. 31°, on June 24, gone the next day. The polarity was opposite that of single or preceding spots of the present cycle, so the confirmation of our 1912 result (or reversed polarity) is all right, as far as it goes. I am eagerly waiting for another new spot cycle, and hope it will come soon." As the new sunspot cycle developed, he was delighted to discover that "the expected magnetic reversal was completely confirmed."

By 1925, as a result of the enthusiastic work of his colleagues in years when he himself was able to do little of the actual observation, Hale was able to publish a paper (with Seth Nicholson) on Law of the Sun-Spot Polarity.[49] In it they reviewed the history of this important discovery and showed how the observation of about 2,200 sunspot groups had led to the establishing of a definite law of sunspot polarity—a law that was to be called after Hale and one that was to assume growing importance as knowledge of the sun's nature increased.

Day after day they had recorded the polarities of spots in the new 11½ year cycle that appeared in high latitudes after a minimum of solar activity. From this record it became apparent that these spots were of opposite polarity in the northern and southern hemispheres. As the cycle progressed, they observed that the mean latitude of the spots in each hemisphere of the sun steadily decreased, but their polarity remained unchanged. Finally, with the beginning of a new 11½-year cycle, they had seen that the high-latitude spots in this cycle began to develop more than a year before the last low-latitude spots of the preceding cycle had ceased to appear, and were of opposite magnetic polarity.

Therefore they concluded: "Thus while the well known 11½ year interval correctly represents the periodic variation in the number or total area of sunspots, the full sunspot period, corresponding to the interval between the successive appearances in high latitudes of spots of the same magnetic polarity is between 22 and 23 years. This may be called the 'magnetic sunspot period' to distinguish it from the 11½ year frequency period."

Looking back and tracing the history of his results on the polarity of spots, Hale wrote in *The Telescope* for 1936: "After sufficient solar and laboratory results had been obtained to demonstrate the existence, strength, and general characteristics of the strong magnetic fields in several of the largest spots then visible, other phases of the problem were attacked. A bar magnet has two poles, north and south. In general, the magnetic fields of the sun-spots observed north and south of the equator were found to be opposite in polarity. As some exceptions were noted, however, a general study of polarities was begun. This led to some curious results.

"Anyone who will examine the early observations of Galileo, Scheiner, and their successors will notice the wide variety in the size and form of sun-spots. They often appear at first as single spots but soon develop into elongated groups, containing large and small members. Many of these groups are double, comprising two large spots, with or without companion, or one large group, followed or (less often) preceded by a train of smaller spots. Such spots were found to be almost invariably bipolar, consisting of two spots or groups of spots having opposite magnetic poles. In the great majority of cases the preceding (western) spots of such pairs are opposite in polarity in the northern and southern hemispheres of the sun.

"It was thus a comparatively simple matter to develop a scheme

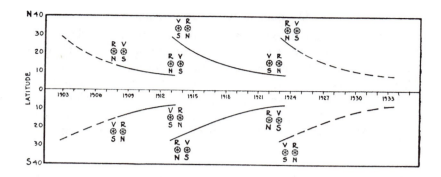

*The law of sunspot polarity. The curves show the latitudes in which
sunspots are most numerous at different times during a sunspot cycle.
The letters N and S indicate the polarities in groups of spots and show
the reversal of polarities in successive cycles.* (Courtesy of Mount Wilson
and Palomar Observatories)

of magnetic classification, which has since been applied to the many
thousands of sunspots observed at Mount Wilson."

By 1936, too, it had been possible to observe the spots through
another cycle. In 1934, at the beginning of that cycle, the characteristic reversal of polarity had been observed. "The complete magnetic
cycle thus lasts between 22 and 23 years and is about twice that of
the better-known sun-spot frequency cycle. It points toward some
general solution, applying to the sun and doubtless to many other
stars, but still not beyond the empirical stage." [50]

Still, therefore, the cause of this "magnetic" period remained puzzling. In 1935, Hale wrote to Charles G. Abbot at the Smithsonian
Institution: "The 23-year sun-spot period is one of the most mysterious phenomena I know about. It is hard to imagine that the *magnetism* of the spots can have any appreciable influence at the
distance of the earth, partly because most of them are bipolar, they
are of opposite sign in the two hemispheres, and the intensity of the
field decreases very rapidly upward from the lowest levels at which
we can observe it. But as we have now completed a full 23-year
cycle, with reversals of polarity exactly as we expected them, there
can be no doubt, I think, about the validity of the period. . . ." [51]

In 1965, thirty years later still, after observation of another complete cycle, Harold Babcock confirmed the validity and the importance of this remarkable law "which shows that the long known

cycle of solar spottedness does indeed occur in a period of about twenty-two years—not eleven years as previously supposed." Moreover, as a result of the development of the solar magnetograph, Harold Babcock suggests in a footnote, "A decade of observation with the solar magnetograph has led to a proposed explanation of the Solar Cycle. Energy derived from differential solar rotation keeps it going. Principal observed features including the 22-year period fit the reasoning." [52]

Meanwhile, in 1922, at the beginning of the spot cycle that was to confirm his earlier observations, Hale was forced to leave Pasadena and Mount Wilson far behind. For some time Dr. McBride had been insisting that he take a year off. Hoping to avoid another breakdown like that of 1910, Hale had tried to get at the root of his difficulties. He had written to Simon Flexner and to Dr. W. W. Keen, describing his symptoms. They were sympathetic. But no one seemed able to reach the core of the trouble. He himself had made numerous attempts at self-diagnosis. Yet he continued to suffer from terrible headaches and from profound depression. Finally, therefore, after weeks of coaxing, he gave in. Soon he was deep in guidebooks, planning "bully" excursions, including one to Egypt, and urging his brother Will to go along.

On June 24, 1922, he sailed on the *Olympic* with his wife and young Bill, who had just graduated from Princeton. In England they stayed with the Newalls at Madingley Rise. After some weeks there they visited William Hale's old friend James Ellsworth, in Switzerland. Ellsworth's strange medieval castle, Schloss Lenzburg, was a weird and wonderful place. It was filled with clocks of every sort, striking the hour, the half hour, and the quarter—all in different keys. On Hale's first night there, as he listened to the chimes echoing through the cavernous stone halls, he found sleep impossible. The next morning he tried to persuade the doughty Ellsworth to remove the clocks from his bedroom. Ellsworth refused.

Yet, as he learned the history of the castle, Hale was charmed. Here the Emperor Frederick Barbarossa had lived. Here, too, in the twelfth century, the Abbot of Marseilles had been imprisoned in a dungeon whose walls were still adorned with a headman's ax and implements of torture. One night, after wandering through the haunted passageways, Hale lay awake, unable to sleep. As he thought of the screams of the tortured, these lines formed in his head:

> Strangled in irons and shaken with fears,
> Naught for his eyes but rack and blade,
> Naught for his ears but moans and tears,
> Dripping their way through the stories deep-laid,
> The abbot crouched in his dungeon cell. . . .

The verses flowed on and on, until at last he fell asleep in his canopied bed.[53]

Shortly after his visit to Schloss Lenzburg, Hale spent a day climbing the Riffle Alp. The exertion proved too great, and he had to spend several weeks in bed in Geneva. Nevertheless, despite his ill health, he pushed himself to attend the meetings of the Committee on Intellectual Cooperation of the League of Nations, to which he had been elected by the council of the league as the American representative. This was a great honor, and Hale found himself one of a brilliant group that included some of the intellectual leaders of the world: Gilbert Murray, Henri Bergson, Marie Curie, H. A. Lorentz, and Paul Painlevé. Albert Einstein was also a member, but did not come to the meeting. "Ostensibly he is going to Japan, but actually (it was rumored) he feared assassination." Ordinarily Madame Curie had refused all invitations to membership on committees. This, as Eve Curie was to point out, was "to be her only infidelity to research." With the other members of the group she hoped that this committee might help the league to create a climate of world peace. Gilbert Murray stressed that "the prime object of the league committee was the establishment of friendly relations between the intellectuals of the Allies and Central Powers." But they would also try to work for specific aims, such as the international coordination of bibliography, the standardization of scientific symbols and work on the Tables of Constants. At first Hale feared this committee might attempt to displace the embryonic International Research Council on which he had spent so much time and effort. This, in fact, was one of his chief reasons for wanting to attend the meetings. He was glad to find that these fears were unfounded.

They met at the Hotel Beaurivage on August 1, 1922. Hale was charmed by Gilbert Murray and also by the philosopher Henri

Bergson. "I can easily see," he wrote, "how his lectures in Paris draw great and fashionable throngs, though it is safe to say that few of his fair auditors understand any of his philosophy." Mme. Curie, he noted, "made a very good and sane member of the committee." He was also delighted by Ruffini, an Italian senator, "a most jovial and agreeable man." But he feared the Belgian Jules Destrée.

Soon, however, much as he enjoyed the stimulation of the discussion and the interchange of ideas on a variety of topics, Hale decided that it was too much for him. He did not want to continue to serve on a committee unless he could really work for it. Therefore he resigned, and arranged to have Millikan (already his alternate) appointed in his place. Millikan was to serve on the committee for ten years, and called it "the predecessor of UNESCO."

After the meetings Hale traveled on from Switzerland toward the south of France and Spain. Everywhere he found new delights. But the weather was cold and dreary, and wherever he went he had to sit huddled in his sweater and overcoat. McBride had advised him to stay in one place "until Hell freezes over." At which he commented, "to judge from the temperature of my feet the ice must be forming over the Pit." [54] At last he journeyed on to Granada and Seville, where it was rumored the sun was shining. Here he was joined by Evelina and Bill, who had been sight-seeing on their own.

In December they arrived in Egypt just as the great tomb of King Tutankhamen was being opened.[55] At Roda on the Nile they met the Breasteds and their son Charles. They anchored their dahabeahs in midstream and had dinner together. That night Hale and Breasted plunged into talk of archaeological exploration. Hale had devised a new tool that he thought might be useful in excavation, especially in the silt of the Nile. It was a simple-sounding device, like an ordinary carpenter's augur with an extensible shaft. He urged Breasted to try it out near Luxor. In turn Breasted told of the magnificent finds in King Tut's tomb and the wonder of its uncovering. Some days later, at the excavation site, they looked on in amazement at the magnificent golden objects as they were brought out of the tomb.

Perhaps it was this last excitement that again sent Hale off balance. He found himself back at Luxor, able only to sit quietly and rest while looking out across the Nile at the great yellow cliffs "pierced with the doorways of rifled tombs." [56]

As he improved, however, he decided to write some articles on

Egypt. The discovery of King Tut's tomb had fired the public imagination, and the time was ripe for the advancement of Egyptian exploration, especially of Breasted's work. Therefore, whenever the excruciating pains in his head permitted, Hale spent his time writing. He finished two articles for *Scribner's*, one on "Recent Discoveries in Egypt"; the other "The Work of an American Orientalist." [57] The response was enthusiastic, and Breasted was delighted with the account of his work, and grateful for Hale's support. Later, too, through the intervention of the Research Council and the State Department, Hale was able to obtain a year's delay in the application of the Egyptian regulations that would have meant the end of excavation for foreign archaeologists.

Meanwhile, when he was still in Egypt and at the nadir of discouragement, a letter arrived from John C. Merriam, president of the Carnegie Institution. "We need you," Merriam wrote, "for so much in science and research that I am hoping to see you come back to this country not merely rested but restored to your normal ability for hard work." [58]

Merriam's letter heightened Hale's sense of desolation. It also brought him face to face with a problem he had long evaded. Finally, on March 29, 1922, after days of indecision, he wrote his letter of resignation as director of the Mount Wilson Observatory. "The real question," he told Merriam, "is to determine how I can accomplish the greatest possible amount of effective work during the rest of my life." [59]

He analyzed the illnesses and other difficulties that had filled a large share of the years since his first nervous attack in 1908. He even went back to his childhood, when he suffered from typhoid, dysentery, and colitis. He emphasized that most of his recent illnesses had been closely connected with the state of his head, "and hence with the environment."

His logical successor was Walter Adams, who had always taken over in his absence. Now he suggested that if Adams were willing, he (Hale) might continue as honorary director in charge of policy, while Adams would be in charge of operations. "I have never," he admitted, "been skillful in carrying on steady and continuous work. In fact, I am a born adventurer, with a roving disposition that constantly urges me toward new enterprises, to accomplish which I have always been willing to take long chances. This leads me to encourage any promising schemes by members of the staff, which interest me practically as much as projects of my own." [60]

A month passed after Hale had mailed this letter. Then, after consulting Adams, Merriam cabled his approval of Hale's plan—subject to the approval of the Carnegie Executive Committee. Soon this too came, together with a personal letter from Merriam in which he expressed the committee's profound regret. He also praised Hale's creation of the observatory and its staff, which had contributed enormously to "our knowledge of astronomy." [61]

In the month that had intervened between his resignation and its acceptance, Hale had gone to Florence, where he stayed at the Villa Palmieri, another of the homes of the tall, thin, gray-haired James Ellsworth, whom he had visited in Switzerland. Breasted joined him there after Evelina Hale and Bill had left for home. In this villa Boccacio had written the *Decameron*, and Marie Antoinette and Queen Victoria had been guests; on the hill above, Dante had dreamed. It was an idyllic place, and the two friends reveled in its beauties. They spent countless hours in the garden, talking or dreaming their own dreams. Slowly Hale found his spirits reviving. Neither he nor Breasted ever forgot that visit. Often, in later, more harrowing days, they wished themselves back in that paradise, where nightingales sang and roses bloomed.[62]

It was during their stay in Florence, too, that Ellsworth expressed interest in Galileo and in his ancient instruments preserved in the Science Museum there. With the help of the Italian astronomer Giorgio Abetti, it was arranged that Galileo's original telescopes should be taken from the museum and attached to Amici's telescope at Arcetri. So far as they knew, no one since Galileo's time had looked at the skies with his telescopes. They gazed at the mountains on the moon, at Saturn, at the four satellites of Jupiter, and the sun, and afterward Hale wrote, "It is wonderfully interesting to be able to see with his eyes." [63]

Hale had, however, one ulterior motive. For many years he had been helping Abetti to find the funds to erect a 60-foot solar tower. He had contributed all he could from the William E. Hale Fund but this was not enough. Possibly now the quixotic Ellsworth might be interested in this project. He might be more interested still if he could in return receive something of value that was unique. Such an object was a curious clock that stood in the dome of the old observatory. It was of no scientific value, but if it could be presented to

Ellsworth at the proper moment he might be induced to finance Abetti's telescope. "As usual," Breasted noted, "Hale's strategy succeeded. We were on the point of leaving the dome when Mr. Ellsworth announced that he would contribute the money for the completion of the new observatory on one condition—that he should be given the curious clock he now saw before him. Hale and Abetti exchanged glances, and the latter, swallowing hard, assured Mr. Ellsworth that the clock was his for the asking.

"What a genius Hale is," Breasted exclaimed, "and what an evening this had been!" [64]

CHAPTER SEVENTEEN

Life in the Hale
Solar Laboratory

✳ *1923-1930*

IN THE FALL of 1923, Hale moved out of his office on Santa Barbara Street and tried to accept his narrow proscribed life. For a long time he found it hard to settle down to any real work. His uselessness, he felt, was becoming more and more apparent. Even a year after his return he was reporting to Harry Goodwin his lack of energy, his inability to accomplish anything. The only thing he could do successfully, he said, was to sit in the sun and read.

His reading in these dark days covered a wide range—everything from his boyhood idol, Harry Castlemon, to "The Turn of the Screw" by his favorite Henry James. When Harry sent him a "bloody pirate tale" he enjoyed it hugely.[1]

As his health improved, he turned to more serious works—a life of Rayleigh, the memoirs of Cassini, Haskins' *Studies in the History of Medieval Science*, and *From Immigrant to Inventor*, Michael Pupin's autobiography. The more he read, the more his interest in the past grew—"a sure sign of old age."

As his interest in the history of science increased, he continued to buy valuable old books as well as to read current ones. When *The History of Science and the New Humanism* by George Sarton came out, he wrote, congratulating Sarton on his breadth of view: "I have been very impatient with the growing public tendency to class science and the humanities as antagonists and you have done a most valuable service in pointing out so clearly that no such antagonism exists, except in the minds of those who are incompetent or unwilling to recognize the truth." [2]

Again and again throughout his life he had considered ways of narrowing the gap between science and literature. Again and again he had written about it. Thus, in an article on "Everyman's London" he was to write: "No one with an open mind who has seen the marvelous variety and beauty of microscopic life, or formed a clear image of the celestial universe or looked down the long vista of evolution can deny the existence of a 'literary' quality in science which some great poets have felt and partly expressed. Nor should anyone who has shared the delights of creative scientific research fail to realize their close analogy with the pleasures of creation in the various forms of art. But how many schools of any kind succeed in making clear to boys or girls of literary trend that science is anything more than a mass of forbidding formulae? And, on the other side, how many schools of technology succeed in really interesting their students, for the most part narrow specialists, in the wide field of art or literature?" [3]

With his interest in science and the humanities, it was natural that Hale should be strongly attracted to the writings of George Sarton. Actually, their friendship had begun some years before, when the Belgian scientist, on his arrival in America, had been seeking support for his monumental plans for the development of the study of the history of science. He had written to Hale outlining his dreams. At the time Hale could do little, but later he was able to help in gaining support from the Carnegie Institution. From the time, too, that Sarton founded *Isis*, the first journal in the world to be devoted to the history of science, Hale became one of its strongest supporters. He was also a charter member of the History of Science Society.

At first, after his resignation as director of Mount Wilson, Hale tried to carry on his work at his own house, Hermosa Vista. He borrowed a coelostat from Mount Wilson, set it up in the back yard, and under conditions as primitive as he had known in the earliest days at Kenwood, began experimenting on the sun with a scheme he had first tested many years before. He set up a 4¼-inch telescope Mr. Hooker had given him and began to make tests of the seeing. As he worked, he felt he had gone back forty years to 1883, when he had mounted his 4-inch telescope on the roof of 3989 Drexel. Now he mounted a temporary horizontal telescope in the garden.[4] Though he had photographed the bright and dark flocculi on the sun's face, he wanted to be able to observe them visually. The idea for a new instrument had long been, as he said, "in my notebook."

He called it a "spectrohelioscope." [5] It seemed to him to offer great possibilities. Soon he was making plans for the building of a permanent solar observatory, and searching for a site away from Mount Wilson, away from the family, where he could work undisturbed.

The property he finally chose was on a corner of the original Huntington grounds. He spent days drawing plans for a "solar laboratory" that he would finance himself. It would include a spectroheliograph, a library where he could work, and a basement where he could have his shop and "lab." As the plans grew, his old enthusiasm for observatory building returned. With all the zest of an amateur, he wrote to Newall in Cambridge, England: "I shall have a small laboratory and a little shop with a lathe that would make your eyes bulge out! I also have my eyes on jewelers' and dentists' tools for work in my shop. If you and I could only join forces, supported by Beckford, what a *bully* time we could have working together." [6]

As soon as the plans were finished, construction began. As with every observatory he had ever built, the delays seemed endless. "It has dragged in every way." Long before the upper part of the building was finished he moved into the basement—a shop just a bit smaller than his first one in the yard at 3989 Drexel. Here he set up the ancient Kenwood lathe, a milling machine, a circular saw and a grinder. Later he added a drill press. With these fine tools he felt he could do almost any kind of work. Before the concrete was dry, he was down there experimenting. [7]

In this new "Solar Laboratory" Hale became in truth, as Gano Dunn called him, the "Priest of the Sun." The building itself was dedicated to the sun's worship. For a position over the fireplace in the library Lee Lawrie designed a bas-relief of Ikhnaton (the King of Egypt and worshiper of the sun god, Aton) rising in his chariot toward the life-giving sun. [8] Hale wrote to him in delight: "The likeness is perfect, almost too perfect for dry eyes. He is merely resting, planning his greatest flight. . . ." When the bas-relief arrived, it took its place across the room from the bust of Nefertiti, "the image of eternal youth." Above the fireplace on one side was a beautiful Persian astrolabe, and opposite it an old armillary sphere. Above the front door appeared another tribute to Ikhnaton symbolized by the sun's rays that ended in hands that grasped the symbol of life—a copy from a Theban tomb. This tribute was vividly expressed in the "Hymn to Aton" that Breasted had translated and given to his old friend—to Hale's everlasting joy:

Thou risest in beauty in the horizon of the sky,
O living Aton, beginning of life!
When thou risest in the eastern horizon,
Thou fillest every land with thy beauty.
Thou art beautiful, great, glittering, high above
    every land.
For thy rays encompass the lands
Even all thou hast made. . . .
Though thou art far away, thy rays are upon earth:
Thou art on high, thy footprints are the day.

From the moment the Solar Laboratory was finished in March, 1925, the exquisite little building became the center of Hale's life. Here he spent hour after hour in the library in his armchair, with a bookrest on one side and a writing table with a shelf underneath on the other. Here he could read and write to his heart's content. When it was cold, he could enjoy the blazing fire. "It is far better than nothing, though I hate to be compelled to live like a hermit."

Soon after he moved in, Evelina Hale wrote to Margaret, "Father is at work every day, early and late, and is as happy over it as a small boy with a new tool chest." By the early months of 1926, he was again deep in solar investigation. A coelostat mirror was mounted in the dome. The 12-inch object glass that he had originally used at Kenwood and had now borrowed from Yerkes formed a two-inch image of the sun on the spectrograph slit in his basement laboratory.[9]

Before long he was ready to make a decisive test with the oscillating slit he had worked out in the Hermosa Vista garden two years earlier. The day he chose was glorious. The sun was covered with numerous large and active spot groups. He set his second slit on the H alpha line in the first-order spectrum, and found he could see the bright and dark flocculi on the sun's disk almost as well as he could photograph them with the spectroheliograph. "In forty years," he wrote exuberantly to Gano Dunn, "I have never observed such interesting and remarkable phenomena with a device which is absurdly simple and is not yet perfected."[10]

Later he explained the "simple" principle of his new spectrohelioscope: "Cut a slit about a hundredth of an inch wide in a piece of cardboard and hold it between the eye and an electric-lamp bulb. If the eye is not too near the slit, only a small part of the incandescent filament can be seen. Oscillate the slit and the entire filament becomes visible. To obtain a steady image, free from flicker, the slit should pass before the eye many times a second."

This principle, as Hale was to point out, could be used to render the forms of the prominences visible. "What is needed is some device between the oscillating slit and the eye which will cut off all light except that due to hydrogen. The spectroscope is such a device."

Many years before, in 1870, Hale's old friend Charles A. Young had foreseen the possibilities of such a method when he attached a pair of oscillating slits to his spectroscope. With these slits he had been able to see the dim outline of the prominence forms at the sun's limb, or edge. But the oscillation of the slits had caused his telescope to vibrate, and he finally abandoned the scheme in favor of the wide-slit method. With this device, however, he was still able to see the bare outline of objects of unusual intensity only on rare occasions. "Nevertheless," Hale noted, "the credit for building the first spectrohelioscope and applying it to the observation of prominences at the limb belongs to Prof. Young, who was also the first to photograph the forms of single prominences through an open slit."

As Hale worked with his new device, the days passed swiftly. He soon discovered he could trace the prominences all the way around the sun's limb. Often they overlapped onto the disk and he could observe their dark structural forms clearly. Sometimes he could even see the most delicate small dark structures all over the disk. Several times he observed the dark hydrogen structures running into spots that he had first photographed with the spectroheliograph in 1908. Now, watching them with the spectrohelioscope, he could see them changing in an extraordinary way. Though his tiny observation window was only two millimeters in diameter, what he was able to see far surpassed anything he had ever seen before.

At times he could watch the dark absorbing gases as they descended into the spots at velocities that, from the shifts in the spectral lines, he estimated at hundreds of kilometers a second. Once on a day in January, 1926, a tremendous eruptive outbreak suddenly appeared in a large spot group. Hale watched in wonder the brilliant jets and the flaming cloud masses, the extraordinary changes in form. Up to this time, astronomers had known that these great solar phenomena occur from time to time, but they had lacked the means of observing them regularly. Years before, their existence had been

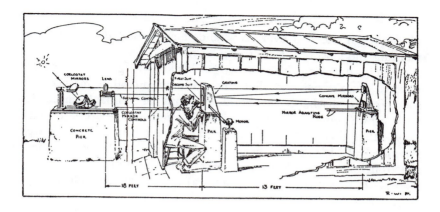

General view of a simple horizontal telescope and spectrohelioscope. From George Ellery Hale, Signals from the Stars *(New York, 1931).*

detected spectroscopically, and Hale had photographed them with his spectroheliograph. But this was not the same thing as being able to watch them directly, wherever and whenever they occurred on the sun's surface. A magnificent new realm of observation had been opened.

Soon, too, he was delighted to receive a letter from Norway that provided earthly confirmation of the effect of the tremendous outbreak he had observed on the sun. A letter arrived from Carl Störmer, a Norwegian who had been studying the phenomena of aurorae, or northern lights, for many years. He wrote: "I have been most fascinated by a remarkable aurora here the 26th of January of exceedingly *red* color, like the aurora in 1870. I should like to know from which active part of the sun this aurora was coming."

"By a piece of rare good fortune," Hale could note, "I was able to answer this question with some chance of certainty." On January 24, 1926, while testing his spectrohelioscope between 11 h. 40 m. and 12 h. 15 m. Pacific Standard Time, he had observed a bright eruption near a great sunspot at about 22° north latitude, which was then close to the sun's central meridian. "Its form changed rapidly, and in this and other respects it was evidently an exceptional object." [11]

On January 25th another eruption (now called a "flare") occurred in the same region. It was "the most brilliant and remarkable solar phenomenon" he had ever seen, and it continued throughout the morning and most of the afternoon.

From his own experience, as well as his knowledge of other observations, Hale concluded that this eruption probably marked the source of the discharge that produced both the brilliant aurora described by Störmer and the intense magnetic storm recorded on January 26th and 27th at Greenwich and elsewhere. Again, too, he was reminded of his observation made on July 15, 1892, with his new spectroheliograph at Kenwood, of a similar outburst, which was likewise followed by a brilliant auroral display and a violent magnetic storm.

Now, in 1926, he immediately wrote to friends all over the world to share his latest successes with the spectrohelioscope. "The thing is so full of previously unexpected possibilities," he told Goodwin, "that I have been in a ferment and I shall not be able to settle down into a quiet state of mind until several of the new ideas, at least, have been adequately tested." "Of course," he added, "the effect on my head has been like that of a powerful champagne, and I have frequently had to lie up for repairs." [12] Yet, in general, as he told his twelve-year old nephew George II (as he was called), "I never had a bullier time, not since the 3989 days when we used to sally forth on Saturday mornings into the shop behind the house." [13]

Then, in order to explain what he was doing, he gave a vivid description of the workings of the spectrohelioscope: "Imagine yourself on the 'Nautilus,' and suppose that all fish were naturally invisible. But you have contrived peculiar windows, through one of which you can see all the sharks, through another nothing but whales, through another only the octopi or pusses.

"My 'Nautilus' is hard aground, and I must look out through a sea of air instead of water. Beyond the air stands the sun, with wild and fantastic beasts roaming over its surface. They are of enormous size, sometimes reaching up to heights of four hundred thousand miles. And they are fearfully hot, made of hydrogen, helium and calcium, in the form of gas so thin that it is only about a thousandth part as dense as the air you are breathing.

"Naturally," he explained, "when you look at the brilliant surface of the sun with a telescope you do not see these beasts, because they are so thin that they don't cut out any appreciable fraction of the light. But if you had a window through which you could see nothing but things made of hot hydrogen, they would surely come into view.

"I have made such a window and have been looking through it at these marvelous beasts much of the time since I came home. Their antics are astonishing, and though I knew something about them before (because I had a method of photographing them without seeing them), it is quite a different matter to *see* them diverting themselves like monsters of the deep."

Weeks passed, and the work became, if possible, even more absorbing. Once he had to go east, but as soon as he arrived home he rushed back to his laboratory. Almost immediately he was rewarded by a sight he had seen only once before in eighteen years of observation. This was the sudden engulfment of a great prominence in a solar vortex. The date was June 26, 1926. With his new toy he was able to watch the phenomenon "in all its remarkable phases." He tried to determine the radial velocity (the motion in the line of sight, which had never before been measured) as the gas was sucked into the vortex. Unfortunately, the special measuring device he had designed was not yet finished, and he could make only a rough determination.[14] Later, when this ingenious device was ready, accurate observation became possible; results poured in so thick and fast he had difficulty keeping up with them.

In November he wrote to his old friend Charles Abbot that these latest solar adventures, "though pursued much less strenuously through the fogs and smoke of the San Gabriel Valley," were proving quite enough for "an old and decrepit codger" like himself. "In fact," he declared, "the spectrohelioscope has opened a little gold mine, chiefly through the power of 'tuning in.'"[15] Just as in radio, he would lose the message unless he tuned to the right wavelength. Yet, in a few weeks, this tuning device had taught him more about the true nature of spots and the regions surrounding them than he had learned in years of study of spectrograms. "Needless to say," he wrote to Turner, "this is very good sport."[16]

But one observer, working in one place, was not enough. What was needed was a string of spectrohelioscopes scattered at strategic points around the world. Hale now set to work to design an inexpensive instrument that could be easily used anywhere. A chain of observers, well distributed in longitude, might be able to record enough solar outbursts to clear up uncertainties about their relationship with such terrestrial phenomena as aurorae and magnetic storms. Moreover, this was a shining opportunity to show the possibilities of small telescopes, and to encourage amateur observers the world over.

From the beginning of his astronomical career Hale had liked best the title of "amateur" for himself. Now he felt he had again joined that select band with Sir William Huggins, who, like Faraday and Darwin, had "worked with an earnestness and devotion that would have made him a great investigator in any country and in any age." Soon after he moved into his new "lab.," a heartwarming letter arrived from H. H. Turner in England. "Vale atque ave! Valedictory words of the warmest possible to the retiring Director of the Mount Wilson Observatory which has revolutionized astronomy; and hail to the new 'amateur' with his modest equipment! How splendid to hear that you are carrying into practice your oft repeated and encouraging advice to the man of modest means—by erecting a 6-inch at the door of the biggest and finest observatory the world ever saw! All good wishes (in all confidence) for its success!" [17]

Throughout his life Hale had tried to do everything in his power to advance amateur interest in science, and especially, of course, in astronomy. He wrote articles, he gave talks, he wrote endless letters that would encourage others as Brashear had once encouraged him. In April, 1916, in an article on "Work for the Amateur Astronomer," he had described the amateur as one who "is in fact a true lover of knowledge for its own sake, one who works because he cannot help it, swept on by a passion for research which he attempts neither to explain nor to curb, and enthusiasm which carries him over obstacles too high to be surmounted by the perfunctory student or the man without zeal. To the sane enthusiast, whether his talents be large or small, great advances are possible. An impelling interest, even if backed by only a very slender stock of knowledge, may accomplish more than all the learning of the schools."

Looking back on the joy of discovery in his own childhood, he wrote that, as he had pushed with larger and larger telescopes into the depths of space, he had often "been forced to confess that the astronomer never beholds sights more wonderful than those which a drop of ditch-water, on the stage of the cheapest microscope, will afford to any boy."

Twelve years later, in 1928, Hale turned again to the amateur. His subject was "How To Build a Small Solar Observatory." In it he described the building of an inexpensive solar telescope and spectroscope that could be turned into a spectrohelioscope or spectrohelio-graph for visual and photographic observations of the solar atmos-phere.[18]

Thus, when a young Hindu, an amateur working in India, wrote for advice on building a spectrohelioscope, all the necessary infor-mation, including a set of blueprints, was at hand. Hale sent these along, and continued to send encouraging letters over a period of years.

Another characteristic case of an eager amateur was that of a boy named Jack Garrison, who lived in Indianapolis. Jack wrote first in 1919 when he was ten. "I like astronomy very much and have been reading for about four years." He asked for books and pamphlets, also for information on the work being done in the "Observeotory." "If you do this for me I will be very glad." In reply Hale sent him a copy of his book *Ten Years' Work of a Mountain Observatory*. Jack replied enthusiastically, "I want you to know I'm as tickled as Tom Sawer [*sic*] over his pulled tooth."

At intervals the correspondence continued. Before long, Jack was writing with amazing confidence, "I think I have a beginning for a famous astronomer." Seven years after the correspondence began, he commented, "Taking it all in all I am very optimistic as to my future as an astrophysisist [*sic*]." Soon Hale was helping him with a check as down payment on a college registration fee. Jack wrote gratefully, "I cannot express my appreciation for what I realize is a sacrifice on your part except to make every possible benefit from it." [19] (In the end, Garrison did not go into astronomy, though he did go into science. He carried on work in the field of acoustics, and became associated with the United States Gypsum Company.)

Soon after his "lab." was finished, Hale began to think of taking up another "amateur" scheme that took him back in spirit to his boy-hood. In 1927, he wrote to T. H. Morgan to thank him for his monu-mental treatise on genetics: "I have read a lot of it and am anxious to repeat for my own delectation some of the remarkable experi-ments you describe. Of course I don't know anything about the subject, but as an amateur microscopist since about 1881, when I began chasing rotifers and other beautiful infusoria, I am still under the fascination of the microscopic world. My ancient Beck micro-scope, given me about 1883 by my father as a prize for studying infusoria, smiles down on me as I write here before the fire, and a marvelous new Leitz 'Universal' stand, with all manner of attachments for biological and polariscopic work, awaits an opportunity to show me some of your phenomena. Like Dunn, I want every new microscope that appears on the market, and a couple of extra magazine articles should pay for the Leitz extravagance." [20]

Then, thinking of another experiment he had long been dreaming of, he asked, "By the way, have any of your phenomena been observed in a very powerful magnetic field (I mean something like 15,000 gauss or so?) I don't suppose there is any reason to try the effect of either magnetic or electric fields, but I have got one of the most powerful magnets in existence downstairs (for work in connection with solar magnetism) and would like to try it in this way just for fun."

As he pondered the enticing question of what would happen to infusoria subjected to the effect of a strong magnetic field, he returned to the reading of his old books on microscopy—"all," as he told Newall, "by English authors and all redolent of memories of glorious muddy ditches, where my brother and sister helped me to break the ice in the spring in order to entrap the unwary Cyclops or Branchipus in our collecting bottles. Colonies of Vorticelli and rare rotifera are as fascinating to remember as they were to gaze at in the old days."

In the pond in Henry Huntington's Japanese garden he was sure the most marvelous infusoria were "dying to leave their habitat" and mount the stage of the new microscope he had ordered from England "for my special delight." He could hardly wait to get at them. "Actually," he acknowledged, "all this longing for the pleasures of my youth is doubtless a sign of incapacity for more vigorous fare, but I get a lot of fun out of it just the same."

For his new infusorial experiments, he set up an apparatus that contained a strong magnet, his new binocular microscope and a tank where the tiny animals could live. He wanted rotifera that could be viewed with a power of 100, and he wanted to watch them first under normal conditions. Then, when they had calmed down, he would switch on the field. Under the magnetic influence he felt that something drastic ought to happen; they ought to blow up or explode or be killed in some other way. Or, possibly, they might go on living—but a different, more active life. He wondered if they might not even acquire paramagnetic properties. After observing all these things visually, he hoped to record them photographically.[21] Unfortunately, alluring as these experiments were, he never really had a chance to follow them up—to discover what would actually happen. About the time he got his apparatus set up, his microscopic plans were shattered by the 200-inch telescope, which was soon to absorb all his limited energies.

Still, all through this time, he never forgot his duties as honorary director of Mount Wilson. He retained charge of policy and research and the development of new instruments and methods. He followed eagerly the revolutionary work of the 1920's—a period that was in a way one of consolidation after the completion of the 60-inch and 100-inch telescopes. He took the greatest delight in discussing their research programs with members of the staff. He rejoiced when Adams performed his epoch-making and extremely difficult feat of observing the spectrum and measuring the line displacement in the companion of Sirius.

Sirius, as anyone who knows the skies at all is aware, is one of the brightest stars in the sky. Early astronomers used it to determine time. But they soon discovered that it was rather a poor clock; sometimes it would gain over a period of years; then it would lose. The cause turned out to be a small dark star around which it moves in an elliptic orbit. This star could not be seen, but the evidence of its existence was clearly apparent in the force it was exerting on Sirius. Years later the tiny companion was observed by Alvan Clark when he was testing the famous 18-inch object-glass that Hale knew as a boy at Dearborn.

In 1914, Adams discovered that this tiny companion was actually a white-hot star; in size it lay somewhere between the earth and Uranus—which for a star is very small. Yet, while it was small, it had a mass not much less than that of the sun (actually ⅘ of that mass.) It must, therefore, be extremely dense. It turned out that it was 53,000 times as dense as water; a density far greater than anything known on earth: a ton of such material could be squeezed into a matchbox.

This result was so extraordinary that Adams decided to test it in another way. Einstein in his theory of gravitation had suggested that all the lines in the spectrum of a star when compared with the corresponding terrestrial lines will be slightly displaced toward the red end of the spectrum. This theory had been tested on the eclipsed sun. Now Adams tested it on this tiny massive star where the effect should be much larger. With extremely fine instrumentation and the extraordinary observational skill for which he was famous, he discovered the large shift in the spectral lines that he had predicted, which was also very close to what Einstein had foreseen. So, too, the prediction that matter can exist that is thousands of times denser than any terrestrial substance was confirmed.

The news of this "amazing" discovery was picked up by newspapers the world over. The *Daily Mail* of London headed its column, "Great Stellar Giant Found. Star Formed of a Substance 10,000 Times Heavier than Iron." When the reporter interviewed Herbert Dingle of the Imperial College of Science and Technology, Dingle noted that previously no one had any idea that there were any stars of such an extraordinary character. "It is all a mystery," he said. He pointed out, too, that so heavy is the substance of these "White Dwarfs" "that a child would not be able to lift a piece of it of the size of a penny. A lump of it sufficiently small to fill a pint jug would weigh considerably over 22 tons. If it were possible for a lump a foot square to fall on the earth it would do enormous damage."

Hale, elated by these and other spectacular advances, wrote to John Merriam, president of the Carnegie Institution. He described the epoch-making advances made under Adams's direction. He told of Adams's own studies of the companion of Sirius; Seares' investigation of the nature and structure of our galactic system; Pettit's and Nicholson's work on the measurement of stellar temperatures with instruments that were so delicate that, if there were no intervening air to absorb the light, and the earth's curvature did not interfere, it would be possible to measure a candle's heat in New York with the help of the 100-inch telescope. He told, too, of Hubble's determination of the distances and dimensions of spiral nebulae, which would lead him to identify them as "island universes beyond our own galaxy." All these observations were, of course, a result of the development of the superb instruments to which Hale had devoted such a large share of his life. They were, though he did not say so, a brilliant vindication of that life.

In writing to Merriam, Hale was not only reporting results. He was also urging the crying need for an increase in the astronomers' salaries, as well as additional support for instruments of new and untried design. Such gambling experiments, he argued, are the essence of scientific progress. From his own experience, he was sure all the "sporting chances" he had taken had paid off. He pointed to the Snow telescope; the first tower telescope (a pioneer in the field);

Michelson's 20-foot interferometer; even the great 100-inch—all of which had been leaps in the dark. He declared, "We have been gambling for high stakes ever since the Observatory was started, and most of our principal discoveries and advances could not have been made otherwise." [22] Even in cases where the experiments did not succeed, he argued that much could be learned from failure.

Yet, if Merriam could see the value in the work Hale described, there was one of Hale's pet projects in which apparently he could see little value. This was the plan for the establishment of an observatory in the Southern Hemisphere. Hale had repeatedly urged that a Mount Wilson expedition to the Transvaal or some other good site in South America, lasting at least five years and equipped with a 60-inch telescope, would be immensely rewarding, especially in the study of galactic structure.[23] His pleas were eloquent, but Merriam and the Carnegie trustees turned him down—repeatedly. When questioned about his opposition at this time, (1925) Merriam stated only that when the need became vital the money would be found. As long as Hale lived that time never came. It was one of the great disappointments of his life. He would have been elated if he could have known that today, forty years later, the Mount Wilson and Palomar astronomers are planning a large telescope in the Southern Hemisphere. By this time, however, it is no longer a pioneer in the field.

Often in these years when he was unable to carry on his own research, Hale found he could write. In addition to his own scientific articles, he wrote others for *Harper's Magazine* and *Scribner's*. Some of these later appeared in the form of small books which, he hoped, would reach a large audience. In 1915, in *Ten Years' Work of a Mountain Observatory*, he had told the story of the founding and building of Mount Wilson. In 1922, *The New Heavens* appeared. In it he described the work of Galileo and the Herschels, and traced their discoveries down to modern times. He showed how modern methods (in which he himself had played a leading role) revolutionized astronomy, and greatly enlarged the astronomer's outlook.

He also wrote on "Giant Stars" and "Cosmic Crucibles." And he repeated the familiar refrain that thirty years before had seemed so revolutionary—and was still not so widely followed in principle as he felt it should be—that an observatory should serve as a physical laboratory.[24]

The more he wrote, the more he found he enjoyed it—though he never found it easy. His next book he called *The Depths of the Universe*. After it came *Beyond the Milky Way* and *Signals from the Stars*. All these were written when his health would permit and when "more serious work was out of the question."

He was delighted to find that these articles and books of a popular nature were welcomed by a wide audience. When the article "Cosmic Crucibles" appeared, Breasted wrote: "It makes oriental investigation seem very trivial and insignificant. What a marvelous prospect, to delve, with an able chemist at one elbow and a brilliant physicist at the other, into these celestial crucibles of which you paint such fascinating pictures. I think if I had it to do over again I should follow astronomy as my job." [25] (Similarly, Hale often said that if he had not become an astronomer, he would have liked to be an archaeologist.)

There were times, however, in these days when he found little joy in anything. And his depression was deepened by news of the death of old friends. In 1924, the former president of the Carnegie Institution, Robert Woodward, who had been his chief support during the early Mount Wilson years, died in Washington. When Woodward's will was disclosed, Hale was profoundly touched to learn that Woodward had left the majority of his professional books to Mount Wilson, or, in the event they should refuse the gift, to the California Institute, because, "it is my desire that these books be placed where they may aid in the development, jointly, of the two establishments named, whose organization and growth at Pasadena are due so largely to the enterprise and altruism of my friend and colleague, Professor George Ellery Hale." At a time when he was feeling so useless, such tokens of esteem meant much. He was grateful, too, when the noted Italian astronomer E. M. Antoniadi wrote, "Were I to have your photograph, I would frame it and place it in my studio as that of the foremost astronomer of our time."

Sometimes, with seemingly endless hours for reflection, he would ponder on various subjects, including religion and its meaning. Once, long before, when Evelina Hale had urged him to go to church "for the sake of the children," he had tried to explain his opposition to dogma that had grown out of the stifling religious atmosphere of his childhood. He had written then, "As for going to church, I understand what you mean, and will try to help you. Of course you must see that it is hard—really impossible—for me to reason one way through the week, and another way on Sunday. My creed is truth, wherever it may lead, and I believe that no creed is finer than this. . . ." In an effort to compromise, he had suggested that the children might gain good from a sufficiently liberal church. "But the Truth will not permit us to teach them beliefs that we do not hold ourselves. Probably the best way is to have them learn that there is a fine underlying idea which they should value, but which does not require them to believe the many absurd doctrines of the church." [26]

Over the years his search for that truth had continued. Now, as he jotted down his ideas, he wrote of the outlook of primitive man that was still shared by those who clung to his limitations and therefore to a creed that remains fixed, "in spite of the growth of knowledge." In contrast, the attitude adopted by the rational thinker (and that which reflected Hale's own feeling) comprises "an open-minded inquiry into the phenomena which express the operation of nature's laws." "Every new fact observed and every underlying law formulated enlarges the known scope of the Creator's powers." [27]

Yet while his friend Millikan continued to write and speak extensively on the relation between science and religion, Hale himself never published anything in this field. Yet once, in answer to a letter from T. H. Morgan, he wrote of his inability to accept any mystical explanation of the universe. "I never had any use for philosophical or metaphysical modes of approach to problems of science, except in so far as they may suggest clear-cut hypotheses subject to actual test by observation or experiment. This modern talk about physics vanishing into thin air is poppy cock. Physics never professed to explain infinity nor to deal with 'ultimate' questions. It still remains quite competent to widen effective knowledge, in spite of the mystics." [28]

During these years the Hales usually spent the summers in Santa Barbara. Mrs. Hooker and her daughter, Marian, and Maude Thomas now lived there, and Hale delighted in their company as well as in the beautiful surroundings. But even there he could not keep his restless mind from scheming. One day in the summer of 1924, he decided to try out an idea that had long fascinated him. He had often wondered if the solar seeing would not be good near the water with a light breeze blowing in from the Pacific. One morning as the fog was breaking, he started down to the beach with a 4¼-

inch telescope under his arm. He mounted it on a high bank a few yards from the water. Through the fog the sun appeared "very sharp." About twelve the fog disappeared. The result was all he had hoped, and more. "By Jove," he exclaimed in a letter to Goodwin, "it panned out as I had hoped!" The seeing was nearly perfect; he could see "the minute details of the granulation and the fine structure of the one large spot exquisitely sharp." [29] He began to dream of expeditions to the seashore for direct visual observations of spot structure and spot changes; of observations of eruptions and of hydrogen flocculi with his spectrohelioscope, as well as studies of prominences with a spectroheliograph.

All this kind of thing was intensely stimulating, but it was not the best thing for his head. Soon he found himself more keyed up than ever. In the same year, 1924, immediately after the National Academy dedication, he went to Johns Hopkins for a complete diagnosis under Dr. Llewellys Barker. There he saw the great psychiatrist Adolf Meyer. Meyer's notes recorded during their interview refer to the various depressed periods the patient had suffered—"some of them very severe." In a letter to Dr. Barker, Meyer called this "an anxiety type of depression." He did not, however, consider the attacks "manic-depressive." Hale's was, he noted, a case of over-anxiety, overintensity—a product of our over-restless civilization. In his notes he concluded, "Patient insists on determination to arrange his life." [30]

For a while after this, that life moved along fairly smoothly, though the patient continued to feel frustrated by his inability to work as he once had. He wrote to Goodwin, "My bursted old head is the bane of my existence, always preventing me from doing what I want. When I got back, after a period of apparent improvement, the very sight of my new spectrograph being assembled would set it boiling for the day." [31]

For his wife, of course, life under these conditions was hard. It was especially so because of her unceasing concern. One day in the summer of 1926 she wrote to Margaret: "Father is coming out of his shell somewhat, so I hope to have more good times." She added that Bill had gone and that she missed him and "the bit of life he brought into the quiet of the house."

While Hale recognized her difficulties, he could do little about it. Once when she was on a trip to Panama, he wrote: "I have been such a burden to you and I know that the only thing for you to do was to take such a trip and to stay away as long as you can find rest and relaxation." [32]

Increasingly, then, Evelina Hale made her own life and carved out a place for herself in the community. As time passed, she became more active in civic affairs—with the Visiting Nurse Society, the Pasadena Hospital, the Coleman Chamber Music Society, the local symphony orchestra. She was also an ardent member of the Garden Club. But her greatest love was the Pasadena Playhouse, to which she devoted a great deal of time, and she remained always one of its most ardent supporters. Her work was considered so valuable and the "need for her counsel" so impelling that when she retired she was voted an honorary member in this and in many other organizations.

Meanwhile Hale was turning to another project that had grown out of the National Research Council, and for a while it would absorb a great deal of his time. This was a program for pure research to be supported by industry. At first he called it the National Research Endowment. Later it bore the simpler title The National Research Fund. A definite plan was first proposed in 1922. But for several years little was accomplished. Finally, in May, 1925, Hale called a conference in New York to discuss ways of educating the public to the need for pure research and of persuading industry that it should turn back each year a certain proportion of its income for its support. By 1925 the stock market was booming, and there seemed every reason to hope for success. The first and most important step was to interest the leaders of industry. First, therefore, he invited Andrew Mellon, then Secretary of the Treasury, to an organizational meeting with Gano Dunn and J. J. Carty, vice-president of the American Telephone and Telegraph Company. He also asked Herbert Hoover, then Secretary of Commerce, and such leading scientists as William Welch, T. H. Morgan, and Vernon Kellogg of the Research Council.

From the first, everyone seemed interested, but no one was eager to take a leading part. In that fall of 1925 Hale set out to get academy support. Yet he confessed to Goodwin, "I have suffered a bit from cold feet lately—it is so easy to talk of raising large sums and so hard actually to do it." [33]

Hoover was being considered as chairman of the fund, but Hale had his qualms. He confided to Goodwin, "I am also perplexed about Hoover who is ready and anxious to raise a million a year from the oil men for petroleum research, but still remains an enigma to me as regards his attitude toward pure science." [34] Still, after a long, reassuring talk with Hoover, he asked him to accept the chairmanship of the fund. Hoover immediately said "Yes," and began scheming on ways to raise funds. Long afterward, in the light of the actual circumstances, Hale would have been amazed to find a statement in Hoover's Autobiography, in which he took entire credit for creation of the fund: "As an aid to research in abstract science, I created in 1926 a committee under my chairmanship to seek financial support from industry to be given to the universities and other centers of pure research." [35] Ironically, too, while he mentioned the names of the committee members, he did not even include the name of Hale. While Hale was never one to push his own claims, he would have wondered at this distortion of fact. And Welch, who presented the fund proposal at the academy meeting in the fall of 1925, would have shared Hale's amazement. After that meeting Welch telegraphed, "This job done." This, however, was only the beginning. No penny of the twenty millions proposed for the fund had been collected.

A few days later a trustees' meeting was held with Hoover, Welch, Carty, Millikan, Pritchett, and Root. Plans were made to tackle the big corporations. But it was a stupendous job, and Hale had misgivings even about his own first assignment—a meeting with the president of Sears, Roebuck in Chicago, Julius Rosenwald, whom he had known in Washington during the war. He went, armed with a letter from Hoover. But he was far from sure that this letter would help his cause. His doubts were enforced when, the night before the interview, he showed the letter to his brother, who considered it a very poor start. At this, as Hale told Goodwin afterward, his feet, already cold, got colder.[36] They got colder still the following day when Rosenwald's secretary informed him that he could not see the great man until the Monday after Christmas. He was forced to wait at Will's house in Winnetka until the appointed day and hour. By that time he was beside himself with anxiety and fear of failure. With "real trepidation" he arrived at Rosenwald's "remote lair" on the West Side.

Rosenwald received him with unexpected cordiality. Soon, however, Hale saw he was in for a hard time. Rosenwald declared emphatically that he did not believe in endowments *at all*. In fact, he said, he was a well-known crank on the subject; then he plunged into his ideas on the evils of setting up perpetual funds for an unknown future.

Hale waited for this explosion to die down; then, quickly changing the subject, he turned to a topic he knew Rosenwald favored— the industrial museum in Chicago to which he had just promised $3,000,000. An entertaining discussion of museums followed, and Rosenwald was soon presenting his guest with a book on the subject that, as Hale told Goodwin, he would treasure "as a memento of an uphill climb."

After a while, when he thought they had discussed museums enough, Hale carefully brought the conversation back to his own fund, and presented Hoover's letter. The letter suggested a gift in annual installments, a scheme that Rosenwald fortunately commended. He had in fact prepared the way by saying that a man would rather give *ten thousand* a year for several years than set up a capital sum of the total amount. With this, Hale began to breathe a bit more freely. "I saw my finish and sparred for wind." He expressed complete agreement with Rosenwald's point of view, then adroitly turned the conversation to Hoover and his personal merits.

Then, smiling inwardly, he slowly built up an imposing picture of the tremendous scale of the fund, endeavoring, as he told Goodwin, "to make ten thousand a year look like thirty cents," and at the same time "spurring" his courage to make his demand.[37] Finally he got it out—$100,000 a year for ten years! There was a slight pause in which he thought his heart would stop beating. Then Rosenwald said that of course he could not decide that day. Hale assured him he could take his time. Yet, afraid of losing his fish, he cautiously mentioned the trustees' meeting the following week, and spoke of the great advantage of having an offer to start with. "The fact was," Hale confessed, "I didn't dare leave him, so I sparred again for wind." Again there was a pause. Then Rosenwald remarked that he could not think of being alone in making such a gift. As Hale sat in suspense on the edge of his chair, his heart pounding, Rosenwald suggested that he might agree to be one of five to give $100,000 a year cash for ten years. Hale, overjoyed, took him up on it "without losing a second and got away before anything else could happen."

"Never," he told Goodwin, "did I go through a tougher job or feel greater relief when I got into the open air. The temperature was about zero or not far above, but my feet were no longer cold as I made for the elevated on my way back to town!" [38]

If all the other campaigns had been equally successful, the fund's future would have been assured. Unfortunately, not everyone had Hale's faith in the plan, or at least his magic touch in conveying his faith to others. Unfortunately, too, Hoover's interests were scattered in a thousand directions, and as far as Hale could tell, Hoover spent little time on the fund. Most of the other offers of contributions came in without his help. The American Telephone and Telegraph Company, inspired by Walter Gifford's interest, offered twice the amount Rosenwald offered, *if*—and it was a big if—the endowment could succeed in getting a total of $20,000,000, or $2,000,000 a year for ten years. All the promises were, in fact, contingent promises.

By 1927, the prospects of ultimate success were dimming. In July, Calvin Coolidge issued his famous pronouncement, "I do not choose to run." The way to Hoover's nomination for President of the United States was cleared. After this, as he became increasingly absorbed in politics, he seemed to forget the National Research Fund entirely. Finally Hale went to New York to see what he could do. He found Hoover angry over the delays in the fund campaign. "Hell!" he cried. "If you fellows don't do something right away there won't be any Fund." [39] He seemed to forget that, as chairman, he too had some responsibility. After seeing Hoover, Hale dashed over to see Carty, and "succeeded in stirring him up." He promised to do all he could to get the additional gifts needed to guarantee success. Hale, intensely relieved, returned to the club to bed. But it had been a strenuous day, and he commented sadly, "This is no life for me."

In January, 1928, an article by Hale had appeared in *Harper's*.[40] In it he dramatized the plight of the research worker—often an underpaid member of a college faculty—who, unless he could have a mind untroubled by financial worries, was in no position to make a contribution to scientific research. Soon after this, Carty took over Hoover's position as chairman. Yet already the financial structure in which Coolidge and Mellon had placed such implicit faith was tottering. The apparently inexhaustible supply of unsound money and easy credit had done its dangerous work. By 1929, the situation became desperate. In June, Carty wrote a long letter to the steel industries in a last attempt to save the fund. He described again the importance of pure research, and chose as an example the work on the constitution of matter at Mount Wilson. On its summit, he noted, astronomers were observing the process of evolution in the sun and stars and nebulae—"but what is most astonishing and most astonishing for the future of the steel industry, the evolution of the chemical elements themselves." He pointed out that in the research laboratories at the mountain's foot chemists and physicists were trying to reproduce, with increasing prospects of success, conditions observed for the first time in the depths of the universe. He even looked into the future toward a time when science might learn how to take the elements apart "and put them together in new and most remarkable combinations." Therefore he urged that research such as that at Mount Wilson be multiplied a thousand-fold in the hope of discovering new sources of energy and of using those already available. "No inquiries could go more deeply to the foundations of industrial progress than these, and none could hold forth such promise of continued prosperity and of such advancement in all human affairs." [41]

In answer to this fervent plea, the steel industries (not including U.S. Steel, which had already committed itself) promised $100,000 on the usual condition—that the total fund be subscribed.[42] By this time the National Electric Light Association had offered $300,000; and George Eastman had promised $100,000 a year for five years, on condition that the entire amount be obtained by January 1, 1930.[43]

In a last desperate attempt to overcome Eastman's condition, Hale appealed to Max Mason at the Rockefeller Foundation. "If we don't put over this scheme this autumn it will surely vanish into thin air." Already there was danger that many of the original offers would be rescinded. One was that of Rosenwald. If this should happen, Hale vowed that he himself, despite his "many weaknesses," would go out to find a donor to replace him.[44] By 1931, however, with the country in the depths of depression, the prospect of raising new funds, and so of saving the old ones, became increasingly remote.[45] In the end, Hale and Carty and all those who had worked to push the fund through had to admit defeat.

Yet the idea was good, and in later years industry itself would recognize the advantages of promoting basic research on a broad scale. In time, too, the government would assume a leading role in the support of pure research. But in 1929 the plan fell victim to the times as well as to the lack of vision of the majority of industrial leaders. Not until the mid 1930's did Hale give up hope entirely. But

CHAPTER EIGHTEEN

The Henry E. Huntington Library and Art Gallery

✳ *1906-1927*

ALL HIS LIFE Hale was a scientist first, but a scientist with a profound interest in the humanities, and a firm belief in the interrelationship of science and art. As a result one of his most important—and most difficult—jobs had little or nothing to do with science. Also, in contrast to so many of his other tasks, his basic goal in this instance was to prevail upon a millionaire to spend his fortune in the advancement of research in the humanities.

The millionaire was Henry Edwards Huntington. In 1906, the year of their first significant meeting, Huntington had been a member of the Southern California community for four years; already his electric-railway empire, with its famous red cars, was spreading octopus-like out of Los Angeles. He had been born in Oneonta, New York, in 1850 into a family of modest circumstances. He had begun his working life as a clerk in a country store. But in the mid-nineties his uncle Collis Huntington, the railroad tycoon, had sent him to San Francisco to take charge of the railroad-owned streetcar system.[1] "No better boy than Ed ever lived," was his uncle's pronouncement. From San Francisco he had moved on to Los Angeles, and before long he was profiting enormously not only from his traction interests but also from vast real-estate holdings that bordered his spreading trolley lines.

Long afterward, his landscape gardener, William Hertrich, was to describe some of the habits of this man with whom Hale was destined to deal over a long period. He was, in many ways, a man of austere tastes. He rose at six, breakfasted at seven, and in general led a simple, abstemious life. He was thrifty, sometimes even stingy. Hertrich tells the story of the evening he was playing cards with the second Mrs. Huntington. Suddenly he noticed a light burning unnecessarily down the hall. He left the game, walked down the hall, and switched off the light. On his return, Mrs. Huntington said, "Edward, I simply can't understand why you walk to the end of that long hall to save a few pennies, when the next minute you'll turn around and pay as much as ten thousand dollars for an old book." His answer was this, "My dear, if I did not save the pennies, I couldn't buy the ten thousand dollar book." [2]

Yet, at times, Huntington's spending could be on the most lavish scale. Like so many of his millionaire contemporaries, he had been convinced by the noted art dealer Joseph Duveen that he could best achieve immortality by means of a magnificent collection of paintings. Great paintings, Duveen explained, were hard to find, and therefore bound to be extremely costly. The more Duveen charged for a painting, the more valuable Huntington considered it.

Over the years Hale must have felt that he had some sort of affinity for traction magnates. Inevitably he must have compared Huntington with Yerkes. Both men had made their fortunes in streetcars. Both had used methods that were ruthless—in Yerkes's case, at least, positively dishonest. Both chose to lavish huge sums on the acquisition of great paintings. Yet Yerkes had a passionate appreciation of art. Doubtless his purchases were motivated by a desire to impress a society reluctant to accept him. Yet he loved beautiful paintings, as he loved beautiful women, and he bought what he himself enjoyed. In Huntington's case, prestige apparently dominated over love. After he had fallen under Duveen's spell, he followed the art dealer's advice unquestioningly.

Hale's first meeting with Huntington had occurred at the foot of the Mount Wilson trail soon after his arrival in California. [3] At that time their conversation dealt mainly with the question of building an electric railway up the mountain. There was little chance for other discussion. But a chance did come on October 3, 1906, at the Hotel Maryland in Pasadena. The city was playing host to the head of "one of the greatest, if not the greatest, trolley systems in the world under a single management," and the biggest and finest dinner possible had been planned. Myriads of lights gleamed in the banquet hall. The tables were lighted with candelabra. On the head table rode a floral trolley car, four feet long. Each guest also had a small tin trolley car, its roof shoved partly back, the interior filled with salted almonds. Even the ice-cream was in trolley-car form. [4]

Evelina was still in the sanitarium, and Hale went to the banquet alone. To his surprise he was seated next to Huntington at the head table. The talk soon turned to books and book collecting. Hale listened with interest as Huntington described his rare books and pictures that were now in his Fifth Avenue mansion in New York. Long afterward, in his autobiographical notes, Hale recalled that night. Because Huntington "was uncertain what ultimate disposition to make of them," he asked Hale's opinion. "The question was whether they should go after his death to some institution in New York or possibly to San Marino (adjoining Pasadena) where he was to build a large house." In turn, Hale told him something of his own plans "for developing a research center in Pasadena, and pointed out how admirably the creation of an exceptional library would harmonize with this scheme." As the talk went on, Huntington "listened with interest, but remained undecided." Hale commented, "Recognizing his independent character and his natural desire to choose for himself, I did not press the matter, but waited in the hope that another favorable opportunity might present itself." [5] That opportunity did not present itself for six years.

In the interim, in 1910, Huntington retired from business. After this, his interest in book collecting grew, and soon he was going at it in a wholesale manner. Before long he was startling the world by his purchase of three or four of the finest book collections in existence. In 1911 he bought the tremendous Church collection of early Americana and early Elizabethan literature for what was rumored to be over a million dollars. If he wanted any book or any collection badly enough, he was willing to pay any price for it.

Yet, as the years passed, Hale became impressed by Huntington's genuine love of books. "His affection for his books and manuscripts was so intense that I have often seen tears come into his eyes as he talked of them, and read aloud some favorite passage. . . ."

After that memorable dinner in 1906, Huntington and Hale may have met and talked occasionally, but it was not until the spring of 1912 that a new and different project gave Hale the opportunity he had been hoping for.

On an evening toward the end of May, 1912, a small and congenial group met for dinner at the house of his old friend Charles Holder on Bellefontaine Street. The group included Scherer, and James Culbertson of Throop, Theodore Coleman, a board member of the old Throop; Ernest Batchelder, the architect and artist; and Hale. After dinner, in Holder's study, over coffee and brandy they talked of Throop, then of the Coleman Chamber Music concerts that Alice Coleman had inaugurated some years earlier. After this, as the talk turned to the need for support of music and art in Pasadena, Hale suddenly exclaimed, "Why don't we found a Music and Art Association?" Everyone was enthusiastic, and another meeting was planned.[6]

This meeting was held on May 28th at Hermosa Vista. That night Hale saw that, as father of the idea, he was expected to implement it. Against his will he was elected president. The purpose of the association, as he put it in a resolution, was this: "To promote the interests of music, painting, sculpture, architecture and the kindred arts in Pasadena. It should foster and stimulate the cultural activities of the community and coordinate the work of individuals and groups interested in music and art."

On the following Saturday evening an organizational meeting was held, and a Board of Directors that included Henry Huntington was elected. Moreover, Huntington appeared at the meeting, accompanied by his daughter-in-law, Mrs. Howard Huntington, and his best friend and attorney, George Patton, who was manager of the Huntington Land and Improvement Company, in charge of the Huntington properties in Southern California. In addition to the original group, Dr. McBride and a few other leading lights were present.[7] After this whenever he was in Pasadena, the tall, handsome Huntington came to the meetings. His balding head with high forehead, his large nose and flowing moustache became a familiar sight at Hermosa Vista. Before long he was joined by his bride, Arabella Huntington.

In 1906, after thirty-three years of marriage, Henry Huntington and his first wife had been divorced. Seven years later, in an astonishing move, he married his Aunt Arabella, widow of Collis P. Huntington. Yet, if it was a curious match, it was a happy one. In 1914 they moved into the beautiful new mansion Huntington had built on the old Barth Shorb estate in San Marino, which bordered on Pasadena. Two months later they came together to the annual meeting of the Music and Art Association, to hear the lecture on "Art and Environment" that, as Hale said, he had the "temerity to give."

Even before this, apparently after an earlier meeting, Hale had the even greater temerity to write to Arabella Huntington. The letter was dated February 14, 1914.[8] Evidently he had decided that a good way to reach Huntington was through his bride. While married to Collis, she had filled their huge Fifth Avenue mansion with fabulous paintings—under Duveen's tutelage. More important, however, from Hale's viewpoint, her son Archer had founded the Hispanic Museum in New York. He was a scholar in his own right, with an interest in historical research, especially in the fields of Spanish and Arabic.

In his letter Hale spoke first of Archer's work, of which he was sure Mrs. Huntington was proud. Then he turned to his own international activities as foreign secretary of the National Academy, and mentioned his plans for a book on the advance of American scholarship in all fields—particularly those in which her son was prominent —and asked for additional information on his accomplishments.

The letter was a tactical masterpiece, and it led to a meeting with Archer himself in New York on April 16th. In his diary Hale notes for that day: Archer Huntington 12:30; Carnegie 7; H. E. Huntington 4:00. (This, incidentally, was the day of his memorable interview with Carnegie on the National Academy Building.) [9]

When Hale arrived at Archer Huntington's office at 15 West Eighty-first Street, he found a very tall, strikingly handsome, charming man with a rugged physique who well merited his mother's praises. He found, too, a poet as well as a scholar with a profound interest in history, literature, and the arts. Among other things Hale discussed the exhibit of comparative art he was planning for the Art Association. He wanted, he explained, to begin the exhibit with the cave dwellers and carry it through Egypt, Assyria, Crete, early Greece, and so on up through the ages. Now he was looking for an advisory board. Would Archer Huntington be willing to accept the position as consultant in Spanish and Moorish (perhaps all of Mohammedan) art? Archer agreed, and said he wished they could go on talking all day. "I see in him," said Hale, "a future ally of great power." [10]

At four o'clock that same afternoon, Hale dashed over to the Hispanic Museum to meet Henry Huntington, who had agreed to show off the museum. This arrangement gave Hale the unexpected chance to discuss again the idea he had broached eight years earlier at the dinner at the Hotel Maryland [11]—the establishing of a great research library in the humanities in Pasadena. As they were driving back from the museum, far uptown, Huntington announced that he had made a will leaving his San Marino estate and all his collections to the Board of Supervisors of Los Angeles County. Hale, distressed by what he considered a fantastic scheme, sat dumbfounded and said nothing. The next day, however, after lying awake the entire night, he decided to take his life in his hands, and warn Henry Huntington on the dangers of such political control.[12] In a long and deeply felt letter, he congratulated Huntington on his intention of making his books and pictures available for the public benefit, and "on the splendid prospects which your generous action offers to the development of art in America."

Yet, under the Board of Supervisors, with its inevitable political outlook, he could foresee only ultimate disaster. Therefore, as he thought of the potential dangers in Huntington's plan, the ideas he had long been turning over in his mind came pouring out. He proposed instead "a body of trustees, selected because of their knowledge, taste and experience." This, he noted "would seem to me preferable to the Board of Supervisors." Then, perhaps more on impulse than as a result of reflection, he suggested that such a body of trustees already existed in the Pasadena Music and Art Association. To the Executive Board of this body, Mrs. Huntington and Archer could be added, so that H.E. would be assured that control was in safe hands.[13]

Then he turned to the details of his plan. It was not enough, as Huntington had suggested, to educate the public. It was just as important, if not more so, to establish a center in the humanities where scholars from home and abroad could come for study and research. "The books would then serve not merely as rare and interesting curiosities, glanced at by many, but appreciated by few, but as a continual source of literary and historical study, which would make them known and valued by thousands otherwise ignorant of their existence." At the end of a long letter Hale concluded diplomatically, "I am sure you realize that my object is not to interfere with any plans of your own, but to offer suggestions that may not have occurred to you." Huntington's answer came three days later, "Some of your suggestions are most excellent and I will take them under consideration." [14]

Three weeks later, back in Pasadena, unable to contain his enthusiasm over this glorious scheme, Hale wrote again. He marked his letter *Personal*.[15] "The powerful attractions of your pictures and library have fired my imagination, and set in motion a new train of ideas. I cannot help feeling that with such rich and uniquely valuable material, it would be a very easy matter to make your collection of real international importance, without greater expenditure than you may already contemplate."

Then he described his own experience in starting international projects: the founding of the *Astrophysical Journal;* the organization of the International Union for Cooperation in Solar Research, "now the leading international body in astronomy." It was easy, he noted, to gain such cooperation, provided it offered the chance "for attractive and important work in which men of all nations could take part." Therefore: "It would be easy to organize an advisory international board of the greatest living scholars and artists whose advice and assistance could be obtained whenever required."

He called Huntington's library the richest in the world, his painting collection one of the finest. Yet he wondered if methods of increasing the attraction of these collections should not be considered. His ideas of possible methods then came flooding out—suggestions for a school of art and architecture, for buildings that would be "absolutely unique specimens of art, so perfect as to command the attention of artists and scholars everywhere." He thought of his recent journey to Greece in the winter of 1913. And the vision of a Greek temple, perfect in every detail, built of Pentellic marble, rising on the Huntington grounds, shone in his mind's eye. Yet, however beautiful, this was not enough. In the Huntington of the future, he saw the possibility of a research library without par—a place to which scholars would flock in droves.

Afterward Hale sent a copy of his letter to James Breasted.[16] Breasted, impressed by his friend's flaming interest in research in the humanities, answered: "The great service which you have rendered natural science and the discoveries you have made have always deeply impressed my imagination. I find myself wondering whether you are not on the threshold of a new career which will render even greater service to research, both in natural science and in the vast field of humanities—a service such as no administrator of scientific research in any land or any age has ever yet been able to render. It is a most inspiring vision, and I am loath to believe that it is merely a dream." [17]

Yet Breasted felt these ideas would require repetition before Hale succeeded in "landing the virus" under Huntington's skin: "But once you have roused his imagination the vision of further possibilities seem boundless. I do hope," he concluded prophetically, "that this dream of an ideal project on the Pacific may be but the initial chapter in the spiritual development of the Pacific coast and that this letter of yours may some day be regarded as the historic germ out of which it has all grown." [18]

Indeed, repetition, or something stronger, apparently was needed; no answer to Hale's second letter came. Huntington, he knew, had gone to his newly acquired château in France, but this could not explain a silence of months. He feared that Huntington had considered his suggestions presumptuous.

Hale's letter had gone out in May, 1914. It was not until the beginning of October that an answer finally came. It was brief, but promising. "Your letter of May 11th reached me as I was sailing for Europe and during the summer I have given the suggestions some thought. I am not ready to reply, but it is quite possible that you have planted a seed." [19]

"*It is quite possible that you have planted a seed.*" In those few words Hale saw glowing promise.

After this, Hale made a point of seeing Huntington whenever he got the chance. In New York that December he went to call.[20] He found Huntington surrounded by his books, many of which he proudly showed to his visitor. Yet, in the months that followed, Hale continued to be perplexed. What could he do to convince Huntington of the importance of his research plan?

He talked with everyone whose opinion might possibly have weight. One of these was Henry M. Robinson, a director of the Art Association and member of the Throop Board of Trustees. Another was George Patton, the attorney, whose influence on Huntington was great. Now, as the long campaign accelerated, Patton was to prove a strong ally.

In 1916, while Hale waited for a chance to talk with Huntington in person, he drew up in rough outline, but with the infinite care for detail that characterized all his planning, a scheme for what he now called the "Pasadena Institute for Arts, Letters and Science" (with the name "Huntington" ultimately to be substituted for that of "Pasadena"). Again he elaborated his ideas on the character of the governing board and on the nonresident associate fellows. He described once more his basic ideas for an organization that would include research departments in science as well as in arts and letters, in which creative work of a high order could be carried on. "I think," he exclaimed in a letter to Henry Robinson, "some such plan, which would be wholly beyond the County Authorities, might appeal to Mr. Huntington." [21]

In March, 1916, the elusive Huntington (who had been ill in New York) finally appeared. He agreed to see Hale in San Marino. After waiting so long, Hale started out apprehensively. But he was delighted to find Huntington in good form and unusually interested in his plans. Huntington promised to consider them in detail. He examined with interest the Year Book of the Carnegie Institution, and noted especially the strength and diversity of its Board of Trustees. As Hale talked, he felt that Mrs. Huntington, who was also present, was sympathetic. He went home elated.[22]

Ten years had passed since the dinner at the Hotel Maryland. Now, at last, Huntington showed signs of real interest in the idea for a research library. Hale was, of course, eager to pin the matter down. Moreover, Huntington was, as he said himself, an old man with little time left. Therefore Hale hoped Huntington would share his own sense of urgency. "If the library could be built, and the institution definitely established, you could plan and work out the organization to your own satisfaction during your lifetime, and enjoy the pleasure of accelerating very largely the intellectual development." [23]

After a long delay Huntington finally replied to a letter in which Hale had enclosed the outline for what he was now calling "A Concrete Plan." "The mode of organization," Huntington said, "is in line with my ideas." Then he added, "It will probably be some time (at least until times are better) before I take up this matter actively, as you of course can understand." [24]

Even if Hale could not understand, there was little he could do about it! Yet barely a week passed before he was off on a new track. On a trip to New York, he called on Archer Huntington, who proved as charming as ever. He called Hale's scheme superb, and promised to do everything he could to get his stepfather to carry it out. He even approved the suggestion for a department of Egyptology with Breasted at its head. "Banzai!" Hale exclaimed. "It is a pity he cannot talk with Henry Huntington at once." [25]

Soon, however, these new hopes were dashed. Archer had evidently changed his outlook in a way that Hale found quite baffling.[26] Later, Ernest Batchelder said that Archer was violently opposed to moving the collection to California.[27] But why, then, Hale wondered, had he not said so before?

This was in the spring of 1916. By this time Hale was in Washington, absorbed in his plans for the Research Council. Therefore he had little time to brood over this latest development. Yet he did not forget his dream. Even before the war's end, he had a brief meeting with Henry Huntington, in March, 1918. That June, back in Pasadena, he took time to address the Music and Art Association on "The Huntington Library." He described, from his own experience, the pleasures of book hunting, which had carried him into all sorts of out-of-the-way places—a sport that in its seeking for solitary pleasure he compared to angling. He spoke of Huntington's rise to power in the book world, his luck in obtaining the superlative rarities in the Church, the Hoe, and the Huth collections—"the crown jewels of English literature." He described him as the owner of the "finest private library that had ever been brought together." Afterward he sent Huntington a copy of the talk.[28]

Some time after this he learned that Huntington was going ahead with his library plans and had chosen Myron Hunt as architect. Even with this welcome news, however, Hale was doubtful that his larger scheme would ever materialize.

Then, one day late in August, 1919, when he was sitting on the veranda at Hermosa Vista, he was surprised to see the tall, white-moustached Huntington coming up the path, leaning on his cane. Without any preliminaries, Huntington announced that he was forming a Board of Trustees for the Huntington Library, and wanted to know if Hale would serve. (With this exception, the board was to be made up of Huntington's family and close associates.) An organizational meeting was planned for the following week. At that time, on August 30, 1919, a trust indenture was drawn up and signed by the board members. "I am looking forward to some very interesting developments," Hale noted. Yet, if the outlook was promising, he was still fearful of Huntington's real aims.

After this, nothing happened for many months. During the gloomy depression days of 1921, Huntington, objecting to the high costs of construction, stopped work on the library.[29] Early in February, 1922, Hale decided to call. He found Huntington eagerly awaiting the arrival of his latest treasure, the "Blue Boy." [30] All that afternoon Huntington talked of his latest acquisitions, which had been increasing at a fantastic rate. And his rate of buying pictures was matched by his purchase of high-priced rare books and incunabula from the famous book dealer A. S. W. Rosenbach. The more he heard, the more worried Hale became. There was as yet no sign of any adequate endowment for this growing collection or for the unfinished library. Under Duveen's ingenious prodding, and under Rosenbach's equally forceful tactics, Huntington was obviously dispensing his fortune at an appalling rate. Yet, at the moment, there was nothing Hale could do to divert Huntington's thoughts into other channels.

Hale therefore turned to an entirely different project, which nevertheless was related to the cultural growth of Pasadena and his dreams for this "Athens of the West." It was also a natural result of his lifelong interest in city planning that grew out of his father's friendship with D. H. Burnham, the architect, who had died in 1912. In 1921, Mrs. Burnham had sent him a copy of the life of her husband by Charles Moore. As he read, Hale thought of the Rookery that Burnham had designed for his father. He recalled his visits to Burnham's drafting rooms in Chicago where he had pored over the blueprints for the Columbian Exposition. He thought of the happy days he had spent with Burnham exploring Paris in 1910. He thought, too, of the city plans for Washington and Chicago on which the architect had worked, and he began to dream of a similar plan for Pasadena.

At this time Hiram Wadsworth, a man of vision, was mayor of Pasadena. Hale lent him the Burnham biography, and afterwards, as he talked about a systematic city plan, Wadsworth quickly saw the possibilities. He agreed that to make the plan a success, it must be directed by those who not only understood engineering needs but who also had artistic vision. He agreed, too, on the need for a city planning commission to study Pasadena's needs for highways, parks, and public buildings. Soon afterward they met with the City Directors in the dining room of the Hotel Green. Different members spoke on the building needs of the city—library, art museum, and so on. Of those who spoke, Ernest Batchelder (a leading advocate of the plan) said that Hale was the most eloquent. "He gave a wonderful talk that night." There were many there who had never seen Hale before; "all were profoundly impressed." [31] When B. O. Kendall urged that the city plan go north and south on Raymond Avenue, and suggested that the civic center be built there, Hale countered by saying that this plan was too small. Never, he emphasized characteristically, do a small thing when you can carry out a big one. Had not Burnham also said, "Make no small plans"?

After this, the plans grew rapidly. Hale, entranced by this latest project, spent all his spare time drawing plans for a civic center around which he visualized the city's growth. On one side he pictured the city hall; on another, a new library. On the third side he saw a great auditorium rising; and on the fourth he envisioned an art museum. With a rough sketch of these ideas in hand, he set out for Chicago to see Edward Bennett of the firm of Bennett and Parsons, who for eight years had been closely associated with Burnham.

Soon the architects arrived in Pasadena. They were flown over the city. Then they drew up a beautiful set of blueprints. Not a word had been said in the papers; nothing had been published about the plans. There were, said Batchelder, not a dozen people who knew about them. Then, in a blaze of glory, a bond issue to cover the cost of the entire plan was launched. It "went off with a bang." This, Batchelder recalled, was typical of George Hale's way of doing things. He did not like to "dribble things out."

With the plan approved, it was decided to have a contest for the design of the city hall, the auditorium, and the library. The jury included three architects and two laymen—George Hale and Ernest Batchelder. The architect Fitch Haskell was selected to design the auditorium; Myron Hunt was given the library. In the case of the city hall, however, the jury "somehow did not have the final say-so." The least interesting design was that of Bakewell and Brown of San Francisco. Yet in the end this design, with some modification, was chosen.[32]

Then, just as they got their plans under way, an unfortunate event occurred. The Board of City Directors changed, and a group of an entirely different caliber came into power. They got in on a plea of lower taxes. For the first time, Hale and Walter Adams, now the able chairman of the Library Board, found themselves part of a political battle, a new and sometimes amusing, if frustrating, experience.

Under the new administration the building plans suffered. Thus the "gold room" for the auditorium, as a result of the efforts to economize, became a room almost useless for the musical purposes for which it had originally been planned. The size of the stage was so drastically cut that there was barely room for a piano.

Another building that failed to materialize was the art museum. Stuart French, chairman of the Planning Commission, designed the layout for such a museum on the "Carmelita site" at the western end of Pasadena at the head of Orange Grove Avenue. It included Hale's dreams for exhibit halls, a library, and a lecture hall; it also had a place for a school of art and design and a group of studios. There was even an open-air auditorium for concerts. To his romantic imagination these plans seemed essential to the overall plan for the cultural growth of Pasadena. He continued to hope they would be carried out.

One day in 1923, while he was deep in these plans, a letter arrived from his good friend Dr. Robert Freeman, the minister of the Pasadena Presbyterian Church. Freeman had been at a meeting of the Kiwanis Club where the City Plan was being discussed. At one point, he said, in a way that warmed the heart of the recipient, the address became a "eulogium" of George Ellery Hale. "We are all human enough, under the crust," he wrote, "to be pleased at the recognition of our endeavors; and it would have done your heart good to note how these successful young business men responded to references made to your great part in our civic affairs. Because you have been compelled to drop out of things social, and because your instinct is to dodge the limelight, not a few of them may never have met you. But the Mount Wilson Observatory is a synonym of your name; the development of the Institute of Technology is known to have been your dream; the Huntington Library is credited to you, though it is given by the millionaire; Carmelita they would be willing to name Evelinita, if it would please you; and now the city plan is conceived when it is still feasible, because that old head of yours which you want to put in a junk shop has worked overtime instead of loafing. All that, just from the pinched viewpoint of dwellers in a little city on the Pacific Coast, and without regard to your national contribution in the great crisis, or your larger service for the advancement of truth and good will among nations.

"Do you wonder that many of us, who have the sense to do so, envy you what you have accomplished. God bless you, old man, you have a bigger place in many hearts than you will ever credit yourself with having, even though you yourself know you have a goodly number of friends." [33]

At this point, however, Hale's dreams for the Huntington were still far from realization. At every opportunity he continued to push his cause. On July 1, 1924, he found himself on the magnificent Huntington estate in San Marino. The occasion was a trustees' meeting at which Huntington transferred to his library and art gallery all the books and pictures he had bought during the last two years. Their worth, he announced proudly, was over $4,100,000. This was fine news, but Hale was worried about the future. "I hardly think his endowment will be on a sufficient scale to keep up expenditures of two million a year." [34]

Two months later Arabella Huntington died. Huntington himself was suffering from cancer, and this loss, Hale knew, would affect him deeply. If he should die without making adequate endowment provision, Hale feared his entire scheme for a research library would be lost. As months passed without action, his fears increased.

During his crowded years of diverse activity, Hale had tackled many different problems. Each time he thought, as he thought now, that he had never faced a project more difficult or more desirable.

He spent hours talking with Patton and Robinson. Yet still nothing happened. In April, 1925, he wrote an article for *Scribner's* on "The Future Development of the Huntington Library and Art Gallery," hoping to stir Huntington up. That summer, in Santa Barbara, he even worked out a new plan for the library.[35] In its essentials it had changed remarkably little over the years. But, as he had recognized the goals Huntington was willing to contemplate even remotely, it had become more limited in scope. He wrote the plan between some terrible bouts with his "blasted head." Soon after this, Huntington revised his deed of gift so that the library was now called "a free public *research* Library; and he revised the trust deed so that it provided for all kinds of research." Still, as Hale confided to Breasted, "I do not think that Mr. Huntington realizes as yet all that I have in mind." [36]

On October 13, 1925, after reworking his plan once more, Hale sent it to Patton in the hope of getting Huntington's final approval and so of getting them "out of danger." As it finally stood, this charter for the future, the result of years of thought, was an extraordinary document, remarkable for its vision, and so far-reaching that years later many of its provisions had not yet been realized. Two days later Huntington approved the entire scheme.[37] Hale, of course, heard this long-desired news with the greatest glee.

The road to victory had been a long, uphill one with many turns and windings. Huntington, a man of independent mind, had never relished suggestions. He had considered himself quite capable of making his own decisions. During these long years Hale had often become discouraged. But he too was stubborn, and as long as there was the slightest possibility of success, he would never willingly give up.

Still, unsolved problems remained. Huntington had formally approved the research policy, but as long as he conducted the library himself, current policy would no doubt continue.[38] Moreover, there was as yet no definite plan for endowment.

In January, 1926, Max Farrand, Huntington's first choice for director, arrived in Pasadena. Farrand had taught history at Yale. He called Hale's plan "wise and far-sighted," as well as essential and inevitable.[39] By this time, Huntington had undergone a serious operation in New York. Seventy-six years old, he was impatient to get his affairs in order and to see the research policy in operation. By this time, too, other trustees were sharing Hale's fear that Huntington did not yet realize the cost of organizing and running a great research library.

Hale, becoming more and more agitated, urged that something be done before it was too late. Duveen and Rosenbach, who also knew they were racing against time, continued to spend Huntington's money at a whirlwind pace, buying incunabula and pictures, most of which, Hale mourned, "will be of little or no use in our scheme of research." [40] It was tough competition.

After numerous delays, Farrand's study of the collections was completed. Hale sent the study to Elihu Root, hoping his endorsement would carry weight with Huntington. "Library at stake," he wired.[41]

On January 12, 1927, Root's wholehearted endorsement of the plans arrived by telegram. "Think them admirable and full of promise for incalculable usefulness. . . . But do not underestimate the magnitude of the task." He described his work as trustee of the Metropolitan Museum, the New York Public Library, and the Carnegie Institution. In every case "the funds which at first seemed ample, proved too small to do the work properly. . . ."

Hale, overjoyed, replied exuberantly: "Unless I mistake the character of Mr. Huntington, he will not be content to have his Library placed in the inferior class. His difficulty has been to pass from the point of view of the collector of rare and unique objects to that of a patron of research. Most collectors, I believe could not cross this chasm, and I think Mr. Huntington has showed decided breadth in accepting the modifications of his original plan. . . . Whether he has fully realized the necessity for a very large endowment still remains to be seen." [42]

The crucial meeting with Huntington was to have been held on January 14, 1927. Hale was counting on Root's telegram to tip the scales. Because of Huntington's illness, the meeting had to be postponed for a few days.[43] At that time Huntington approved Farrand's plan and promised immediate financial support. On February 28th the news was released to the press.[44] The Huntington Library was called "a great research laboratory."

Letters of congratulation poured in on the original promoter. His brother Will wrote, "You have now founded five great institutions with large endowments. . . . If you had gone into business I am sure you would surpass Henry or John D. by now!" [45]

Three months later Hale was on his way back from New York. On the train, the night before he was to arrive in Pasadena, he heard the news he had so long expected and dreaded. Huntington had undergone an operation he considered minor and had not survived. When he died, some of his plans were incomplete. He had left at least ten millions for endowment. If he had lived, Hale felt he would have given at least five million more.[46] Yet how wonderful it was that all the strenuous work had paid off and guaranteed the Huntington Library's future as a magnificent research center in the humanities and a "scholar's paradise."

Twenty-one years had passed since Hale and Huntington had first discussed books over the toy trolley cars at the Hotel Maryland.

CHAPTER NINETEEN

The Birth of the 200-Inch Telescope

✳ *1921-1931*

IN A POOL of green slime tiny infusoria swam. Through his microscope the boy of fourteen gazed at them in wonder. Years later, on a California mountaintop, the boy, now grown, gazed through the great 100-inch telescope at a distant green nebula floating in a deep blue sky. The sense of wonder, born "of an inborn teasing imagination," remained. The insatiable desire to try new experiments persisted.

George Hale, the man, felt he would have been quite happy to pursue those microscopic bodies all his life. The marvels were as astonishing as anything revealed by his largest telescopes. Yet, with his first glance through a spectroscope—that magic key to the composition of the universe—his life's course had been changed. With the telescope it had led him far from the earth. In his search for "more light," it had carried him far beyond his "typical star," the sun, to the most distant nebulae. In turn this search had inspired him to devise new instruments and methods and to become a beggar in order to build one great observatory after another. The 100-inch, the culmination of a long line of telescopes, would have been achievement enough for most astronomers. But Hale, as always, was not satisfied. "Starlight is falling on every square mile of the earth's surface and the best we can do is to gather up and concentrate the rays that strike an area 100 inches in diameter." [1] He began to dream of a still larger telescope that could penetrate more deeply distant and unexplored realms of the universe. He spoke of his dream to Francis Pease, who had come from Yerkes to join the staff soon after the founding of Mount Wilson. Pease, carried away by the possibilities, began to put some ideas on paper. He had taken his first magnificent photographs of the moon with the 100-inch in 1919. Two years later he was designing a 300-inch.

In March, 1921, Hale was surprised to receive a letter from the Philadelphia surgeon W. W. Keen, in which he exclaimed, "If I were rich enough, you should have a 200″ spy glass." [2] Two years passed. Then, in 1923, Hale wrote an article for *Popular Astronomy* on "The Possibilities of Instrumental Development." [3] In it he wrote, "Looking ahead and speculating on the possibilities of future instruments it may be mentioned that comparative tests of the 60-inch and 100-inch promise well for larger apertures." Yet, when three years later Pease suggested that anything up to a hundred feet could be built, Hale could not share his extravagant optimism. He knew the hazards, as well as the financial difficulties, in the way of building a larger telescope. He told Keen, "Alas, . . . there is no money in sight to pay for it." [4]

Yet, just a week later, on October 1, 1926, H. J. Thorkelson of the General Education Board of the Rockefeller Foundation arrived in Pasadena. After a tour of the Solar Laboratory with Hale he went up the mountain with Walter Adams. There Pease brought out his blueprints for a 300-inch telescope. This was, as Thorkelson noted afterward, a proposal so striking "that I think the astronomers detected my excitement and also my dismay at the probable cost." [5] He reported the incident to Wickliffe Rose, of the International Education Board of the Rockefeller Foundation, and Rose replied immediately, 'This is very interesting. We should watch closely for any possible developments." In this unexpected way another seed was planted. But at the time neither Hale nor Adams could have any knowledge of that planting.

The following year a letter arrived from the editor of *Harper's Magazine*. In it he asked for an article on some astronomical topic. Hale chose "The Possibilities of Large Telescopes." "Like buried treasures," he wrote, "the outposts of the universe have beckoned to the adventurous from immemorial times. Princes and potentates, political or industrial, equally with men of science, have felt the lure of uncharted seas of space, and through their provision of instrumental means the sphere of exploration has rapidly widened. If the cost of gathering celestial treasure exceeds that of searching for the buried chests of a Morgan or a Flint, the expectation of rich return is surely greater and the route no less attractive. . . . Each expedition into remoter space has made new discoveries and brought back permanent additions to our knowledge of the heavens. The latest explorers have worked beyond the boundaries of the Milky Way in the realm of spiral 'island universes,' the first of which lies a million light years from the earth while the farthest is immeasurably remote. As yet we can barely discern a few of the countless suns in the nearest of these spiral systems and begin to trace their resemblance with the stars in the coils of the Milky Way. While much progress has been made, the greatest possibilities still lie in the future." [6]

Here Hale looked back on the development of the telescope, the photographic plate and the spectroscope, and described "the extraordinary harvest" that had been reaped from the use of these and other instruments. But much as had been gained, an even greater future lay ahead "in dealing with the most vital problems of the nature and evolution of stars and the structure of the universe."

The results with the 100-inch, as he had often shown, and as he now showed again, had surpassed all forecasts. These results were indeed a rich treasure trove and a lure to any astronomer to extend the range of exploration into farther space with a much larger telescope.

Hoping, therefore, that some wealthy man might be found to realize the astronomers' dreams, he wrote, "Lick, Yerkes, Hooker, and Carnegie have passed on, but the opportunity remains for some other donor to advance knowledge and to satisfy his own curiosity regarding the nature of the universe and the problems of its unexplored depths."

Early in 1928, a proof of his article arrived from *Harper's*. Hale asked the editor to send a copy to Wickliffe Rose at the Rockefeller Foundation. He also wrote to Rose himself, emphasizing the progress of recent research and the importance of a large telescope to future advances in physics and astronomy.[7] "In fact, the range of celestial temperatures, densities, masses and states of matter so enormously transcends that of the physical laboratory that many of the most fundamental advances in physics depend upon the utilization of these conditions." He concluded his long letter by asking if the General Education Board would consider making a grant to determine how large a mirror it would be feasible to cast. Significantly, he did not ask for the telescope itself.

Rose replied immediately, "It is a matter that interests us. We should be very glad to discuss it with you." [8] A few weeks later Hale went east. On the 14th of March he called on Rose. Rose listened spellbound as his visitor described the glowing possibilities as well as the hazards of building a 200-inch telescope. Because of the turbulence of the earth's atmosphere "which envelops us like an immense ocean, agitated to its lowest depths," the building of a larger telescope, much more sensitive to these effects than a smaller one, remained a lottery.

This last, however, did not seem to disturb Rose, who asked eagerly how much such a telescope would cost. What were the chances of success? What about a 300-inch telescope? Hale pointed out why he considered the jump from 100 to 300 inches too great. It would be safer, he said, to aim at a 200-inch.

After this, the conversation took an unexpected turn. The question that Rose asked, to Hale's surprise, was this, "To whom should the telescope be given?" To Hale the answer seemed obvious. The greatest observatory in the world had been created at Mount Wilson under the Carnegie Institution of Washington. He naturally felt that the new telescope should also be under its jurisdiction. Rose, however, disagreed. The telescope, he insisted, should go to an educational institution, probably the California Institute of Technology. As Hale soon discovered, Rose believed an educational institution was more secure, its policy more permanent, its effect on the community more lasting than that of a research organization. Paradoxically, however, he insisted that the Mount Wilson staff cooperate completely by carrying on the research of the observatory, as a prerequisite for the donation of the telescope to the California Institute.[9]

Yet, if Rose and Hale did not agree on this crucial point, they agreed completely on everything else. Indeed, Hale was astonished by Rose's enthusiasm. "Rose," he said, "is not only very enthusiastic about a very large telescope, but speaks quite seriously of spending as much as fifteen millions for one. . . ."[10] Afterward he wrote to R. G. Aitken at Lick, "An article of mine on large telescopes, shot like an arrow into the blue, seems to have hit a 200″ reflector, and I have been forced to take to the air myself and try to parachute it to earth."[11]

As it happened, he was to spend some time in the air, trying to land his prize. As a result, his quiet life in the Solar Laboratory was destined to be seriously disrupted. But he relished the challenge, and he enjoyed working with Rose, of whom Raymond Fosdick was to write: "He had a bold, fresh mind, with daring imagination and originality, combined with a practical sense of the immediately obtainable. It was adventure, in terms of creative work, that tempted him." [12] He had a favorite saying, "Make the peaks higher." To Rose, as to Hale, the "master builder" (as George Gray so aptly called him), this new project was the kind of challenge both men liked best.

All too soon, however, Hale was to run into the first serious hurdle—in the form of the president of the Carnegie Institution, John C. Merriam. Immediately after his interview with Rose, Hale had tried to get in touch with Merriam. But Merriam, in Mexico, could not be reached. On April 3rd the president, back in Washington, found a copy of the *Harper's* article on his desk. He wrote to Hale, "This is an extremely interesting statement and I have had much pleasure in reading it. . . . Sometime when we have the opportunity I hope that we may touch on this subject in the course of conversation." [13]

This, of course, was the opening Hale had been seeking. But he wanted the "conversation" immediately—not "sometime." Rose was about to retire; action was vital.

Meanwhile Rose, who had been in California with Thorkelson, had arrived back in New York. Enthusiastic over their visit to Mount Wilson, he told Hale he had decided to ask his board for $6,000,000 to build a 200-inch telescope. It would go, he said, to the California Institute.[14]

At this point Hale was hopeful but cautious. "It may all blow up," he said. Rose was sure of success.[15] Yet, before it could become a reality, he had to convince the Rockefeller board of the telescope's importance. Hale had to prove to Merriam and the Carnegie board the advisability of cooperating in Rose's plan. Neither task would be easy; of the two, Hale guessed that his would prove the hardest.

On April 12th, after some hesitation, Merriam agreed to a meeting at the University Club in New York. Nearly a month had passed since Hale's first meeting with Rose. Yet he had had no chance to tell Merriam of Rose's interest or of his desire to give the telescope, not to the Carnegie Institution, but to the California Institute. "It was," as he told Evelina, "a curious task." [16] To his surprise and relief, Merriam, though reluctant, apparently accepted the plan.

On April 16th, in a letter requested by Rose, Hale summed up the arguments for a 200-inch.[17] He listed the outstanding questions waiting to be solved. He outlined the problems from an engineering and optical standpoint, and suggested possible procedure in the building of the telescope—the making of the mirror, the building of a mounting, the choosing of a site. He concluded with a request for $6,000,000. The next day, after another interview with Rose, he left for Pasadena.

Two days after his arrival there, a telegram came from Arthur Noyes, who had been in New York with Millikan, to discuss the plans with Rose. "Sorry to communicate sad news but you should know it promptly. Met Rose today in Cosmos Club. Said he had just talked with Merriam who said he considered expenditure for large instrument at present not justified. Rose said consequently he would drop the matter till all concerned were in agreement. Asked me to tell you when I reached Pasadena." [18]

Hale read the telegram in disbelief. Merriam's actions seemed totally incomprehensible, especially in the light of his apparent agreement to the plan twelve days earlier. Later, Hale learned of his resentment of Rose's plan, which was reflected in his report to Henry Pritchett, a Carnegie trustee: "As I understand Dr. Rose he believes that Mt. Wilson Observatory is eminently fitted to do the work of planning, constructing and operating a two-hundred-inch telescope, but he believes the organization and policy of the Institution inadequate to give proper guarantee for the utilization of such an instrument or the organization of its work in the future. . . ." [19]

Because Merriam could not understand Rose's outlook, he could not accept it. It would be better, he felt, to give up the telescope entirely than to deviate from his principles and beliefs. He argued that the California Institute was not suited in any way to run a great observatory. It did not have equipment, either in staff or in administrative facilities, to develop such a program, "excepting such facilities as may be offered by the Carnegie Institution." It had, moreover, never dreamed of owning such a telescope, or even of having a department of astrophysics. The building of a 200-inch telescope, he declared, was inevitable. But the means now proposed did not warrant the end, and the scheme was not so urgent that "great plans and programs should be thrown aside to bring the plan to instant realization."

Doubtless, because of his sensitivity, the fact that the plan had been worked out without his knowledge was a factor in his violent opposition. Whatever the reason, his verdict to Rose at the Cosmos Club was final: "It would not be in harmony with the plans for the development of the Carnegie Institution to have this gift made to the California Institute." As a result, Rose, who knew that the project would be impossible without the cooperation of Mount Wilson and the backing of the Carnegie Institution, returned to New York convinced the project was dead.

As soon as Noyes's telegram arrived in Pasadena, Hale went to see Adams in great anguish. He called in Henry Robinson, chairman of the California Institute trustees and a wealthy man with whom he had worked on the Huntington Library. All agreed that drastic steps must be taken to save the project. Hale telegraphed to Merriam: "Have received report conversation between you and Rose which Robinson, Adams and I feel must be distorted. Nevertheless am informed it is likely effectively to prevent carrying out greatest project in history of astronomy previously practically assured. We have asked Carty to talk with you as he is in immediate contact by telephone." [20]

That afternoon Hale telephoned to J. J. Carty, who agreed to talk with both Merriam and Rose. Hale also wrote to Elihu Root, chairman of the Carnegie trustees. "At the present moment an emergency exists on which I need your advice more than ever before in the past. It is impossible to explain the whole matter by letter, and I did not feel while in New York that I ought to trouble you with the tentative development that preceded the extraordinary and wholly unexpected news from Washington yesterday." [21]

The following day, after a long telephone talk with Merriam and an interview with Rose, Carty telegraphed encouragingly, "I believe we have encountered merely a temporary setback." [22] The next day, however, he begged Hale to come east immediately.[23] That night, the 29th of April, 1928, Hale packed his bag and left on the Santa Fe Chief for Chicago. By this time he was beside himself. He saw his hopes shattered, his dreams vanishing, the $6,000,000 disappearing forever. If this magnificent opportunity were missed, he could not tell when, if ever, another might appear. On the train east he scribbled notes and queries. "If he [Merriam] intended not to cooperate, why did he not tell me so when I described to him, at dinner with Noyes at the University Club, on April 12, my talks with Rose and my desire to work out a complete plan of cooperation with him and Adams when he comes to Pasadena in June?" [24]

These and other things he could not, and never would, understand. They stemmed from an irreconcilable difference in the nature of the two men. Merriam, cautious in outlook, often elliptical in his thinking, believed everything would work out in time. Someday there would be the southern expedition for which Hale had pleaded for years. Some day there would be a 200-inch telescope. He could see no reason for haste.

Hale, on the other hand, was impatient and direct: "The general interests of scientific research are far more important than the reputation of any single institution. . . . The accomplishment of the work is the main thing, the other question after all is an incidental one." The object was "to get the telescope."

On his journey east, he debated whether he should go first to Washington to talk directly with Merriam or to New York to discuss the problem with Root and Carty. From Chicago he called Carty, who told him to come directly to New York. "Apparently," he said gloomily, "Merriam's opposition has spoiled our case." After a meeting of the Rockefeller board, Rose had called off a lunch with Pritchett and Carty. No one knew exactly what had happened. Rose said only that the whole 200-inch plan had been abandoned.[25] Carty surmised that John D., Jr., might have been present and insisted on avoiding any conflict with the Carnegie Institution. Nevertheless Hale noted in his diary, "Carty refuses to give up and agrees that we should push the fight to the limit."

The following morning Hale arrived in New York on the Twentieth Century. The next day he met with Rose. Rose finally agreed that, if all hands, including Merriam, could agree, the proposal might be revived. The following day Hale, joined by Breasted, called on Simon Flexner, a member of the Rockefeller board. Flexner also advised revival of the project "as soon as Merriam's conversion has been effected." Then, speaking of Merriam's actions, he exclaimed, "Wasn't that the damndest thing!"

At the same time a telegram arrived from Noyes and Millikan, urging the need for conciliating Merriam. "To save his face must propose some modification." [26]

On May 5th, Hale went with Carty to see Elihu Root. Root, now eighty-three, was suffering from angina, and Hale had hesitated about going to him. He felt, however, that the situation warranted the risk. As they came in, Root rose to greet them, standing as erect as a "young West Point graduate." His eyes, looking out from under the silver bang that ran straight across his forehead, were as keen as ever. As always, he was dressed in a pepper-and-salt suit with flatlapelled waistcoat, and underneath, the soft blue shirt with polka-dot tie. And, as always, he wore his high black shoes, mark of his generation.

As Hale told him what had happened, Root leaned back in his chair and puffed on his familiar cigar. He said little; occasionally he nodded agreement. At the end he called the project wonderful, and said the Carnegie Institution should offer complete cooperation. He also confirmed in strong language what Rose had admitted—that, because of the old personal rivalries of Carnegie and Rockefeller, it was absurd to imagine that the Rockefeller board would make such a gift to the Carnegie Institution. He depicted them graphically—"Carnegie as always on the warpath, armed with a tomahawk, and Rockefeller as smooth and compromising in most of his dealings." At the end Root commented with a smile that the Carnegie Institution would furnish the brains while the Rockefellers provided the funds. He promised to do all he could to help.[27]

That night Carty and Hale drafted a new letter to the Rockefeller Foundation, offering the complete cooperation of the Carnegie Institution. Root signed it and sent it to Merriam by way of the pocket of Henry Pritchett, a trustee of the institution. If Merriam refused to sign, Root asked that he come to New York. All agreed that to sign was Merriam's easiest way out. Apparently, however, Merriam, as balky as a Mount Wilson mule, was loath to give in. On May 9th, Pritchett telegraphed, "Merriam goes to New York tomorrow morning and hopes to see Root in the afternoon. He does not agree." [28] As soon as he received this disturbing news, Hale telephoned Carty. He was unable to reach him. So he went to bed but not to sleep. In his diary he wrote: "Bad night. Can think of nothing except possible effect of Merriam's stubbornness on Root." [29] Somehow Merriam had to be headed off. In the morning, still unable to find Carty, he reached Gano Dunn, who dropped everything and went at once to see Elihu Root, Jr. Together with Frederick Keppel, a board member, and Carty (who had been at a Carnegie meeting), they rushed off to the Pennsylvania station to await the arrival of Merriam's train at 1:30 P.M.[30] No record remains of Merriam's reaction when he saw these three powerful members of his board waiting to greet him. He did tell them, however, that he was going to see Rose before visiting Root. Carty advised him to reflect the feeling of the Carnegie Institution rather than his personal views. They did not want him to suffer sole responsibility for the failure of the 200-inch scheme. Merriam promised that no one would have anything to regret.

Meanwhile, at the University Club, Hale waited anxiously to hear whether the dash to intercept Merriam had been successful. He hoped for the best but feared the worst. Finally, at a quarter to four, after what seemed an eternity, the telephone rang. It was Carty. When he described the events at the station, Hale could hardly believe his ears. In fact, he would not believe them until he learned the outcome of Merriam's meeting with Rose.[31]

At five, after another eternity, the telephone rang again. It was Merriam himself. In a very friendly voice he suggested that they dine together that night. Still incredulous, Hale listened as Merriam said he would bring good news. He was then on his way to see Root.

Soon after six Merriam came. As immaculate as always, he looked quite unruffled. Hale, on the other hand, red-eyed from lack of sleep, felt he had aged ten years since he left California ten days before.

They went down to dinner. Merriam told of his interview with Rose and said he thought there was now no obstacle to carrying out the telescope scheme. Hale could only look at his guest in wonder. He could only wonder more when Merriam assured him he had never wanted the money for the Carnegie Institution and would not dream of taking it. Afterward Hale wrote to Evelina, "We had as pleasant a dinner as though nothing had happened." [32]

All that was needed now was a formal agreement for cooperation from the California Institute and the Carnegie Institution. Both agreed. Hale was immensely relieved, and Rose was elated. "The train is on the track," he exclaimed, "and steaming along as though nothing had happened." He was sure the scheme would go through the Rockefeller board without a hitch.

After this, Hale left for California, hoping nothing would happen to bring him back to New York. In Pasadena, on Rose's recommendation and his own responsibility, he returned to his planning for the telescope. He began negotiations with Elihu Thomson of the General Electric Company for a quartz mirror. He wrote down ideas for cooperation between Mount Wilson and the California Institute.

In June, 1928, he would celebrate his sixtieth birthday. Looking ahead, he foresaw the long years of work in design and construction that "some of us may not outlast." "Building a large telescope, as I have found before, is not a rapid job," he noted. But whether he was to live to see the end, or whether the telescope was even to become a reality, he wanted to make sure that "this undertaking may bring advantages of some kind to science." His far-reaching plans included experiment on new devices and methods that "may prove no less useful to other observatories than to our own."

His plans, of course, as they had been twenty-five years earlier, when he founded Mount Wilson, were still a gamble. Six million dollars was a huge sum (especially in a time when money for science was hard to come by); some of the more conservative members of the Rockefeller board might balk. On May 25th, the day before the fateful meeting, he wrote to Carty, "I am holding my breath anent tomorrow." [33] For several days he waited impatiently. No news came. At last Rose wrote: "The International Education Board has given to the Observatory proposal sympathetic consideration." He did not say they had approved it or voted the anticipated $6,000,-000. He said only that he would come to Pasadena to discuss the "next necessary steps."

Meanwhile a more encouraging message came from Frederick Keppel, another Carnegie trustee. "I've been keeping my fingers crossed but I know that all is well at 61 Broadway, and in due course of time the telescope will be a reality. Hooray!" [34]

Still Hale knew nothing of what had really happened. Two weeks dragged by. On June 11, 1928, Rose and Thorkelson arrived in Pasadena. As soon as Rose appeared at the Solar Laboratory, Hale could see that he was worried. The board had approved the telescope scheme, but had done so only on condition that the California Institute first raise the necessary endowment. The $6,000,000 was for construction only—"to provide for an instrument and accessories as complete and perfect as possible." Until assurance of adequate endowment was given, there would be no telescope.

Hale, of course, knew that the institute had no money for such endowment. Again he saw the telescope vanishing.

When the meeting called to consider the problem came to order in the Solar Laboratory, those present were Rose and Thorkelson, Hale and Adams, Henry M. Robinson and Millikan, Noyes and Merriam, who had arrived the day before. (Merriam was in such an amiable mood that it was hard to believe that a short time before he had been so uncooperative.)

As soon as Rose mentioned the endowment condition, Robinson announced that he would "undertake to secure the necessary amount." (He had already told Hale confidentially that he planned to underwrite the entire amount himself.) [35] Everyone, especially Rose and Thorkelson, was astonished. No one could believe that this last obstacle had been overcome so easily. Robinson beamed with pleasure. However, the Rockefeller trustees would not meet again until fall; until then Hale had to await the final vote.

That night he wrote to Carty of "our adventure." He confessed, "A strong desire to get back to the beach, where we spent a few days a week ago is, I must admit, a dominant notion with me." [36] These last weeks had been some of the most strenuous of his life. Yet, considering the fearful strain, he had come through remarkably well. For this he felt he owed his friends a great deal, especially Carty and Root, to whom he wrote, "Once more we are directly indebted to you for saving a large project." [37] As he wrote to Root, he was swept by an emotion far deeper than gratitude. His feeling for the old statesman was unique. In a way Root had taken the place of his father as an unfailing source of comfort and advice in his hours of darkest trouble. Just two days before the crisis with Merriam, he had received from Root a remarkable portrait photograph. In his letter of thanks he tried to express some of his feeling: "If you will permit me to say so, no one since my father's death in 1898 has ever been to me such a source of perfect confidence and invaluable support. As a young man, confronted with difficult problems in securing funds for and organizing the Yerkes Observatory, I appealed to my father in every trying situation. After his death, I felt as though I had lost my one firm and reliable source of aid. Since then I have known intimately many able men, but your advice alone has always impressed me as conclusive and your aid has been beyond all price. You will thus understand why I so greatly value the gift you have made me."

A little over two weeks after the meeting in the Solar Laboratory, on his sixtieth birthday, Hale wrote wistfully, "I wish I were thirty years younger and able to jump into the task as I did at Mt. Wilson, but under the circumstances I must confine my attention to methods of organization and various general matters, though I shall also help all I can in other respects.

"In view of the broad and liberal policy of Dr. Rose and his associates, the opportunity is by far the greatest in the history of astronomy and we must try to make the most of it." [38]

As soon as Harry Goodwin heard this latest news, he sent his warmest congratulations: "It is simply great, George, that you have the almost unlimited means at your disposal to put through this wonderful project." Yet, knowing his friend, he warned, "Remember that you are not thirty years old but sixty and take things *easy*— don't overdo in your enthusiasm—and injure your health—the success of the whole scheme would be in danger. I mean this literally! For Rose and Thorkelson would never have recommended $6,000,000 for a telescope if they did not expect *you* to be the directing mind back of the whole scheme. . . ." [39]

A few months later, after reading a reference to the 200-inch in the January issue of *World's Work*, Goodwin wrote in lighter vein. "The Rockefeller Foundation," the article had said, "will finance the undertaking, which is the conception of Dr. George Ellery Hale, known as the 'grand old Man of Astronomy.'" "Has it come to this?" Harry demanded. "At 60 years to be called that! It is a case for a suit for libel. *Old Man* indeed! I am the Antique—white-haired Dean— damn it!" To which he added more philosophically, "I suppose we might as well get used to accepting the judgment of the rising generation no matter how young we may feel!" [40]

On October 29, 1928, to quell rumors that were spreading about the telescope, Hale issued a preliminary announcement. Everywhere public reaction was quick, enthusiasm great. An interest in the huge telescope was roused that would increase over the long years of its building. Never, perhaps, had there been such widespread fascination over a scientific instrument with a peaceful aim.

Still, in these years, Hale continued to fight ill-health. In 1927, fearing another breakdown, he had gone to Dr. Riggs's sanitarium in Stockbridge on Dr. Welch's recommendation. The regime was quite different from that in Bethel, Maine. There was no wood chopping, no hypnosis. He took walks and practiced weaving "so as to learn something about a new thing." He liked the place, considered the doctors scientific, and was sure he would benefit from the treatment. He had written to his wife, "They have no idea of a sudden or magic cure, and neither of them think I can do nearly all the things I would like so much to do. Dr. Riggs said very emphatically, after going over my life history, that he was surprised my breakdowns had not been more numerous and more severe. Both say I am not a hypochondriac and both agree that I must pick out certain things to do and avoid others." [41] They had expressed the belief that in time, by following their system, he could reduce his head pains and the resulting depression.

During his stay at the sanitarium, Hale had learned that he had been awarded the Franklin Medal by the Franklin Institute. Dr. Riggs advised strongly against wasting energy on attending the presentation ceremony. He also advised against lecturing or accepting the National Academy presidency. He noted that Hale's difficulties were deep seated and could not be expected to vanish quickly. "He says I am very hypersensitive, feeling all emotions keenly, and therefore a marked case of the type that suffers most in these troubles."

"My trouble," he had written to Evelina Hale in what was certainly no news to anyone, "is 'over-mobilization'—the high tension and great interest in many things that you have so often noticed and rightly objected to. This I must try to overcome, but as it was born in me and has been exercised so long I may have to work for a long time before I succeed." [42]

On his return from Stockbridge, Hale had determined to spend "many moons," if necessary, trying to overcome this "besetting sin." He had dedicated himself to a slow life with quiet writing and some observing in his hermit's laboratory. Then, like that proverbial "bolt from the blue" that had struck so often before, he found himself plunged into a new job—the planning of the biggest telescope and the largest observatory in the world. Later he wrote to H. H. Turner, the British astronomer, "The last thing I had in view was to create a new observatory." Now, however, there was nothing to do but to turn to the huge job of organization. Actually he himself must have been amazed at what he was able to accomplish in a period when he had sometimes thought his lifework finished. Despite many ups and downs, often when he felt weak and ill, he pushed himself to keep on. And those few who saw him could only admire his courage. Somehow his eternal passionate drive prevailed over his physical frailty.

Perhaps his good friend Breasted understood his problems and their cause better than anyone, since he himself suffered similar difficulties. So often, too, his enthusiasms had carried him beyond his strength and had led to the nervous illness and depression to which both were prone. Thus Breasted was to write, "You call it Americanitis and I think that about right . . . in your case and

George Ellery Hale. (*From a painting by Seymour Thomas in the National Academy of Sciences*)

The Hale Solar Laboratory, Pasadena. *(Courtesy of Arnold T. Ratzlaff)*

Walter S. Adams at the front door of the Hale Solar Laboratory. *(Courtesy of Edison R. Hoge)*

George E. Hale observing with the spectroheliograph in the Hale Solar Laboratory. *(Courtesy of Mount Wilson and Palomar Observatories)*

The 200-inch mirror, showing honeycomb pattern. Insert shows comparative size of Newton's telescope.

The 200-inch Hale Telescope, showing observer in prime focus observing position with plateholder and guiding eyepiece. *(Photos courtesy of Mount Wilson and Palomar Observatories)*

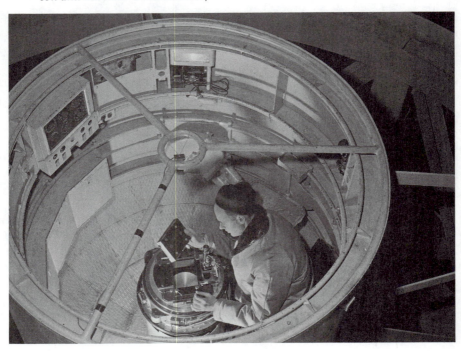

The Hale Telescope. *(Courtesy of Mount Wilson and Palomar Observatories)*

Dedication of the Hale Telescope at Palomar, June 3, 1948. *(Gordon Wallace photo, courtesy of the Los Angeles* Times)

The dome of the Hale Telescope by moonlight. *(Courtesy of Mount Wilson and Palomar Observatories)*

Spiral nebula in *Canes Venatici*, Messier 51, photographed with the 200-inch Hale Telescope. *(Courtesy of Mount Wilson and Palomar Observatories)*

Nebula in *Scutum Sobieski*, Messier 16, photographed with the Hale Telescope. *(Courtesy of Mount Wilson and Palomar Observatories)*

mine we have been trying to carry too many large responsibilities far beyond the borders of our own immediate personal research work." [43] To which Hale, trying to be philosophical, replied: "As you well know, there is never a time for either of us when we want to stop, but it pays to do so once in a while. The great object to hold in view is to get as much accomplished in our remaining years, and this means conservation of energy. My own total output is absurdly low, but I try to stick to a policy that will enable me to keep my head above water." [44] Yet, as they both knew all too well, it was difficult to become a "successful loafer."

This exchange of letters took place in May, 1929. By this time Hale was deep in plans for the 200-inch. An Observatory Council had been formed to direct the planning, construction, and operation. Hale became its chairman, Robinson, Millikan, and Noyes were members; Walter Adams always attended the meetings, and in 1934 was officially added to the group. From the beginning the theme of the council, as of every other committee and individual who worked on the telescope, was cooperation—not only on a national but also on an international scale; not only with astronomers but also with scientists in any field who might have ideas to offer.

On the Advisory Committee, which included members of the observatory and institute staffs, as well as members of other institutions, there were representatives from astronomy, physics, chemistry, mathematics, and aeronautics. The committees on the mirror, the mounting, the site, and the equally important question of auxiliary devices, included men from these and other fields— geophysics, seismology, meteorology. There were experts on optics and photography; there were architects and artists, and structural engineers who had built great bridges and electrical engineers who had perfected the finest electronic devices. All saw "the exceptional nature" of the project, and were glad to have some part in it. All, as Hale said, "gladly put aside any desire for personal reputation and commercial profit, and dropped other work to give their best services."

In 1928, Hale chose those who were to direct the project. From Mount Wilson came the astronomer John Anderson as executive officer; from Vermont the astronomer, architect, explorer, and "father confessor" of amateur telescope makers—Russell Porter. From all over the country help was enlisted to make a mirror and mounting for the great telescope. Great men, famous companies, unknown men—all felt honored to share in the project. It was to be an epic, built by a hundred brains and a thousand hands, guided by a man with profound faith and infectious enthusiasm, who made everyone believe that anything was possible.

The fundamental problem was, of course, the mirror. The chief question was this: "Can a 200-inch mirror be cast successfully?" If so, the telescope would probably succeed. If not, it was obviously doomed to failure. Hale asked the opinion of astronomers and engineers all over the world. Some replied with strong expressions of doubt; others, as it turned out, were too optimistic.

The next questions, on which the first hinged, were these: How is the mirror to be made? Of what material? Of plate glass like the 100-inch, or of quartz or Pyrex; or of some metal, perhaps stainless steel, cast iron, copper alloy coated with beryllium, chromium, or rhodium? Or could obsidian from the Yellowstone cliffs or the Icelandic hills be used?

From England, Holland, and Germany suggestions came. In Eindhoven, Jena, and Newcastle upon Tyne experiments were made, materials tested. After considering countless proposals, Hale concluded that fused quartz would be the ideal material. For twenty-five years he had dreamed of owning a large quartz mirror. His desires had been enforced by that memorable night in 1917 when he had first looked through the 100-inch telescope, and Jupiter had appeared as a seven-headed object, owing to the expansion of the plate-glass mirror with heat. With quartz this would not happen. Its coefficient of expansion is low; its tendency to expand or contract with heat is negligible. If possible, he exclaimed, "we must have quartz."

While negotiations with the Rockefeller Foundation were still going on, he had corresponded with Elihu Thomson of General Electric about a quartz mirror. Thomson, now seventy-five, was eager to take on the job. The largest quartz disk he had ever made was only twenty-two inches. Nevertheless he wrote confidently, "We have little doubt that we can coat a disk with clear quartz of varying diameter." [45] Actually, it was not all that easy. As Gerard Swope, president of the General Electric Company, was to say, this was an "adventure into the unknown." [46]

Hale met with the assistant director of the G.E. laboratory, A. L. Ellis, at the Commodore Hotel in New York. In his hand Ellis was carrying a beautiful piece of clear fused quartz. He spoke eloquently of its possibilities. How much, Hale asked, would a 200-inch mirror of this magnificent material cost? In reply Ellis scribbled down estimates on a piece of brown paper. At the bottom he added up the total for a series of disks from 22 to 300 inches, including a 60-inch and a 100-inch. The total figure was $252,000—a figure Hale would later remember with longing.

From the beginning, the quartz project seemed hounded by a swarm of demons. The difficulties were endless. A year after the signing of the original agreement, the 22-inch disk was still not successfully glazed. The experiments went on, and finally it was successfully coated; a second disk was also cast. These disks arrived in Pasadena, and the first was subjected to violent heating tests "which no glass could possibly bear." The results were all they could hope, and Hale and Anderson were elated. A 200-inch like this would be magnificent. But at the end of 1929, this goal was still far away. Already the cost had almost equaled the original estimate for the entire project. While those responsible for the work were still confident, Hale and those responsible for spending the Rockefeller money were worried. As months passed without tangible result, and the spending went on at the same fantastic rate, the worry turned to fear.

Then, just when the picture appeared darkest, a jubilant telegram arrived from Ellis: "We have laid one more ghost. The first sixty inch mirror blank has been reduced from annealing temperature approximately 1100° C. to room temperature in eight days, an astoundingly short time compared with glass." [47]

Everyone was delighted. "If," said Hale, "it were not for the *fierce* cost, I should feel much encouraged. Their capacity for spending is appalling." However, he added, "They have overcome all sorts of difficulties, and although they are two years behind their original time schedule, some of us may live long enough to see a 200-inch disk, if the money holds out." [48]

As the work moved slowly on, impeded by unforeseen difficulties, Hale realized more forcibly than ever that he might not live to see it finished.

Soon disaster fell again. The 60-inch had been coated with a spray of clear quartz. For nine days it was allowed to anneal. Everything was running smoothly. Then, a few hours before the estimated time was up, someone, overconfident, lifted the top of the oven. The disk cracked. Ellis telegraphed cryptically: "Experimental 60-inch completed fourteenth. Surface quality not good. Cooled too rapidly below two hundred degrees. Started crack." [49]

The result was the great "spiritual depression" described by Harlow Shapley when he visited Lynn, the scene of the quartz experiments, a few days later.[50] The result in Pasadena was equally gloomy, if not more so. Hale called a meeting of the Observatory Council in the Solar Laboratory. It was decided to ask the Corning Glass Works about the making of a Pyrex disk. The casting of a second 60-inch quartz disk was also authorized. "Perhaps," said Hale, "it may be worth while to make a 60" disk in order to get some return for our money, but I don't see how we can go further unless some miracle occurs." [51]

Unfortunately, the miracle never did occur. The work at Lynn went on, with Thomson eternally optimistic and Ellis breaking under the strain. By July, 1931, the second 60-inch was finished and sprayed. It was believed usable, if not perfect. Thomson, sure this was the final hurdle, wrote, "We fully believe that the completion of this 60" disc will point the way most assuredly to the building of the 200" disc of fused quartz." [52]

But Hale could not share his optimism. He was sure that at this rate neither he nor Thomson would live to see a 200-inch quartz disk. On that same day he wrote to Sir Herbert Jackson, in England, "We have just reached the end of this rope and must try another." He described the agonizing results: "Thus you will see that we have been holding our breath ever since I last saw you and we feel the need of oxygen! The financial depression has complicated the situation—we are fortunate in not completely losing the support of our backers. But perhaps you will partially realize why I have not known what 'minnit's gwine to be the next,' as Uncle Remus puts it." [53]

Yet Ellis, like Thomson, was not ready to give up after nearly three years of "arduous struggle." After preliminary inspection of the 60-inch disk, he telegraphed, "We will carry on as usual unless advised to the contrary." [54] Hale, knowing his decision would break their hearts, was forced to reply: "It is evident that you have accomplished an important technical achievement. . . . If we had not already spent the huge sum of $639,000, and if the estimates for the larger mirrors were not so far beyond our means, we should certainly wish to proceed at once with a larger disk." He concluded

"With heartiest appreciation of all your work and very sincere regret that it now appears impossible for us to ask you to continue it." [55]

On October 3, 1931, Hale met Ellis in New York. Together they drove out to Swope's house in Ossining.[56] With Swope was C. E. Eveleth, chief engineer of General Electric. They stated that the further cost of their project would be $1,000,000—probably more. Later, Adams estimated that, at the current rate, the total cost for the mirror alone might run to $2,500,000. Even this, everyone acknowledged, would not assure success. A large part of the money for the entire project, including the telescope mounting and the dome to house it, would thus be gone, and the telescope would still be a dream on paper.

So ended the first "adventure into the unknown."

Hale had seen Swope on October 3, 1931. Five days later he invited O. A. Gage, G. V. McCauley, and J. C. Hostetter of the Corning Glass Works to lunch at the University Club. They discussed the question of a Pyrex disk, the possibility of success, the probable cost.

Actually, this was not the first time that Pyrex had been considered. Some consultants had urged it from the beginning. In the spring of 1928, Arthur L. Day of the Carnegie Institution's Geophysical Laboratory, a vice-president of Corning, had made a plea for the use of Pyrex, though he had emphasized: "The plan is entirely beyond any experience available from here or elsewhere, and courage is an essential asset to be reckoned with. . . . The task is a heroic one for those who may undertake it." [57]

During the three years of the quartz project, experiments on low-silica Pyrex continued. An improved form of Pyrex was produced, with less chance of crystallization, or of becoming turbid and brittle. All promised well.

Meanwhile, two other developments at Mount Wilson promised to increase greatly the efficiency of a 200-inch. The first was a corrector lens developed by F. E. Ross, which Anderson called the most important improvement in the telescope since Dollond's discovery of the achromatic lens. "The beauty of it," said Hale, "is that it is applicable not merely to the 200-inch (for which it will be invaluable) but to all other reflectors in use." It corrected the distortion of star images at the edge of the photographic plate, and, as Ross himself pointed out, it "should prove enormously useful for both the 100″ and the 60″ for the fine astrometric work involved in getting the internal motions in spirals (nebulae), and relative proper motions of faint and bright stars, which at present are vitiated by coma." As soon as the new corrector lens was applied to the 60-inch, his predictions came true. It worked magnificently. Images that had formerly appeared as daggers out toward the edge of the plate now became brilliant points of light. Thus the field was enlarged without changing the focus. Hale, delighted, sent some enlargements to Max Mason at the Rockefeller Foundation. One of these—a photograph of the Andromeda nebula—was taken with an hour's exposure on a night of poor seeing; as a result it did not show what the corrector might do under good conditions; but it compared well with a two-hour exposure of the same object, taken without the corrector on a night of good seeing.

"The whole Mt. Wilson staff," he wrote, "is greatly delighted over this splendid advance, which can be applied to any existing reflecting telescope, and will therefore increase the output of many observatories." [58]

The other invention—a new form of spectrograph objective developed by W. B. Rayton of Bausch and Lomb—held out equal promise. It promised to double the range of the 100-inch. By March, 1931, Milton Humason at Mount Wilson was getting some magnificent results with this new objective. With a 13½-hour exposure he obtained the spectrum of a nebula in Leo (an exceedingly faint object of the 17th magnitude, 105 million light-years away). The spectrum showed a shift toward the red, which indicated that the nebula was receding at a velocity of 19,700 km./sec., and so upheld Edwin Hubble's conclusion that on the average the velocity of these remote objects is directly proportional to their distance.

As soon as Hale heard the news of this observation, which had been made possible by the new objective, he sent Milton Humason his hearty congratulations and thanks: "I was delighted to hear of the immense velocity of the small nebula. I must confess that I don't see how you kept this 17th magnitude object on the slit during 13½ hours." [59] He also sent a print of the spectrum to the great English physicist Sir Joseph Larmor. "I do hope," he wrote, "the 200" telescope (if we can ever get a suitable mirror disk!) will permit the test to be carried much further. I am also curious to know what you believe to be the true interpretation of the remarkable phenomenon." [60]

This invention and the results obtained with it emphasized the importance of getting the 200-inch into operation. With its much greater light-gathering power it could not penetrate far deeper into space to test the theory of an expanding universe. As Hale wrote to Max Mason, "Either the entire universe is flying apart or a more fundamental physical law must be elucidated to account for these extraordinary phenomena. . . . Whatever we may think about the curvature of space, we shall thus have with the 200″ the only possible means of testing the theories of Einstein, de Sitter and others. Whether the increasing red shift of the lines with distance means actual motion or not is one of the subjects that greatly interests Einstein, who is ready to accept any explanation that proves to be best supported by all the available evidence."

This comment doubtless stemmed from a talk that Hale had recently had with Einstein, who was spending several months as research associate at the institute. His first visit to the Solar Laboratory came in January, 1930. It was followed by others. He was especially interested in Hale's experiments on the sun's general field. If the mystery of solar magnetism could be solved, he felt it might help in the development of his theory of gravity and electromagnetism.

Hale found the author of relativity theory a genial soul, "very simple and agreeable," with a distaste for notoriety. Will Hale was enchanted by this contact with genius. "It would certainly be more or less astonishing to see such a fellow walk in at the door. One might about as soon expect Sir Isaac himself."

Fortunately, when Einstein came to stay in Pasadena he was able to stay in the new Athenaeum on the institute campus. For many years Hale had dreamed of having a center where distinguished guests—scientists, writers, and artists—could meet, and where visiting luminaries to Mount Wilson, the Huntington, or the institute could stay while they were in California. He had also hoped for a place where national and international fellows, as well as graduate students, could live. Such a center was an integral part of his dream of making Pasadena "a center of advanced study and research." Soon after his dream was realized, Einstein and his wife arrived to spend the winter, and the new Athenaeum became their home.

In January, 1932, Arthur L. Day of the Geophysical Laboratory arrived for consultation on the Pyrex mirror. It had been decided to proceed by steps—from a 30-inch to a 60-inch to a 120-inch, and so finally to a 200-inch. The cost had been estimated, hopefully, to be between $150,000 and $300,000. The major obstacle was the enormous weight. A solid Pyrex disk 200 inches in diameter would weigh about 42 tons, would take about nine years to anneal safely, and would be almost impossible to mount. Yet Dr. George McCauley of the Physical Laboratory at Corning, who had been chosen to direct the project, was confident of success.

Francis Pease, the Mount Wilson astronomer, drew a design for a ribbed disk that would cut the weight without sacrificing necessary rigidity. It was studded with bumps and cores, somewhat similar in design to a waffle iron. These cores, as it turned out, proved the most difficult part of the casting. When the molten glass was poured, the heat attacked them; the cores broke loose and bobbed to the top. This caused trouble with the 60-inch. It was almost fatal to the 200-inch. Another difficulty came in the pouring of the Pyrex, which chilled and congealed before it could flow into all the corners of the mold. To keep the Pyrex fluid, it proved necessary to heat the molds and house them in special igloos.

As the work progressed at Corning, Hale followed each step anxiously. In general the program ran far more smoothly than the quartz fiasco. Despite inevitable difficulties, the results from the beginning were encouraging; those with the 60-inch were epoch-making. The ribbed structure, Day reported, "came through splendidly." It was, he noted, an historical event in the annals of astronomical science.[61]

Meanwhile, in Pasadena, the new optical shop on the institute campus was "under roof." The instrument shop and the Astrophysical Laboratory were finished. Four years had passed since the project began. Everything was going well, and Hale was encouraged. Yet the end, he knew, was not in sight. "I doubt," he wrote to Goodwin in 1932, "If we shall get the 120″ before late next year and the 200″ by the end of 1934. Large telescopes, as I have learned before, are secular phenomena. But fortunately the Corning estimates of cost do not increase beyond their original figures, in the frightful way the G.E. estimates (and bills) did."[62]

CHAPTER TWENTY

Final Triumph

✳ *1931-1938*

AS LONG AS THINGS ran more or less smoothly in these difficult years, Hale remained in the quiet of the Solar Laboratory. Ill health continued to plague him, and he was forced to refuse many attractive opportunities in the countless directions of his interests. At times, however, the refusals themselves were refused. This was the case with his brainchild the International Research Council, which had just been renamed the International Council of Scientific Unions. On July 13, 1931, a cable from Brussels informed him that he had been elected president. Fearful of the effect on his head of a scientific meeting, he had cabled his refusal. But his protestations were useless. In the end he gave in, deciding that, if his health did not improve, he could resign later.

Fourteen years had passed since he had first discussed an International Research Council with scientists in Europe and America. Since that time several highly successful international unions—in astronomy, geodesy, chemistry, geophysics—had been organized. But, as he had watched from the sidelines, he had not felt that the council was doing all it could to promote international cooperation. Unless a more vital policy were followed, he feared the organization might collapse. If the presidency gave him the chance to carry out a few of his original ideas for cooperative researches, he would be satisfied.

A preliminary council meeting was to be held in London on May 18th. On his way to the meeting he wrote to Sir Henry Lyons, the secretary general of the council, asking that the meeting be held at teatime, instead of in the evening. "In order to be able to accomplish anything on the following day," he wrote, "I spend all of my evenings reading quietly at home, as the stimulus of interesting talk of any kind stirs up my head so violently that it takes me a long time to get over the effects." He felt obliged to explain further, "I have an irresistible tendency to talk with all such men as our group might include, and the list of attractive subjects is a long and varied one. Hence I have had to abandon most of my old activities here and have become an almost invisible hermit, much to my own disgust." [1]

The tone of this letter must have roused misgivings, but the meeting went better than anyone expected. "Your coming over made all the difference," Sir Henry said. "The work of the council has definitely been pushed up on a higher plane, and a more active one." [2]

The principal topic of discussion was how fuller effect might be given to the purposes of the council. In light of the success of the committee on the study of Solar and Terrestrial Relationships, Hale proposed the formation of additional special research committees, such as one on "Instruments and Methods of Research," a subject close to his heart. In scores of cases, as he pointed out with many examples, principles perfectly familiar in one branch of science are wholly unused in other fields. He urged that members from each union be nominated to this committee to provide greater chance for interchange of ideas and development of new methods. "You see," he wrote to H. H. Donaldson, "I believe that our particular gardens are still too tightly enclosed, and wish to make more openings for pollen obtainable from other gardens in all parts of the world." [3]

The pace at the meetings was strenuous, and inevitably he found himself getting tired. Yet he was having a fine time, and he felt that his efforts were rewarded. For three years he continued those efforts until the summer of 1934, when a meeting was to be held in Brussels. As president, he knew he should be there. But he feared the result, as his health had become increasingly precarious. Therefore he abandoned the idea. Frederick Seares of the Mount Wilson staff went in his place.

By this time the shadow of Hitler was looming over Germany, forcing many of her best scientists, like Einstein, into exile. The chances of international cooperation in which Germany would continue to play a role were becoming remote. Yet Hale was still hopeful. As he resigned the office of president, he declared, "My belief in the future possibilities of the Council and the Unions is unbounded, and I am confident that great progress may be expected from their united efforts." [4]

Eleven years later, after World War II, when Hale himself was no longer alive, the International Council would become the sponsor of the International Geophysical Year, a vast cooperative program that would cross the boundaries of individual sciences and nations. He would have been delighted.

So Hale returned to his "hermitage" in the Solar Laboratory. He passed many contented hours sitting in the garden in front of his laboratory, reading or writing, or gazing at the yellow roses and the oranges slowly ripening. Some of his letters were to his six grand-daughters, Margaret's children, who lived on a ranch on the Rogue River in Oregon. Occasionally he would compose bits of doggerel for them.

Now, too, Bill and his wife, the former Margaret Brackenridge, and their children lived next door in the house they had built on the old tennis court. Often these children would come over to play, to delight "Papoo," as they called him, with their small games and amusements. Often he told them the same tales he had told his own children—of the Great Cave, of the watermelon, of jolly "Rigama-role." But, best of all, they liked their visits to the Solar Laboratory, always a weird and wonderful place, with its great pit leading down, down, where "Papoo" claimed an old man with a peaked cap lived and guarded the spectroheliograph mirror. A narrow winding ladder led to the mysterious place, and occasionally he let them go down that ladder—a rather scary but always exciting performance. There they could push buttons "to make the wheels go round"— those electric buttons that turned the dome, focused the sun's image, and started the machinery in the shop.[5]

These children, of course, knew little of their grandfather's fame. They knew nothing of the honors that poured in on him during these years except when they took the tangible form of a gleaming gold medal that even then seemed more of a toy than a symbol of honor. One such medal arrived from the Holland Society of New York in 1931. Another—the Copley Medal—came from the Royal Society of London in 1932. This was, in fact, the highest award of that great society. According to the citation, it was given for Hale's work on solar magnetism; it called his discovery of magnetic fields in sunspots the greatest such discovery in three hundred years. Hale was overwhelmed. He was even more overwhelmed when Newall wrote to ask for biographical material to be used in the series "Scientific Worthies." He felt he was getting into a class where he did not belong. "It would be in keeping," he told Margaret, "to depart from this world soon after receiving it." However, he said, "I have no such intention." [6] He still had a lot of work to do, especially on the sun's magnetic field. He hoped, therefore, to avoid the example of those friends who had received this talisman, then "quickly passed into the west."

"If you know a good beard maker, capable of turning out one at least the length of Johnny Runkle's," he wrote to Goodwin, "please let me know at once. For what is a 'Worthy' without a long gray beard.

"Luckily," he noted, "my old friend Newall, whom I know better than anyone else in Europe, is to write the article, and he has a good sense of humor. For to put me in such a list, would draw a smile from the deepest pessimist." [7]

In answer to Newall's request for biographical material, he wrote down some "autobiographical notes" in pencil. In them he summarizes his accomplishments, but gives little of his personal history. Nor does he attempt to explain what special qualities made his manifold accomplishments possible. He omits entirely any mention of the intense and unusual powers of concentration that his brother and sister had noted in his childhood. He does, however, refer to his lifelong desire to make rapid progress, so rapid indeed that he continued to want things "yesterday"—always. He adds, "I soon learned from experience how to devote many years, if necessary, to the long task of overcoming the many difficulties presented by some cherished project. . . . Furthermore, though deeply absorbed in every phase of my work, I constantly aimed to develop and improve my tools and apparatus." [8]

He realized that many people might think he had been more interested in organization than in personal research. To correct any such impression he spoke of his refusal of the secretaryship of the Smithsonian and the Carnegie presidency. "It should be clear where my chief interests lie."

Always, especially in his building of large observatories, his chief desire had been for "more light," his chief goal, the expansion of opportunities for research.

Years before, on his first visit to Rowland, he had seen photographs of the solar spectrum on a scale he had never imagined possible. He had begun to dream then of photographing the brighter stars on a scale equally great. With the long fixed spectrograph at the coudé focus of the 100-inch telescope, that dream had been approached. With the completion of the 200-inch, he was sure it would be realized.

He had only one regret. His years of illness and his activity in other fields had forced him to give up such "large and intensely attractive tasks" as the complete study and interpretation of the sunspot spectrum. Nevertheless he realized that his contribution to the growth of astrophysics in particular, and of science in general, was far greater than if he had spent all his life on his favorite star, the sun. As Walter Adams was to write, "It is perhaps symbolic of this man of great gifts and wide horizons that he who had devoted his life to the study of the nearest star should find his last deepest interest in an instrument destined to meet the remotest objects of our physical universe."

The Copley Medal, as it happened, arrived just at a time when he was struggling to solve the still-unsolved problem of the sun's magnetic field in the Solar Laboratory. Years before it had appeared that the sun had a field of its own. Now, as he tried to check the results, it seemed uncertain. The quantities, small and extremely difficult to measure, were subject to the difficult factor of "personal equation." For months, lengthening into years, the experiments went on.

Yet, by 1934, he was able to write to Margaret: "I am at last making better progress with my work here, or at least *seem* to be. But the sun is a difficult customer, and the job of 4 years duration (not to speak of the same kind of work 20 years ago) is the hardest I ever tackled." [9]

A year later he was still having trouble; but with a new and promising method for measuring the plates devised by Theodore Dunham, he hoped for more positive results. These new results, however, were still inconclusive. It was a baffling business, not to be solved in his lifetime. In one of his last communications for publication in *The Telescope* in 1936, he concluded that the existence and polarity of a general field had been confirmed but that its exact numerical value and the possibility of any variation in intensity would have to await further investigation. [10]

If Hale could have lived until 1952, he would have been elated by the success of Horace Babcock (the son of Harold Babcock) in the development of a highly sensitive electronic instrument, called a magnetograph, that would make it possible to determine the sun's general magnetic field. When the instrument was perfected, Harold

Babcock began routine daily observation in the Hale Solar Laboratory. The results were published jointly by the two Babcocks. Later, as Harold Babcock writes, "Horace Babcock developed a synthesis, based on a decade of study, which connects the observed results with the differential rotation of the sun and the 22-year period of sunspot phenomena and the puzzling changes of polarity first observed by Hale. A special study was made of magnetic field near the poles of rotation. This is the 'general magnetic field.' Its average intensity is about 1 gauss, far below the sensitivity of Hale's equipment. The polarity of this weak field was found to reverse near the peak of the solar activity cycle then (that is, about 1958)."

Moreover, magnetism has been found to be a force not only in sunspots and the sun itself but also in other stars and even in interplanetary space. Thus, looking back on the significance of Hale's work, Harold Babcock writes: "The significance of scientific discoveries may be gauged by the researches of later workers. Thus, a new category of stars was found by Horace Babcock about 1947, each of which has a strong general magnetic field. These stars are representative of a much larger number. The strongest stellar field yet measured is 34,000 gauss.

"In his writings Hale dwelt on the broad lines of future progress. The structure of the universe (at that time meaning by universe the local galaxy), the structure of matter, the evolution of the stars, etc.—progress in these directions has confirmed Hale's perception.

"Recent theories of galactic structure include magnetic fields of intensity only about 10^{-5} times that of the earth. Such fields are of undoubted importance in star formation, and probably in the spiral structure of galaxies." [11]

During those long months in 1932, Hale had been defeated not only by the problem of the sun's general field but also by continuing ill health. He suffered from severe nosebleeds and dizziness. His blood pressure soared. Torturing nightmares disturbed his sleep. In the spring of 1932 he found himself back at the Riggs sanitarium; again he improved. But in January, 1933, he was overcome by a wave of pain and depression that knocked him completely out of the running for days. After this, his illness progressed in waves. "I seem to be good for nothing lately," he said. Yet he kept on working whenever he could.

That summer he had hoped to get away for a rest. But for weeks the tantalizing problem of the sun's magnetic field kept him in Pasadena. "When an expected result refuses to come after long effort," he wrote to George Sarton, "one's nerves may be unduly affected." [12] In Santa Barbara he tried to do some writing—an article on Galileo, tributes to his old friends Albert Michelson [13] and Dr. William Welch of Johns Hopkins, who had died, leaving gaps in his life that could not be filled. Gradually he relaxed, as much perhaps as was ever possible in one so highly keyed. The aches in the back of his head diminished, and he slept better. But it was not a happy existence, and he champed against it.

In general, in these days when he wished he could tackle a thousand things and instead could do less and less, he tried to take his infirmities philosophically. Nevertheless he continued to analyze himself and his history, and continued, too, to catch at any thread that might offer a clue to his trouble. He discussed his condition with his old friend Jim Scherer. Afterward Scherer wrote, "It distresses me to learn you still have to suffer so much discomfort! How any man could accomplish what you've done with such handicaps is beyond me!" [14]

Often in these dark days Hale felt the desperate need to escape. In the fall of 1933, on the spur of the moment, he suggested a trip abroad. After a month in England with his wife, visiting Newall at Madingley Rise, motoring in the country "in perfect sunshine," he came home refreshed, ready again to tackle the "pesky" sun.

But the effect of this interlude did not last long. Soon his familiar difficulties returned. Evelina Hale, exhausted from looking after him, pondered the uselessness of her life. "I would not mind going to the desert myself," she wrote once. "Instead, I go on doing the same senseless things." And again, "I am sick of things here and long to get away. . . . Foolish yes, but so." [15] In her boredom any change was welcome. Life obviously did not grow any easier for husband or wife.

Despite his many difficulties, Hale continued to follow eagerly and even to share in every step of the 200-inch project. In 1933, on his return from Europe, he dashed up to Corning. In June the 120-inch disk had been successfully cast; [16] now it was in the annealing oven, slowly cooling. George McCauley, director of research at

Corning, and his associates welcomed him warmly and showed him over the plant.[17] He saw everything, from the great igloo built for the 200-inch pouring to the giant ladles that looked like something out of Gulliver's Travels and made him feel like a Lilliputian. In December the 120-inch was removed from the oven. It was examined and found satisfactory. "The point of doubt as to the possibility of getting a satisfactory 200-inch Pyrex disc has passed."

Three months later the great day for the pouring of the 200-inch —Sunday, March 25, 1934—was set. Hale longed to be there, but in his place he sent Adams and Pease. On their return, he listened eagerly as they described the great event, which was attended by hordes of curious sightseers and reporters, as well as by noted scientists.[18]

The pouring began at 8:40 A.M. Hour after hour it went on without mishap. Then, suddenly, one of the cores broke loose. McCauley yelled at the chief ladler, Charlie Wilson, to stop the ladles. He looked through the peephole into the igloo; the core was floating on the surface of the molten liquid. The steel bolt holding it had melted. As McCauley watched, there was a bubbling in the glass and another core shot up, then another.

There was ominous quiet. McCauley gave orders and his assistants rushed up with metal tongs. The igloo doors were opened. They shoved the tongs in the blinding fire and tried to fish out the cores. With the intense heat they could not see. McCauley suggested breaking the cores into small pieces. The men struck at the cores but soon gave up. The heat of the glass threatened to melt the ends of the bars. No one could tell the extent of the damage.

That night the disk was hoisted into the annealer, and the period of waiting began. It was cooled at ten times the ordinary rate, and when it was lifted out of the annealer, tests were made. The results were promising. In Arthur Day's words, "the disc remained intact. Accordingly," he wrote to Hale, "this is the assurance we give you that a 200″ glass can be made." [19]

This news was reassuring. But there was still doubt about the success of the first disk. Therefore plans were made for a second pouring on December 2, 1934. Everything went beautifully. The pouring was a "perfect success." The long process of annealing began.[20] Yet, while it was cooling, the Chemung River, swollen by spring rains, flooded its banks. The Corning officials, watching anxiously, debated whether the annealer could be moved to higher ground. The flooding water penetrated the ground floor of the factory; it rose higher than it had in seventeen years. The annealer was on the second floor, but the electrical apparatus that controlled it was in the basement. The flood swelled toward the transformer and the wires that carried current to the annealer.

The entire Corning staff worked desperately to throw a sandbag barrier around the apparatus. For a day and a night the work went on; but it was useless. The water gained and the current had to be cut. For seventy-two hours it remained off while the electrical equipment was moved to higher ground. Power was finally restored; yet no one could tell the extent of damage to the disk: the annealing had another three months to run.[21]

The Observatory Council could only go ahead with the planning, hoping the mirror had been saved.

Ever since 1928, even before the money for the telescope had been voted, Hale had been pondering the question of a site. His first thought had been of the beautiful mountain Hussey had described so glowingly twenty-five years earlier—that "hanging garden above the arid lands, with bursting springs of water" and "forests that a king might covet"—Palomar Mountain.[22] The disadvantages of 1903—the remoteness from civilization, the lack of neighboring cities—had become advantages. Around Mount Wilson, Los Angeles had grown into a sprawling octopus with streetlights and flashing signs that interfered seriously with observation. Even Pasadena had become a neon-lighted city.

In 1928, Hale had gone down to Palomar to see this miraculous mountain for himself. John Anderson and Henry Robinson had gone with him. The mountain road was almost impassable. The "Nate Harrison Grade" was a winding, rocky trail, blocked by snow in winter, rutted by rain in spring, worse in many ways than the first Mount Wilson trail.

But the day of the expedition was beautiful. They could see for miles in every direction. Toward the coast lay San Diego; to the south ran the Mexican border; to the east lay the Colorado Desert and the Salton Sea. To the north, Lake Elsinore gleamed in the sun; beyond rose the towering peaks of Old Baldy, San Jacinto, and San Gorgonio, with Mount Wilson eighty miles away as the crow flies.

Palomar Mountain was covered with forests of fir and cedar, spruce and oak, which alternated with grassy valleys and cool,

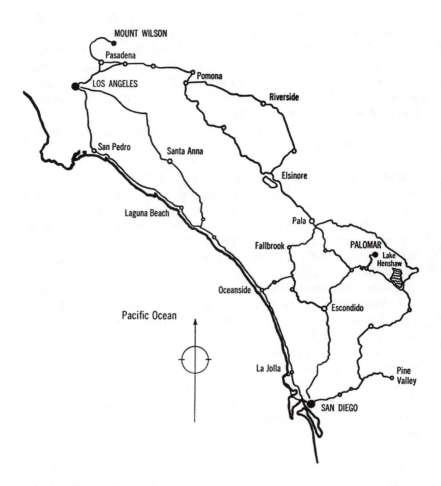

The Mount Wilson-Palomar area.

densely shaded canyons. Among the great rocks grew giant ferns, some as high as eight feet. In the canyons streams flowed; ice-cold springs were abundant.

They spent the night on the mountain. It was still. The stars shone with a steady light. In the distance the lights of coastal cities could be seen. At the foot of the mountain the dim lights from the Indian reservation at Pala gleamed like tiny fireflies. From such lights they knew there was nothing to fear.

Everyone was enthusiastic. Yet, before a final decision, other sites had to be investigated. For five years the search continued. A hundred different sites were tested—from the high Arizona plateau to Catalina Island. One site lay too close to the San Andreas fault. Another, after long observation, proved too windy. The choice was finally narrowed to Table Mountain and Palomar. The balance of opinion favored Palomar.[23]

Next the question arose: Where on Palomar should the telecope be mounted? To answer this question Hale, with Adams and Anderson, made another trip to Palomar. Dan Tracey, the forest ranger on Mount Wilson, joined them. On March 8, 1934, a little over two weeks before the first casting of the 200-inch disk, they drove up the "Nate Harrison Grade." From one end of the mountain to the other they went, looking for favorable sites. From these, a site below the High Point, almost at the center of the great plateau, at an altitude of 5,500 feet, was chosen. This, as it turned out, was destined to be the last trip Hale was ever to make to Palomar. Often in the dark days ahead he would remember it with longing.

The final deal for the Palomar Mountain property was signed by lamplight on a stormy night in the rancher William Beech's weather-beaten cabin. Representatives of the California Institute, the landowners and the San Diego County Board of Supervisors, who had agreed to build a road up the mountain, were present. The agreements were signed at 3:00 A.M. on September 21, 1934. Thus, said a reporter in the San Diego paper, "with the closing of the deal Southern California was assured a scientific institution which will comprise one of the wonders of the world."

Some months later, work was started on the "Highway to the Stars." In August, 1935, Hale received an invitation to a "Luncheon and Ground-Breaking Ceremony" from the Board of Supervisors and the Road Department of San Diego County. He was not well enough to go. But others went and had a fine time. Anderson wrote afterward, "Too bad you didn't feel up to participating and taking the spade in hand yourself—as should have been done—for if any *one* person is the father of the 200"—you are he." [24]

Meanwhile plans were being made for the mounting, which someday would make the journey up that road. The problems of construction were nearly as great as those with the mirror. As Hale had pointed out, the mounting for such a mammoth instrument, while composed of ordinary elements, had some extraordinary requirements. Bigger bridges and larger battleships had been built, but none demanded such unique optical precision and accuracy of movement. The mounting had to be as strong as a bridge, yet as delicate as a watch. To support the mirror perfectly, it had to be more rigid than any bridge. Yet, unlike any bridge, more like a mammoth gun, it had to be able to move on its axis in extremely slow motion—if necessary, a tiny fraction of an inch at a time. There could be no lost motion. The telescope must respond instantly to the astronomers' demands. It must be so perfectly controlled that the touch of a child's finger could move it. Clyde McDowell, a navy captain with wide experience in construction, and a man of tremendous drive and energy, was put in charge of engineering and construction.[25] He was given a leave of absence from the navy. To build the mounting, the Westinghouse Electric and Manufacturing Company was chosen.[26]

All these plans were being made while the 200-inch disk was slowly cooling. But actual work could not begin until the fate of this second disk was known. At last it was taken from the annealer; from surface tests it appeared satisfactory.[27] But final testing would have to await its arrival in California. On March 26, 1936, McCauley telegraphed to Hale: "Our work of loading the 200-inch disc was completed at noon March 25, and the N.Y.C.R.R. started moving the train early on the morning of March 26. This by chance is just two years and a day since the casting of the first 200" blank."

All over the country excitement rose as the special train, loaded with its precious cargo, pulled out of the Corning station and began its long journey across the country. Thousands of people watched by the tracks. At the rate of twenty-five miles an hour it moved during the daylight hours, an advance scout clearing the way, an automatic recorder registering the slightest vibration.

Hale, waiting anxiously on the West Coast, heard of its safe arrival, first in San Bernardino, then on Good Friday in Pasadena. On Easter Sunday, in the Optical Shop, the mirror crate was opened. He telegraphed McCauley: "Disc arrived Friday. Moved into Optical Shop Saturday. Steel case opened Sunday. Disc now standing on a wooden rack. No sign of injury in transit has been detected hitherto. Hearty congratulations. . . ." [28]

The New York Times announced triumphantly:

GIANT EYE ENDS RAIL TRIP ACROSS CONTINENT:
PLACED IN SHOP AT PASADENA FOR POLISHING.

The following day Hale wrote to his old college friend Harry Goodwin: "The 200″ is now in the optical shop, apparently uninjured by the long trip. The railroads surely deserve much credit for their great care every inch of the way from Corning to Pasadena. There has been an astonishing furor about the whole thing, and thousands of people went out to the way station east of Pasadena when the car was unloaded and stayed there for hours, many following the big truck to the optical shop. I haven't seen the disc yet, as I have to keep out of crowds."

"To an old timer like myself," he commented when he finally got the chance to see the great mirror, "it is difficult to realize, when looking at the new disc, that the aperture of its central hole is equal to that of the 40″ Yerkes refractor. No other scale gauge could be more striking to me, as I recall so vividly the arrival of the 40″ objective at the Yerkes Observatory in 1897. . . ." [29]

Nearly forty years had passed since that day in 1897. Over fifty years had passed since the dedication of his first little shop at Kenwood—that little wooden shack he had built with the help of Martha and Will.

In these fifty years the once embryonic science of astrophysics had gone far. In 1886, Robert Ball, writing in The Story of the Heavens, had emphasized the importance of making good drawings of the planets and nebulae so that, if changes occurred in these bodies, they could be identified. Ball did mention the spectroscope, "a new department of science," which was just beginning to reveal the physical nature of planets, stars, and nebulae. But he wrote little of the photographic plate, still in a primitive state. By 1886, William Huggins had shown that the Orion Nebula is gaseous in nature, at least in part. But astronomers still had no way of telling the distance of this or any other nebula. They considered the problem of determining such distances insurmountable. They must lie, they thought, at "appalling" distances; their light, they knew, must have left them centuries before. But beyond this, nothing was known. They estimated that there were about three or four thousand nebulae in the sky, and concluded that the entire universe of stars was contained in the Milky Way. When they considered the possibility of penetrating farther into space, the majority shared Ball's view: "We have reached a point where man's intellect begins to fail to yield him any more light, and where his imagination has succumbed in the endeavor to realize even the knowledge he has gained." [30]

In 1886, Ball could not foresee that man was on the threshold of discovering a vast universe far beyond the reach of his wildest imagination. Nor could he foresee that this discovery would be due in large part to the development of new instruments and methods in which Hale would play a preeminent role.

By 1936, astronomers contemplated a universe in which the Milky Way was only one of millions of galaxies. They considered its diameter to be about 100,000 light-years; they knew there were other galaxies far greater. The most distant of these, according to Hubble's estimate at the time, was 500 million light-years away. Looking at such a nebula at this distance was comparable to looking at a candle with the unaided eye at a distance of 8,575 miles.

Yes, in 1936, the universe was known to be a far vaster place than when Hale had begun work fifty years earlier, and man had shrunk in relative size even more in that age when Copernicus and Galileo had removed him from the center of the solar system.

Yet as Hale looked back, he also looked forward into that great future when the 200-inch would be finished and when from Palomar it would start on its long journey into space. With its estimated light-gathering power of 1,000,000 human eyes, it would penetrate at least 1,000 million light-years. What would it discover in those far reaches which no human eye will ever see, where no human being may ever go? Would light in those almost inconceivable reaches of space continue to behave as it does in more neighborly regions? Would the nebulae still recede from one another? Would the universe, therefore, still show evidence of expansion?

As he looked down on that great mirror in the Optical Shop, seeking answers to such questions, he gazed into a future in which man, on this tiny earth moving around a little star, would occupy a role so subordinate that in it man's arrogance could have no place.

Twenty-eight years later, Ira S. Bowen, director of the Mount Wilson and Palomar Observatories, was to look back also to trace the developments since the 200-inch Hale telescope began regular observation nearly a decade and a half before. He was to ask "to what extent the opportunities for discoveries in the fields enumerated by Hale have been realized" by that telescope and its supporting instruments. These would be some of his answers to that question.

On the structure of the universe: With the 200-inch the steps in the determination of the distances of the Andromeda Nebula had been reexamined; the magnitude scale had been recalibrated; the luminosities of the Cepheid variables (which provide a means of determining the distances of remote nebulae) had been redetermined; and the magnitudes of the Cepheids in the nebula itself carefully studied. The results showed that the Andromeda Nebula is actually 2,200,000 light-years away from us, or three times farther than Hubble had estimated its distance with the limited powers of the 100-inch. This meant "that Andromeda has a diameter three times, a luminosity nine times, and a mass three times greater than the earlier estimate." [31]

The same reevaluation had, of course, proved necessary for more distant galaxies. It appeared that Hubble's values for these objects would have to be increased by a factor of 4 to 7.

In 1928, when Hale wrote his *Harper's* article, he had tried to look into the future to predict the results with the 200-inch. "Most interesting of all," he noted, "will be the interpretation of the apparent outward flight of these spirals, which seem to be receding from us at almost incredible velocities, increasing with their distance." [32] If he could have lived to see the 200-inch at work, he would have been astounded to learn that the velocity of recession of some of the most distant objects in the universe is four times as large as any that Humason had observed with the 100-inch. He would also have been fascinated that the way to this discovery had been paved by the entirely new technique of radio astronomy, which had been discovered but which was still in embryo at the time of his death. The "radio stars" found in this way are intensely bright; some of them are at least 100 times as luminous as the Andromeda nebula; they must lie at much greater distances than normal galaxies. The determination of the exact distance for the most distant of these objects is difficult. "It can be said, however," Bowen writes, "that the light from this radio star left on its journey to us when the universe was only half its present age." [33] This and other stars like it, therefore, offer a promising key to the understanding of the origin and evolution of that universe.

In 1928, Hale suggested that one of the chief problems to be solved by the 200-inch would be the one that had puzzled him throughout his life and had been the impelling force in his founding of one great observatory after another—the evolution of the stars. This problem was still a puzzling one in 1964. Yet, as a result of advances in nuclear physics, already in progress at the time of Hale's death, a great deal more would be learned about the structure of matter, and therefore about the source of energy radiated in the sun and stars. The results obtained in this way would make it possible to determine the true ages of stars in different parts of our own galaxy and of neighboring galaxies in a way that was previously quite impossible. "Ages extending from a million years up to 12 billion years for the oldest globular clusters have been found," Bowen notes.

"Clearly," he concludes, "the 200-inch has performed much as Hale had hoped and predicted. Most of the programs he listed have made great steps forward. Moreover, major breakthroughs have occurred along lines of which Hale had no inkling. If he were with us today, I believe he would be content with the results of the great adventure he started." [34]

In 1936, all this, of course, lay far in the future. Yet with the safe arrival of the 200-inch mirror, the first great step that would make all these results possible had been taken. With that arrival, a tremendous load was lifted from Hale's mind. Somehow now he felt that this, the last of his great adventures, would be successful.

Still, he doubted more seriously than ever that he would live to see the telescope mounted. His energies had decreased to such a point that he rarely left the narrow path between his house and the Solar Laboratory. He was forced to refuse all invitations, even one from Harlow Shapley, director of the Harvard Observatory, to a special symposium in his honor. He had, in fact, urged that this honor be conferred on someone else. "Old and battered fossils may retain a certain antiquarian interest, but in the midst of recent revolutionary advances, they are rapidly outclassed." [35] He was, however, deeply moved. He wrote to Harry Goodwin, his lifelong friend, "I really couldn't keep tears out of my eyes when I read that *you* above all others, were to speak." [36]

On the appointed day, the Harvard Observatory Library was crowded with staff and friends. Shapley, in his introductory speech, said he had planned the symposium with two thoughts in mind. The first was "the recognition of Hale's remarkable contributions to science and to the techniques and equipment of astronomy." The second was to call the attention of the younger astronomers "to the great debt we all owe to one man for the commendable position of astronomy in America at the present time." [37]

Eight other speakers evaluated Hale's contributions to the development of science in the United States and in the rest of the world. They included Harry Goodwin, Robert King, Willard Gerrish, Gerard Kuiper, E. B. Wilson, Arthur L. Day, Theodore Dunham, and George McCauley. At the end of the meeting McCauley showed a dramatic series of pictures of the casting of the 200-inch and its loading for transportation to Pasadena.

The day of the meeting, an editorial appeared in the New York *Times*. It was headed "Priest of the Sun," and in it Hale was called a "Zoroaster of our time." The editor wrote: "From the first Hale thought on an epic scale. If American observatories are the finest in the world, if their instruments are the largest and most interesting, it is because of him." [38]

The editorial spoke also of Hale's great human qualities: "Perhaps in imbuing them [his disciples] with his own enthusiasm he has done work as fine as any with the spectroheliograph. There is hardly an American observatory that cannot produce at least one human discovery of George Ellery Hale."

Goodwin sent Hale a copy of this editorial together with his own account of the ceremony. "I only wish you might have been at the meeting, old boy, but knowing your *excessive* modesty, it is perhaps as well that we had you with us in spirit and with your wonderful likeness, life size, on the screen before us instead." [39] It was indeed a bit like seeing one's obituary before one is actually dead. Nevertheless Hale was grateful. When he tried to send his thanks, the "whirligus," as he called it, got him. A week later he tried again. "Now I dare the Irish faeries, whether from the 'old sod,' the Catholic city of Boston, or any other see of Papal domain, to interfere." [40]

Each day he continued to go to the Solar Lab. In the basement, the experiments on the sun's magnetic field went on, yet somehow he could not find the initiative to join in. He had an idea for a moving-picture attachment, which he had first proposed many years earlier. Robert McMath in Michigan had devised a scheme for taking moving pictures of the sun. Hale thought his device might work better. He was never able to try it out.

On the 29th of June, 1936, he celebrated his sixty-eighth birthday. The observatory staff sent flowers. "It means more to me than I can tell you," he wrote gratefully, "especially during a long period of depression and uselessness." To many members of that staff, he was in fact almost a mythical being. Their visits to the Solar Laboratory were restricted, and in time were abandoned entirely, with the exception of Adams and one or two others. There were some, indeed, who never saw him after his retirement in 1924. One of these was Elizabeth Connor, the librarian. Just before his retirement he gave her letters of introduction to the leading British libraries, and urged her to visit Salisbury "without fail." "There I must walk across the meadows south of the Cathedral until I had a certain view; then in London, at the National Gallery, I must look for Constable's painting of the Cathedral, viewed from that particular spot, which I did." [41] On her return she visited Hale in the Solar Laboratory. After that he would call about books or references he needed from the library; with his remarkable visual memory he could usually tell her exactly where to look, even to the exact page. Yet, that visit of 1924 was the last time she saw him in person.

In a way, the stringent regime of a hermit's life into which he withdrew more and more was beneficial to science. In the quiet of the laboratory he had time for contemplation and for the development of ideas that is impossible for those who are constantly dashing around the country to one meeting or another, always talking, rarely meditating. His ideas and feelings were reflected in a flood of letters amazing for their length as well as for their quantity—on every conceivable subject.

Still, he would have greatly preferred being out in the world. As the days dragged on, they often became increasingly lonely. The sense of loneliness deepened when on December 2, 1935, news came of the death of James Breasted. The next summer Breasted's son, Charles, came to Santa Barbara and arranged a special showing of *The Human Adventure* at the Fox Theater. It was an account of Breasted's life and work. He invited Hale and a few close friends. The film opened with James Breasted presenting the film. Hale looked at his friend on the screen, and felt for a while that he had returned to earth. All through the film he sat as in a trance, living over the long years of their friendship. At the end, when they came out into the dazzling sunlight, Charles saw tears in his eyes. He also saw a faraway look—a look of detachment from everything that was going on around them. To Hale himself it seemed that he had gone through an almost mystical experience.[42]

Nor was Breasted the only friend who departed. In 1937, Arthur Noyes died of pneumonia after a long period of suffering from cancer. In his will he left a bequest for the founding of the "George E. Hale Research Fellowships in Radiation Chemistry."

So to loneliness was added profound sorrow. Often, then, Hale longed for escape. In April he had what he called "a queer but profound attack." Yet he refused to give up his plans for a trip to England in the summer of 1936. En route by way of Quebec he had what was later diagnosed as a stroke. He stayed for a while with his brother in Chicago, then returned to Las Encinas sanitarium in Pasadena. On February 1, 1937, he wrote to John Evershed in England, describing his illness, then added, "I intend to be all right before long."

During his stay in the sanitarium in Pasadena, his wife decided it would be too difficult to maintain the big house, and so she moved to an apartment at 360 Grove Street. When he arrived at the apartment, George Hale felt completely lost, and never did become accustomed to the unfamiliar place. He tried to spend more time at the Solar Laboratory. One day he stayed too long, and confessed he had done too much. After that he was forced to stay away entirely; he felt that his life had ended.

Still, whenever Adams came to see him, he listened with passionate interest to the reports of latest developments on the 200-inch. The Westinghouse engineers, using designs for a yoke mounting worked out by engineers of the California Institute of Technology, had built the tube, the yoke, the bearing parts. By April, 1937, the 55-foot, 125-ton tube, as long as a large pleasure cruiser and weighing as much as an ancient schooner, had been completed. On April

30th a celebration, to which two hundred scientific and industrial leaders came, was held. Among those in attendance was Albert Einstein.

Down on Palomar the dome to house this monster was slowly rising, a landmark on the skyline, visible from miles around. Designed by Russell Porter, it was 137 feet in diameter and 135 feet high—just about the size of the Pantheon in Rome. In his mind's eye Hale could envisage that dome on the beautiful mountain; he would have given his soul to see it. In May, 1937, he wrote hopefully, "In another month I may be able to go down to see the work in progress on the mountain." [43] Night and day he dreamed of such a visit. But it was not to be. His arteriosclerosis was progressive and soon it became necessary that he be cared for at Las Encinas sanitarium.

All his life he had fought against illness. Usually he had won, if only a partial victory, and had accomplished stupendous feats despite his breakdowns. But this was different. There was no longer anything he could do about it.

The year 1938 was full of promise for the telescope. Everything was going well, and everyone had hopes of ultimate success. The huge horseshoe bearing, the largest ever made, was finished. Each section of the horseshoe weighed over 100,000 pounds; the total was about 320,000 pounds. That summer the tube and mounting, too large to be shipped by land, started on their long sea journey to California. When mounted, the entire assembly, including the mirror and pedestals, would weigh over 500 tons—five times as much as Columbus's ship the *Santa Maria*.

But if 1938 was a year of promise, it was also a year of tragic loss. By the time the mounting reached the end of its journey, many of those who had prepared its way had gone on a far greater journey. One of these was the thirty-nine-year-old Sinclair Smith, who had helped to design the final telescope drive. Another was Francis Pease, the architect of the blueprints for a 300-inch telescope. The third was the originator of the entire project.

One day, early in that same year of 1938, Hale looked up at the blue sky and murmured: "It is a beautiful day. The sun is shining, and they are working on Palomar." [44] A few days later, on the 21st of February, at one thirty in the afternoon, George Ellery Hale died.

Afterward his niece Ruddy, Will's daughter, wrote to her father, "I loved him more than any other relative although I saw him more seldom, but of course his loving powers and social talent were as extraordinary as his scientific powers." Then she added, "I do wish they would dedicate the 200" telescope to him, for he certainly would deserve the honor." [45]

The New York *Times* echoed her feeling in a leading editorial: "Some fitting monument should be erected to his memory. If the two hundred inch mirror could be called the 'Hale mirror,' or better still, if the whole plant which is to be erected on Palomar Mountain could be called the 'Hale Observatory' both astronomers and the public would be perpetually reminded of their debt to one of the most eminent men of science this country ever produced." [46]

Naturally, no such thought had ever entered Hale's head. He would have been amazed if he could have returned to Palomar ten years later to the dedication, when the 200-inch was named the Hale Telescope—in honor of George Ellery Hale, "whose vision and leadership made it a reality."

PUBLICATIONS OF GEORGE ELLERY HALE

In addition to over 450 articles, Hale published the following books:

The Study of Stellar Evolution; An Account of Some Recent Methods of Astrophysical Research. Decennial publication of the University of Chicago, Ser. 2, Vol. 10. Chicago: University of Chicago Press, 1908.

Ten Years' Work of a Mountain Observatory. Carnegie Institution, Pub. No. 235. Washington, D.C.: Carnegie Institution of Washington, 1915.

The New Heavens. New York: Charles Scribner's Sons, 1922.

The Depths of the Universe. New York: Charles Scribner's Sons, 1924.

Beyond the Milky Way. New York: Charles Scribner's Sons, 1926.

Signals from the Stars. New York: Charles Scribner's Sons, 1931.

MEDALS AND DECORATIONS

1894   Janssen Medal, Paris Academy of Sciences
1902   Rumford Medal, American Academy of Arts and Sciences
1904   Gold Medal, Royal Astronomical Society
        Draper Medal, National Academy of Sciences
1908   Silva Medal, Sociedad Astrónomica de México
1916   Bruce Medal, Astronomical Society of the Pacific
1917   Janssen Medal, Astronomical Society of France
1920   Commander of the Order of Leopold II of Belgium
        Galileo Medal, University of Florence
        Commander of the Order of the Crown of Italy
1921   Actonian Prize, Royal Institution of Great Britain
1926   Elliott Cresson Medal, Franklin Institute
1927   Franklin Medal of the Franklin Institute
        Arthur Noble Medal for Civic Service (Pasadena)
1931   Gold Medal of the Holland Society of New York
1932   Copley Medal of the Royal Society of London
1935   Frederic Ives Medal

HONORARY DEGREES RECEIVED

1897   Sc.D. University of Pittsburgh
1904   LL.D. Beloit College
1905   Sc.D. Yale University
1907   D.Sc. University of Manchester
1909   D.Sc. University of Oxford
1910   Ph.D. University of Berlin
1911   Sc.D. University of Cambridge
1912   LL.D. Throop College of Technology (now California Institute of
        Technology)
1916   Sc.D. University of Chicago
1917   LL.D. Princeton University
        D.Sc. Columbia University
1921   Sc.D. Harvard University

KEY TO ABBREVIATIONS IN NOTES

E.C.H.—Evelina Conklin Hale, wife of G.E.H.
G.E.H.—George Ellery Hale
G.E.H. II—George Ellery Hale, son of W. B. Hale
M.H.H.—Martha Hale Harts, sister of G.E.H.
W.B.H.—William Browne Hale, brother of G.E.H.
W.E.H. I—William Ellery Hale, father of G.E.H.
W.E.H. II—William Ellery Hale, son of G.E.H.
M.B.H.—Mary Browne Hale, mother of G.E.H.
M.H.S.—Margaret Hale Scherer, daughter of G.E.H.

A.A.A.S.—American Association for the Advancement of Science
Pub. A.A.S.—Publications of the American Astronomical Society
Am. Assoc. Proc.—Proceedings of the American Association for the Advance-
  ment of Science
Am. Astron. Soc.—American Astronomical Society
Ap. J.—Astrophysical Journal
Astron. and Astrophys.—Astronomy and Astrophysics
B.A.A.S.—British Association for the Advancement of Science
C.I.W.—Carnegie Institution of Washington
C.I. Yearbook—Carnegie Institution Yearbook
H.C.O.—Harvard College Observatory
I.R.C.—International Research Council
M.I.T.—Massachusetts Institute of Technology
N.A.S.—National Academy of Sciences
N.R.C.—National Research Council
P.A.S.P.—Publications of the Astronomical Society of the Pacific
Pop. Astron.—Popular Astronomy
Proc. R. Soc. Ser A.—Proceedings of the Royal Society, Series A
Proc. Royal Institution—Proceedings of the Royal Institution
R.A.S.—Royal Astronomical Society

Chapter One

1. G.E.H., autobiographical notes. Also W.B.H. on G.E.H., "Childhood and Youth," unpublished.

2. Genealogical records and biographical notices of W.E.H. Note on ancestors of G.E.H., written by W.B.H., July, 1938.

3. Contemporary newspaper notices.

4. G.E.H., "Some Personal Recollections," unpublished.

5. W.E.H. and the hydraulic elevator business, described in biographical notices and in newspaper notices at his death, November 16, 1898.

6. G.E.H., "Some Personal Recollections," *op. cit.*

7. Reminiscences of their childhood on this and the following pages based on G.E.H., autobiographical notes, and W.B.H.'s vivid account, *op. cit.*

8. Story of making electric bell, told by Alicia Mosgrove as G.E.H. told it to her.

9. Recollection of Ada Browne, G.E.H.'s step-grandmother, recorded in letter to G.E.H., April 16, 1916.

10. M.B.H. to her mother, Mrs. Philemon Scranton, December 12, 1883.

11. G.E.H., "Some Personal Recollections," *op. cit.*

12. Account of building shop based on G.E.H.'s autobiographical notes, on W.B.H.'s account, and memories of Martha Hale Harts, as told to author in Madison, Conn., 1949. Also G.E.H., "Work for the Amateur Astronomer," *P.A.S.P.*, XXVIII (April, 1916), 9–10.

13. G.E.H., autobiographical notes.

14. *Ibid.*

15. Based on memories of Mary Browne Hale of her childhood. Also on genealogical records in the Hartford State Library, and accounts of Erastus Scranton, a Yale graduate of 1802, in the Yale Library. The region around Burlington and its background were also explored by the author.

16. Memories of a classmate of G.E.H. at the Allen Academy, as told to the author in Chicago.

17. Burton Holmes reminiscences, published in *The Griffith Observer*, September, 1947, p. 106.

18. M.B.H. to Ada Browne, January 15, 1884.

19. Burton Holmes, *op. cit.*, 1947.

20. Incident recounted by W.B.H. in his notes on G.E.H., "Childhood and Youth," *op. cit.*

21. G.E.H., "Reminiscences of Madison," unpublished. Also recollections of Martha Hale Harts, whom the author visited in the family homestead, in Madison, Conn. Also Jane Bushnell Shepherd, *From New Haven to Madison in the Sixties* (New Haven, Conn.: Tuttle, Morehouse & Taylor Co., 1931).

22. G.E.H., "Reminiscences of Madison," unpublished.

23. G.E.H., autobiographical notes.

24. *Ibid.*, and memories of W.B.H., *op. cit.*

25. G.E.H.'s notebooks, now in the Hale Solar Laboratory.

26. Account given by W.B.H.

27. G.E.H., autobiographical notes.

28. *Ibid.*

29. Memories of Robert Hale, son of W.E.H.'s brother, Stedman, when he visited the Hale Solar Laboratory.

30. W.B.H., *op. cit.*

31. W.B.H., *ibid.*

32. Accounts of the Rookery given in histories of Chicago, e.g., Frank A. Randall, *History of Chicago Buildings* (Urbana, Ill.: University of Illinois Press, 1949), pp. 5, 18, 65, 112, 152.

33. G.E.H., "Work for the Amateur Astronomer," *op. cit.*, 1916.

34. *Ibid.*

35. *Cassell's Book of Sports and Pastimes* (London: Cassell, Peter, Galpin & Co., 1881).

36. *Ibid.*

37. G.E.H., "Work for the Amateur Astronomer," *op. cit.*, 1916. "How to Build a Small Solar Observatory," *P.A.S.P.*, XL (1928), 285–302.

38. *Ibid.*

39. G.E.H. to Miss Mary Caulfield, Phoebe Brashear Club, Pittsburgh, Pa., November 12, 1915.

40. Account of meeting, told by John A. Brashear, at dedication of the Kenwood Observatory, June, 1891.

41. Samuel P. Langley, *The New Astronomy* (New York: The Century Co., 1884), p. 3.

42. *Ibid.*, pp. 6–7.

43. R. S. Ball, *The Story of the Heavens* (London: Cassell & Co. Ltd., 1887), pp. 26–27.

44. G.E.H., autobiographical notes.

45. Burton Holmes on G.E.H. in *The Griffith Observer*, *op cit.*

Chapter Two

1. W.E.H. to G.E.H., September 23, 1886.

2. Letters to Harry Goodwin. These are one of the most prolific sources of material for this biography. Dr. Goodwin gave these letters to the author to read

when she visited him in Brookline, Mass., in 1948. He told the author then that she was free to use any or all of this material in the biography. After his death, his son, Richard Hale Goodwin, gave this invaluable collection to the Huntington Library, San Marino, Calif.

3. Personal reminiscences of H. M. Goodwin, as told to the author.

4. William Conklin to G.E.H., October 11, 1886.

5. Reminiscences of H. M. Goodwin. Also published in *The Technology Review*, M.I.T., April, 1938.

6. Verses on G.E.H. and H. M. Goodwin in *Technique*, the M.I.T. 1890 class yearbook, p. 157.

7. Reminiscences recorded on this and following pages obtained from Hale's classmates Harry Goodwin, George A. Packard, Frank Greenlaw, and Willis R. Whitney.

8. Quote on G.E.H.'s lack of barroom experience, told to the author by W.E.H. II, May 27, 1949.

9. Interview with Willis R. Whitney, Director of Research, General Electric, Schenectady, N.Y.

10. Reminiscences of H. M. Goodwin, *op. cit.*

11. Material for G.E.H.'s life with Bumpus family gleaned from H. M. Goodwin. Background also obtained from H. C. Bumpus, Jr., *Herman Carey Bumpus, Yankee Naturalist* (Minneapolis, Minn.: University of Minnesota Press, 1947).

12. Based on genealogical records of Hale, Browne, and Scranton families. Also W.B.H. on G.E.H. "Childhood and Youth," unpublished, *op. cit.*

13. Description of Hale family mansion on Drexel Blvd., obtained from E.C.H., from local Chicago records, from a personal visit there and from M. H. Scherer, who knew the house as a child.

14. Records in observing notebooks of G.E.H.

15. G.E.H. to H. M. Goodwin during the summer of 1887, especially a letter written July 4, 1887.

16. G.E.H. to H. M. Goodwin, written many years later, April 4, 1936.

17. E. C. Pickering to G.E.H., February 27, 1888.

18. G.E.H., autobiographical notes.

19. *Ibid.*

20. G.E.H. to H. M. Goodwin, June 2, 1888.

21. Reminiscences of G.E.H. and E.C.H. on blizzard of 1888. Also contemporary newspaper accounts.

22. G.E.H. to H. M. Goodwin, August 5, 1889. (Description of the "Kenwood Physical Laboratory" given in letter to R. G. Aitken, December 6, 1916.)

23. G.E.H. to H. M. Goodwin, undated letter, late summer 1889.

24. Henry A. Rowland to G.E.H., May 29, 1886, courtesy Harriette Rowland.

25. Account of meeting with H. A. Rowland, told to the author by W. S. Adams as it was told to him by G.E.H.

26. G.E.H. to H. M. Goodwin, June 6, 1889.

27. G.E.H., autobiographical notes.

28. G.E.H to H. M. Goodwin, June 6, 1889.

29. J. A. Brashear to Edward S. Holden, in files of Lick Observatory, June 19, 1889.

30. G.E.H. to H. M. Goodwin, June, 1889.

31. G.E.H. to H. M. Goodwin, July 1, 1889.

32. Article by G.E.H. on "Spectroscopic Astronomy," in the *Beacon*, Chicago, July, 1889.

33. G.E.H. to H. M. Goodwin, August 5, 1889.

34. Sir Robert Ball, *The Story of the Sun* (London: Cassell and Co. Ltd., 1893), p. 191.

35. G.E.H. description of the spectroheliograph, in *The Study of Stellar Evolution* (Chicago: University of Chicago Press, 1908), pp. 82–83.

36. C. A. Young to G.E.H., as reported to H. M. Goodwin by G.E.H., summer, 1889.

37. G.E.H. to H. M. Goodwin, August 16, 1889.

38. G.E.H. to H. M. Goodwin, August 28, 1889.

39. Record of college activities, in *Technique*, the M.I.T. 1890 class yearbook.

40. Transcript of G.E.H.'s record, obtained from M.I.T.

41. Thesis of G.E.H., published in *Technology Quarterly*, III, No. 4 (November, 1890).

42. Talk given by H. M. Goodwin at celebration in honor of G.E.H. at Harvard College Observatory, April, 1936.

43. Photograph by H. Notman Photo Co., Boylston St., Boston, Mass.

44. G.E.H. to H. M. Goodwin, July 4, 1890, from Hotel Vendome, San Jose, Calif.

45. Mary Hale to her mother, Mrs. Philemon Scranton, June, 1890.

46. G.E.H. to H. M. Goodwin, July 30, 1890.

47. Reminiscences of visit to Lick Observatory, recorded in G.E.H.'s autobiographical notes.

Chapter Three

1. G.E.H. to H. M. Goodwin, November 16, 1890.

2. G.E.H. to H. M. Goodwin, November 22, 1890.

3. Mary Hale to Ada Browne, November 18, 1890.

4. Personal reminiscences of E.C.H., as told to the author, 1947–48.

5. G.E.H. to H. M. Goodwin, January 26, 1891.

6. G.E.H., "The Rapid Development of American Research," unpublished manuscript in Hale Solar Laboratory.

7. E. C. Pickering to W. R. Harper, February 7, 1891.

8. W. R. Harper to Gayton Douglass, February 11, 1891.

9. Gayton Douglass to G.E.H., February 13, 1891.

10. W. R. Harper to G.E.H., May 13, 1891.

11. W. R. Harper to G.E.H., May 19, 1891.

12. G.E.H. to W. R. Harper, May 30, 1891.

13. Lincoln Ellsworth in *Beyond Horizons* (Garden City, N.Y.: Doubleday, Doran & Co., 1937), p. 5.

14. G.E.H. to H. M. Goodwin, May 15, 1891.

15. "Photography and the Invisible Solar Prominences," *Sidereal Messenger*, June, 1891.

16. J. A. Brashear to G.E.H., May 19, 1891.

17. G.E.H. to H. M. Goodwin, November 3, 1890.

18. Speeches given at the Kenwood Dedication, published in the *Daily Inter-Ocean*, and in *Science*, June, 1891.

19. G.E.H. to W.B.H. and M.H.H., July 5, 1891. Description of their prominence observation and coincidental observation by M. Fenyi, of Kalocsa, Hungary, given by G.E.H. in "Recent Results in Solar Prominence Photography," *Astron. and Astrophys.*, XI (1892), 235.

20. William Huggins, in *Publications of the Tulse Hill Observatory*, Vol. I. Quoted by G.E.H. in *The Study of Stellar Evolution* (Chicago: The University of Chicago Press, 1908), pp. 54–55.

21. *Ibid.*

22. W. Huggins, quoted by G.E.H. in "The Work of Sir William Huggins," *Ap. J.*, XXXVII (1910), 145–53—after Huggins' death, May 12, 1910.

23. *Ibid.* Also G.E.H. to Lady Huggins, January 9, 1913.

24. G.E.H. to H. M. Goodwin, August 21, 1891.

25. *Ibid.*

26. *Ibid.*

27. G.E.H., "The Ultra-Violet Spectrum in Prominences," read at Cardiff meeting of B.A.A.S., published in *Astron. and Astrophys.*, XI (1892), 50–59.

28. G.E.H. to H. M. Goodwin, July 10, 1891. On reception in England and on the Continent.

29. *Ibid.*, on Deslandres.

30. G.E.H. to H. F. Newall, on Deslandres, February 9, 1933.

31. J. A. Brashear to G.E.H., October 9, 1891.

32. G.E.H. to J. A. Brashear, October 21, 1891.

33. G.E.H. to C. A. Young, December 28, 1891.

34. G.E.H. to H. M. Goodwin, February 11, 1892.

35. H. H. Turner on G.E.H. when he received the gold medal of the R.A.S., 1904. *Monthly Notices of the R.A.S.*, LXIV, 1904.

36. G.E.H. to H. M. Goodwin, July 17, 1892. Described also in "A Remarkable Solar Disturbance," *Astron. and Astrophys.*, XI (1892), 611–13.

37. G.E.H. to W. W. Payne, November 4, 1891.

38. E. S. Holden to G.E.H., January 18, 1892.

39. R. A. Gregory to G.E.H., January 28, 1892.

40. G.E.H. to H. M. Goodwin, May 6, 1892.

41. Mary Hale to Ada Browne, February 8, 1892.

42. Account largely based on letter from Ferdinand Ellerman to E. C. Pickering on question of observing variable stars, February 25, 1891.

43. W. R. Harper to G.E.H., June 20, 1892.

44. G.E.H., "Some of Michelson's Researches," *P.A.S.P.*, XLIII (1931), 175–85.

45. A. Michelson to W. R. Harper, July 3, 1892.

46. W. R. Harper to G.E.H., July 1, 1892.

47. W.E.H. I to W. R. Harper, July 1, 1892.

48. G.E.H. to H. M. Goodwin, July 17, 1892.

49. T. W. Goodspeed, Secretary, University of Chicago, to G.E.H.

50. G.E.H., "The Rapid Development of American Research," unpublished.

51. G.E.H., "The Spectroheliograph of the Kenwood Astrophysical Observatory, Chicago, and Results Obtained in the Study of the Sun," *Am. Assoc. Proc.*, 1892, pp. 55–56.

52. Account of Yerkes' beginnings and meeting of A.A.A.S. in Rochester, N.Y., given by G.E.H. in address read at a meeting of the Am. Astron. Soc. at Yerkes, 1922.

53. G.E.H. to H. M. Goodwin, September 25, 1892.

54. Based on correspondence of W. R. Harper, T. W. Goodspeed, Charles Hutchinson, and Frederick T. Gates, University of Chicago Archives.

55. G.E.H. to W. R. Harper, September 23, 1892.

Chapter Four

1. Accounts of C. T. Yerkes' decision to build the largest telescope in the world, based on reminiscences of G.E.H., E.C.H., and contemporary newspaper reports.

2. W. R. Harper to Frederick T. Gates, October 10, 1892.

3. S. W. Burnham to E. E. Barnard, October 11, 1892.

4. Quote from W. R. Harper, reported in *Chicago Inter-Ocean*.

5. G.E.H. to J. A. Brashear, October 12, 1892.

6. J. A. Brashear to G.E.H., telegram, October 13, 1892.

7. J. A. Brashear to G.E.H., October 27, 1892.

8. S. W. Burnham to E. E. Barnard, October 25, 1892.

9. G.E.H. to W. R. Warner, October 28, 1892.

10. G.E.H. to Alvan G. Clark, October 28, 1892.

11. W. R. Harper to G.E.H., November 3, 1892.

12. Harriet Monroe, *A Poet's Life, Seventy Years in a Changing World* (New York: Macmillan Co., 1938), p. 117.

13. G.E.H. to C. T. Yerkes, November, 1892.

14. J. A. Brashear to G.E.H., December 30, 1892.

15. W. R. Harper to G.E.H., January 9, 1893.

16. J. A. Brashear to G.E.H., February 2, 1893.

17. G.E.H. on Harper in "The Beginnings of the Yerkes Observatory," written in 1922 for Yerkes meeting of the Am. Astron. Soc.

18. *Ibid.*

19. G.E.H. to H. M. Goodwin, March 5, 1893.

20. G.E.H. to H. M. Goodwin, July 17, 1892.

21. Account of expedition to Pike's Peak based on letters to H. M. Goodwin, on an unpublished account of G.E.H. written March 1, 1934, and on a

description of its scientific work in *Astron. and Astrophys.*, XIII (1894), 662–87.

22. J. E. Keeler's diary, with an account of his observatory at Mayport and of his boyhood, is in the Lick Observatory files.

23. G.E.H. to E. S. Holden, March 23, 1891.

24. Henry Crew, memories of G.E.H. and of the 1893 World's Fair, as told to the author.

25. Description of telescope mounting in *Chicago Tribune*, April 29, 1893.

26. G.E.H. to H. M. Goodwin, September 4, 1893.

27. Description of life in Germany by G.E.H. on this and the following pages is contained in an unpublished account and in his autobiographical notes.

28. G.E.H. to James Hall, a New York newspaperman, July 12, 1895.

29. Quotation on the nature of the astrophysicist, from G.E.H., *The Study of Stellar Evolution* (Chicago: University of Chicago Press, 1908), pp. 11–13.

30. T. C. Chamberlin to G.E.H., January 31, 1893.

31. G.E.H. to Warner and Swasey, April 1, 1893.

32. Journey to Etna recorded in unpublished reminiscences of G.E.H. The scientific aspect is described in "On Some Attempts to Photograph the Solar Corona without an Eclipse," *Astron. and Astrophys.*, XIII (1894), 662–87.

Chapter Five

1. Account of meeting with W. W. Payne, as told to the author by Henry Crew in Evanston, Ill. Also based on account given by G.E.H. to J. E. Keeler after the meeting, September 1, 1894.

2. G.E.H. to W. R. Harper, on his efforts and his hope of obtaining a guarantee fund of $1,000 a year for five years for the *Ap. J.*, September 6, 1894.

3. Results of meeting November 2, 1894, reported in *Ap. J.*, I (1895).

4. G.E.H. to J. E. Keeler, February 8, 1895.

5. J. E. Keeler to G.E.H., February 12, 1895.

6. J. E. Keeler to G.E.H., April 10, 1895.

7. G.E.H. to J. E. Keeler, April 12, 1895.

8. W. R. Harper to G.E.H., December 6, 1894.

9. G.E.H. to W. R. Harper, January 28, 1894.

10. T. J. J. See to W. R. Harper, November 15, 1895.

11. Account of further difficulties with T. J. J. See, based on correspondence between Harper and See: T. J. J. See to Harper, February 3, 1896; Harper to G.E.H., on See's dismissal, March 1, 1896. Also See to Harper, February 12, 1898, and March 4, 1898, in University of Chicago Archives.

12. Karl Runge to G.E.H., May, 1895.

13. G.E.H. to Karl Runge, June 11, 1895.

14. G.E.H. to C. T. Yerkes, October 28, 1895.

15. C. T. Yerkes to G.E.H., October 31, 1895.

16. G.E.H. to C. T. Yerkes, November 11, 1895.

17. G.E.H. to J. E. Keeler, January 23, 1896.

18. J. E. Keeler to G.E.H., January 27, 1896.

19. G.E.H. to H. M. Goodwin, August 13, 1896.

20. Story told to the author by E.C.H.

21. Attempt at corona at Yerkes Observatory, G.E.H. to E. B. Frost, May 6, 1895.

22. G.E.H. to W. R. Harper, January 2, 1897. Also personal memories of E.C.H. on days at Yerkes.

23. Comment of Margaret Brackenridge Hale, wife of W.E.H. II, made to the author.

24. Account of G. W. Ritchey, his background and personality, given by W. S. Adams and by other astronomers who knew him at Yerkes and Mount Wilson.

25. Description of E. E. Barnard, given by W. S. Adams, amplified by biographical memoir of Barnard of N.A.S., and from correspondence in the Lick Observatory files.

26. *Reminiscences and Letters of Robert Ball*, ed. W. Valentine Ball (Boston: Little, Brown & Co., 1915), pp. 342–43.

27. G.E.H. on E. E. Barnard, unpublished notes.

28. G.E.H. to W. R. Harper, April 13, 1897.

29. C. T. Yerkes to G.E.H., April 24, 1897.

30. Account of arrival of 40-inch lenses given in *Williams Bay Observer.*

31. *Chicago Times Herald,* May 20, 1897.

32. Description of workings of telescope given by G.E.H. in *The Study of Stellar Evolution, op. cit.,* pp. 21–23.

33. G.E.H. to C. T. Yerkes, May 31, 1897.

34. *Ibid.*

35. C. T. Yerkes to W. R. Harper, May 24, 1897.

36. Crash of rising floor described by G.E.H. to Charles Young, June 1, 1897, also in *Chicago Tribune,* May 30, 1897.

37. G.E.H. to J. E. Keeler, September 11, 1897.

38. Walter S. Adams, "Some Reminiscences of the Yerkes Observatory, 1898–1904," *Science,* CVI (September 5, 1947), 196–200.

39. H. H. Turner in "Oxford Notebook," published in *The Observatory.*

40. Account of arrival of Yerkes at the observatory, in *The Chicago Record,* October 20, 1897. Complete texts of speeches given at the dedication, published in the *Chicago Evening Post,* October 21, 1897. Keeler's address also published in the *Ap. J.,* 1897.

41. Account of banquet, etc., given in the *Chicago Journal.*

Chapter Six

1. G.E.H. to Stanley C. Reese, February 15, 1904.

2. Philip Fox, "George Ellery Hale," *Popular Astronomy,* XLVI, No. 8 (October, 1938).

3. Reminiscences of W. S. Adams, as told to the author. Based also on unpublished autobiographical account of Adams written for N.A.S.

4. Harlow Shapley on Walter Adams, "A Master of Stellar Spectra," *Sky and Telescope*, July, 1956, p. 401.

5. Karl Runge to G.E.H., after J. E. Keeler's appointment at the Lick Observatory, March 11, 1898.

6. Storrs Barrett to G.E.H., January 31, 1912.

7. F. Ellerman to G.E.H., February 10, 1899.

8. Based on W. S. Adams's reminiscences of life at Yerkes, as told to the author and in W. S. Adams, "Some Reminiscences of the Yerkes Observatory, 1898–1904," *op. cit.*

9. G.E.H. to J. E. Keeler, on his appointment as director at Lick Observatory, and on difficulties of an observatory directorship, March 10 and 12, 1898.

10. Keeler to G.E.H., April 19, 1899 and June 22, 1898.

11. G.E.H. to J. E. Keeler, August 2, 1900. J. E. Keeler died August 12, 1900.

12. Mrs. J. E. Keeler to G.E.H., September 19, 1900.

13. J. A. Brashear to G.E.H., November 25, 1898.

14. J. C. Kapteyn to G.E.H., October 23, 1920.

15. G.E.H. to H. M. Goodwin, November 25, 1898.

16. Stories told by G.E.H. as he wrote them down toward the end of his life. Account also based on memories of M.H.S.

17. G. W. Myers to G.E.H., March 17, 1898.

18. G.E.H. to E. C. Pickering, March 26, 1898.

19. G.E.H. to J. E. Keeler, September 11, 1897.

20. Simon Newcomb to G.E.H., December 24, 1898. Also G.E.H. to Newcomb, January 5, 1899.

21. G.E.H. to S. P. Langley, December 28, 1898.

22. G.E.H. on results with 40-inch, second annual report of the director of the Yerkes Observatory for the year ending June 30, 1899.

23. G.E.H., "On the presence of carbon in the chromosphere," *Ap. J.*, VI (1897), 412–14.

24. Agnes Clerke, *The History of Astronomy in the 19th Century* (London: A. & C. Black, 1902), p. 200.

25. G.E.H. to H. F. Newall, May 27, 1903.

26. Described in *New York Daily Tribune*, December 20, 1903.

27. G.E.H. to W. R. Harper, February 12, 1901.

28. G.E.H. to C. T. Yerkes, February 28, 1901.

29. G.E.H. coelostat destroyed in December, 1902. See "The Snow horizontal Telescope," *Ap. J.*, XVII (July, 1903).

30. Results of observation of Secchi's red stars. G.E.H., "The Spectra of Stars of Secchi's Fourth Type." (With F. Ellerman and J. A. Parkhurst.) Decennial publication, University of Chicago, Ser. 2, 8, pp. 251–385.

31. Otto Struve on G.E.H.'s results: statement made in personal conversation.

32. G.E.H. to N. C. Duner, March 23, 1899.

33. G.E.H. to E. C. Pickering, April 1, 1899.

34. Agnes Clerke, *The History of Astronomy in the 19th Century, op. cit.,* pp. 432–33.

35. G.E.H. to W. R. Harper, September 19, 1898.

36. G.E.H., "Statement Concerning the Reflecting Telescope of the University of Chicago." See also G.E.H. to W. R. Harper, May 9, 1899, and September 19, 1899.

37. W. R. Harper to G.E.H., September 21, 1898.

38. C. B. Scoville to G.E.H., on visit to Durand, December 5, 1899. Also G.E.H. to Scoville, on Griffith, December 13, 1899.

39. G.E.H. to E. F. Nichols, June 14, 1899.

40. G.E.H. to J. S. Ames, June 2, 1900. Also personal memories of H. M. Goodwin and account sent by C. G. Abbot to W. S. Adams, for the use of the author, January 8, 1950.

41. G.E.H. to F. L. O. Wadsworth, November 6, 1900. W.E.H. II born November 6.

42. *Reminiscences and Letters of Sir Robert Ball,* ed. W. Valentine Ball (Boston: Little, Brown & Co., 1915), p. 342.

43. Memories of E.C.H. and M.H.S., as told to the author.

Chapter Seven

1. *Chicago Tribune,* January 10, 1902.

2. G.E.H., autobiographical notes. Also G.E.H. to Mrs. Charles D. Walcott, July 16, 1929

3. J. A. Brashear to E. S. Holden, March 22, 1892, in Allegheny Observatory files.

4. Andrew Carnegie, *The Gospel of Wealth and other Timely Essays* (New York: The Century Co., 1900), p. 25.

5. G.E.H. to J. S. Billings, January 24, 1902.

6. This statement, undated, was originally prepared by G.E.H. in the hope of finding a donor for the 60-inch.

7. G.E.H. to Lewis Boss, August 15, 1902.

8. G.E.H. to Lewis Boss, September 2, 1902.

9. C. D. Walcott to G.E.H., December 2, 1902.

10. "Report of the Committee on Southern and Solar Observatories," *C.I. Yearbook,* No. 2. Account also based on letters in Lick Observatory files.

11. E. C. Pickering to G.E.H., report of Harvard expedition to Mount Wilson, September 21, 1903. S. I. Bailey, *History and Work of the Harvard Observatory,* Harvard Observatory Monograph, No. 4 (New York: McGraw-Hill Book Co., 1931). "History of the Harvard College Observatory, during the Period 1840–1890," by David W. Baker, in *Boston Evening Traveller,* August and September, 1890.

12. "A Study of the Conditions for Solar Research at Mt. Wilson, California," *Contributions from the Solar Observatory, Mt. Wilson,* No. 1, 1903.

13. W. J. Hussey to G.E.H., July 5, 1903.

14. Based on memories of G.E.H. in letter to Mrs. Charles D. Walcott, July 16, 1929, and recollections of W. S. Adams.

15. Harold Babcock, "In 1903," *P.A.S.P.*, LI (February, 1939).

16. G.E.H., diary record.

17. W. J. Hussey to G.E.H., W. W. Campbell. Also report of Hussey to Commission on Observatories to the C.I.W., October 14, 1903.

18. G.E.H. to W. W. Campbell, June 30, 1903.

19. W. J. Hussey to W. W. Campbell, July 1, 1903.

20. W. W. Campbell to Lewis Boss, July 28, 1903.

21. G.E.H. to J. S. Billings, July 6, 1903.

22. G.E.H. to J. S. Billings, November 9, 1903.

23. J. S. Billings to G.E.H., November 28, 1903.

24. W. W. Campbell to Lewis Boss, October 11, 1903, in Lick Observatory files.

25. J. S. Billings to G.E.H., December 11, 1903.

26. G.E.H. to Mrs. Charles D. Walcott, 1929, *op. cit.*

27. G.E.H., diary record.

28. G.E.H. to Mrs. Charles D. Walcott, 1929, *op. cit.* Also recollections of E.C.H.

29. G.E.H. to H. M. Goodwin, December 24, 1903.

30. Horace White's letter, quoted by G.E.H. to H. M. Goodwin, December 24, 1903.

31. C. D. Walcott to Andrew Carnegie, January 4, 1904.

32. A. Carnegie to C. D. Walcott, January 7, 1904.

33. G.E.H. to H. F. Newall, January 9, 1904.

34. Personal recollections of Seward Simons, as told to the author in Oakland, Calif.

35. G.E.H., diary record, January 10, 1904.

36. W. S. Adams to G.E.H., January 28, 1904.

37. S. B. Barrett to G.E.H., June 8, 1904.

38. S. B. Barrett to G.E.H., December 22, 1909.

39. G.E.H., diary record, February 29, 1904.

40. Description of George Jones, given by W. S. Adams and other Mount Wilson astronomers.

41. G.E.H. to Mrs. Charles D. Walcott, 1929, *op. cit.*

42. G.E.H. to Agnes Clerke, March 7, 1904.

43. Account of burros, etc., on Mount Wilson, as told to the author by W. S. Adams.

44. G.E.H. to W. W. Campbell, March 7, 1904.

45. G.E.H. to H. M. Goodwin, May 28, 1904.

Chapter Eight

1. G.E.H., diary entry, January 8, 1904.

2. W. H. Christie to G.E.H., December 17, 1903.

3. G.E.H. to H. H. Turner, account of his work to date, January 17, 1903.

4. H. H. Turner's address, given February, 1904, published in *Monthly Notices of the R.A.S.*, LXIV (1904), 388.

5. W. W. Campbell to G.E.H., December 26, 1903.

6. G.E.H. to Marian Hooker, August 19, 1929. Also "John Daggett Hooker" in *The History of Los Angeles*, by William A. Spalding (Los Angeles: J. R. Finnell & Sons Pub. Co., 1931), II, 177.

7. G.E.H., diary entry, February 3, 1904.

8. Description of house on West Adams St., given to the author by Alicia Mosgrove. House also visited by the author.

9. G.E.H., diary entry, February 4, 1904.

10. G.E.H. to H. M. Goodwin, February 25, 1904.

11. W. S. Adams to G.E.H., February 12, 1904.

12. G.E.H. to W. S. Adams, March 7, 1904.

13. G.E.H. to W. R. Harper, March 7, 1904.

14. Based on memories of W. S. Adams.

15. G.E.H. on work with Wadesboro coelostat, April 10, 1904.

16. Meeting with J. S. Billings in Washington, April 16, 1904, described in letter to Mrs. Charles D. Walcott, *op. cit.*, 1929.

17. Early history of N.A.S. and its aims, given by G.E.H. in "National Academies and the Progress of Research. The First Half Century of the N.A.S.," *Science*, n.s. XXXIX (1913), 189–200.

18. G.E.H. to H. M. Goodwin, May 20, 1902.

19. $10,000 grant made by C.I.W. to bring the Snow telescope to California. The expedition for solar research was officially organized under the joint auspices of the University of Chicago and the C.I.W., with the understanding that the funds granted by the C.I.W. would be used for the construction of piers and buildings and other expenses, while the University of Chicago would furnish the instrumental equipment, and pay the salaries of some of the members of the party.

20. Hale's report of interview with W. R. Harper, April 23, 1904.

21. Robert Curzon, *Visits to Monasteries in the Levant* (1849).

22. Memories of W. S. Adams, told to the author.

23. Recollections of Daisy Turner, sent to W. S. Adams, June 6, 1950.

24. Address of G.E.H. at Congress, September 23, 1904. "International Cooperation in Solar Research," Minutes of the Meeting of Delegates to the Conference on Solar Research, September 23, 1904, *Ap. J.*, XX (December, 1904), 301–12.

25. M. Minnaert in *Vistas in Astronomy*, ed. Arthur Beer (New York: Pergamon Press, 1955), I, 9.

26. Meeting with Billings, described by G.E.H. to Mrs. Charles D. Walcott, *op. cit.*, 1929.

27. Meeting at Mrs. Draper's, *ibid.*

28. G.E.H. to H. M. Goodwin, October 14, 1904.

29. G.E.H. to Mrs. Charles D. Walcott, *op. cit.*

30. G.E.H. to H. M. Goodwin, October 14, 1904.

31. *Ibid.*

32. G.E.H. to W.B.H., November 2, 1904.

33. W.B.H. to G.E.H., November 11, 1904.

34. Horace White to G.E.H., November 28, 1904.

35. G.E.H. to Andrew Carnegie, December 5, 1904.

36. Minutes of the trustees, in C.I.W. files in Washington.

37. G.E.H. to C. D. Walcott, December 13, 1904.

38. Telegram from C. D. Walcott to G.E.H., December 20, 1904. Also memories of G.E.H. recorded in letter to Mrs. Walcott, *op. cit.*, 1929. And G.E.H. to H. M. Goodwin, December 21, 1904.

39. R. S. Woodward to G.E.H., December 20, 1904.

Chapter Nine

1. G.E.H. to H. M. Goodwin, February, 1905.

2. W. W. Campbell to G.E.H., December 26, 1903.

3. Memories of Alicia Mosgrove, told to the author in San Francisco.

4. Memories of M.H.S., recorded by the author in Washington, D.C., and Glenn Dale, Md.

5. Memories of Alicia Mosgrove, *op. cit.*

6. G.E.H. to W. R. Harper, January 7, 1905.

7. W. R. Harper to G.E.H., January 17, 1905.

8. G.E.H. to Charles Hutchinson, February 23, 1905.

9. Memories of George Van Biesbroeck, told to the author, 1949.

10. A. A. Michelson to G.E.H., December 28, 1904.

11. G.E.H. to H. M. Goodwin, February 22, 1905.

12. G.E.H. to E. E. Barnard, April 30, 1906.

13. W. S. Adams on G.E.H.

14. Memories of W. S. Adams and M.H.S.

15. Description of truck, recorded by G.E.H. and W. S. Adams.

16. Memories of W. S. Adams.

17. G.E.H. to R. J. Wallace at Yerkes, March, 1906.

18. Memories of motorcycle, W. S. Adams and M.H.S., as told to the author.

19. H. D. Babcock and W. S. Adams, memories of life on the mountain.

20. Poem written by G.E.H. on a yellow pad, in the Hale Solar Laboratory files.

21. Extracts from correspondence between G.E.H. and C. G. Abbot, in Hale Solar Laboratory files.

22. *Henry Gordon Gale, 1874–1942*, collected and arranged by his wife, Agnes Cook Gale (Chicago: 1943).

23. Description of Snow telescope, in G.E.H., *The Study of Stellar Evolution, op. cit.;* also in *Ap. J.*, XVII (1903), 14.

24. G.E.H. to Dr. George Isham, nephew of Helen Snow, who was instru-

mental in persuading her to make the gift of the Snow telescope, October 27, 1905.

25. S. P. Langley, *The New Astronomy, op. cit.,* 1884, pp. 20 and 24.

26. G.E.H., *The New Heavens* (New York: Charles Scribner's Sons, 1922), pp. 64–65.

27. G.E.H., "The Spectroscopic Laboratory of the Solar Observatory," *Ap. J.,* XXIV (1906), 61–68.

28. G.E.H. to R. S. Woodward, July 24, 1906.

29. G.E.H. to Arthur Noyes, September 24, 1906, and G.E.H. to H. H. Turner, October 3, 1906.

30. W. S. Adams to author on importance of Hale's observation.

31. W. S. Adams, "Sunspots and Stellar Distances," *C.I.W. Publication,* No. 501 (1938), 135–47. *P.A.S.P.,* No. 51 (1939), 133–46.

32. *Ten Years Work of a Mountain Observatory,* pub. No. 235 (Washington, D.C.: Carnegie Institution of Washington, 1915), pp. 20–23.

33. J. Halm to G.E.H., quoted in letter from G.E.H. to R. S. Woodward, March 9, 1908.

34. Sir William Huggins to G.E.H., March 26, 1908.

Chapter Ten

1. G.E.H. to W. W. Campbell, May 6, 1908, and G.E.H. to Sir W. Huggins, May 9, 1908.

2. J. A. Brashear to G.E.H., May, 1908.

3. Account of St. John, by W. S. Adams and H. D. Babcock.

4. G.E.H. to Edwin Frost, June 15, 1908.

5. G.E.H., "Solar Vortices," *Ap. J.,* XXVIII (1908), 100–16.

6. G.E.H. on Thomson's discovery and early work of Rowland and Zeeman in his autobiographical notes. See also "Solar Vortices and Magnetic Fields," *Proc. Royal Institution,* XIX (1909), 615–30.

7. G.E.H. to R. S. Woodward, December 19, 1905.

8. G.E.H. to R. S. Woodward, October 29, 1907. Also "The Tower Telescope of the Mt. Wilson Solar Obs.," *Ap. J.,* XXVII (1908), 204–12.

9. G.E.H. to Andrew Carnegie, November 27, 1907.

10. G.E.H. to J. A. Brashear, June 8, 1908.

11. Reminiscences of C. G. Abbot sent to W. S. Adams, for use of the author, November 8, 1950.

12. G.E.H. to H. M. Goodwin, July 7, 1908.

13. R. S. Woodward to G.E.H., July 29, 1908.

14. G.E.H. to Sir W. Huggins, July 6, 1908.

15. J. C. Kapteyn to G.E.H., September 3, 1908.

16. P. Zeeman to G.E.H., November 25, 1908.

17. W. S. Adams in Biographical Memoir of G.E.H., *Proc. N.A.S.,* XXI (1940), 202.

18. Comment of H. D. Babcock to the author, 1965.

19. V. Bumba and Robert Howard, "Solar Magnetic Fields," *Science*, CXLIX (September 17, 1965), 1332.

20. H. W. Babcock, personal communication, 1965.

21. G.E.H. to Sir Joseph Larmor, February 5, 1934.

22. G.E.H., *Ten Years Work of a Mountain Observatory, op. cit.*, 1915, pp. 30–31.

23. G.E.H. to A. A. Michelson, July 7, 1908.

24. A. Eddington, "Recent Results of Astronomical Research," *Proc. Royal Institution*, March, 1909.

25. Address given by G.E.H., May 14, 1909. "Solar Vortices and Magnetic Fields," *Proc. Royal Institution*, XIX (1909), 615–30.

26. G.E.H., autobiographical notes. Also Minutes of Mount Wilson Toll Road Co.

27. Godfrey Sykes, *A Westerly Trend* (Tuscon: Arizona Pioneers Historical Society, 1944), p. 264.

28. G.E.H. to R. S. Woodward, January 10, 1907.

29. Memories of W. S. Adams, told to the author.

30. G.E.H. to R. S. Woodward, November 27, 1907.

31. Description of optical shop, given by G.E.H. in *The Study of Stellar Evolution* (Chicago: University of Chicago Press, 1908), 223–24.

32. *Los Angeles Examiner*, August 9, 1908.

33. G.E.H. to R. S. Woodward, December 7, 1908.

34. G.E.H. in *The Study of Stellar Evolution, op. cit.*, 1908, pp. 227–28.

35. G.E.H. to Sir W. Huggins, December 20, 1908.

36. G.E.H. to E. B. Frost, June 4, 1908.

37. G.E.H. in *The Study of Stellar Evolution, op. cit.*, 1908, pp. 7–8.

38. G.E.H. on results with 60-inch, 1913, Mount Wilson Solar Observatory Annual Report, *C.I. Yearbook*, No. 12, 1913.

39. Memories of H. D. Babcock.

40. Memories of Hale and Mount Wilson, sent to the author by Giorgio Abetti from Florence, Italy, translated by Mrs. Gerald Kron.

41. G.E.H. on results with 60-inch, 1915, Mount Wilson Solar Observatory Annual Report, *C.I. Yearbook*, No. 14, 1915.

42. H. D. Curtis, Address of the retiring president of the A.S.P., in awarding the Bruce Medal to J. C. Kapteyn, *P.A.S.P.*, XXV, 1913, 15–27.

43. H. D. Babcock, personal communication.

44. *Ten Years Work of a Mountain Observatory, op. cit.*, 1915, pp. 50–51.

Chapter Eleven

1. Based on memories of G.E.H., E.C.H., M.H.S., W.E.H. and his wife, M.B.H., as told to the author; also on a visit to Hermosa Vista.

2. Account of early Pasadena and the beginnings of Throop, based on local histories, such as H. D. Carew, *History of Pasadena in the San Gabriel Valley, Calif.* (Chicago: S. J. Clarke Pub. Co., 1930), pp. 437 ff. Lon F. Chapin, *Thirty*

Years in Pasadena (Los Angeles: Southwest Publishing Co., 1929). C. F. Holder, *All About Pasadena and its Vicinity* (Boston: Lee & Shepard: 1889).

3. January 8, 1904, Throop Executive Committee meeting: "Pres. Edwards reported that Prof. Hale of Chicago has been secured to deliver a lecture in Throop Assembly." The subject was "Evolution of the Stars."

4. Based on Hale's autobiographical notes. Also G.E.H. to H. S. Pritchett, May 8, 1907.

5. S. Hazard Halsted to G.E.H., April 29, 1907.

6. G.E.H. to Simon Newcomb, March 10, 1906.

7. A. A. Noyes to G.E.H., September 15, 1906.

8. G.E.H. to A. A. Noyes, October 4, 1906.

9. "A Plea for the Imaginative Element in a Technical Education," *Technology Quarterly*, M.I.T., 1907.

10. Comment of H. D. Babcock.

11. Memories of G.E.H.'s visit to Boston, given by H. M. Goodwin.

12. G.E.H. to H. M. Goodwin, November 27, 1907.

13. G.E.H. to E. B. Frost, December 3, 1926.

14. J. A. B. Scherer unpublished autobiography, owned by Paul A. Scherer.

15. Visit to Skibo, G.E.H. to R. S. Woodward, June 17, 1907.

16. G.E.H. to Throop Trustees, 1907—quoted by J. A. B. Scherer at his induction ceremony, November 19, 1908.

17. August 7, 1907, "The Board [of Trustees] proceeded to elect by ballot a trustee to fill the vacancy caused by the resignation of Mr. J. S. Cravens. Prof. George E. Hale was unanimously chosen to such position."

18. Quote from J. D. Rockefeller, told to A. Mosgrove by G.E.H.

19. G.E.H. to George V. Wendell, January 13, 1908.

20. G.E.H. to J. A. B. Scherer, May 9, 1908.

21. J. A. B. Scherer unpublished autobiography.

22. J. A. B. Scherer to G.E.H., July 6, 1908.

23. N. Bridge to G.E.H., July 5, 1908.

24. J. A. B. Scherer inducted, November 19, 1908.

25. J. A. B. Scherer, unpublished autobiography.

26. Andrew Carnegie to J. A. B. Scherer, read at Dedication Ceremony, June 8, 1910.

27. On Throop, from *Six Collegiate Decades, the Growth of Higher Education in Southern California* (Los Angeles: Security-First National Bank of Los Angeles, 1929).

Chapter Twelve

1. G.E.H., *The Study of Stellar Evolution* (Chicago: University of Chicago Press, 1908), p. 242.

2. A. Mosgrove, memories of negotiations with Hooker on 100-inch telescope.

3. G.E.H. to J. D. Hooker, July 27, 1906.

4. J. D. Hooker on 100-inch plan, quoted by G.E.H. to R. S. Woodward, September 1, 1906, and G.E.H. to W. W. Campbell, September 29, 1906.

5. G.E.H. to R. S. Woodward, July 13, 1906.

6. R. S. Woodward to G.E.H., December 12, 1906.

7. G.E.H. to R. S. Woodward, September 19, 1906.

8. G.E.H. to R. S. Woodward, December 12, 1908.

9. G.E.H. to H. M. Goodwin, January 7, 1909.

10. G.E.H. to Sir W. Huggins, January 28, 1909.

11. Based on memories of W. S. Adams, A. Mosgrove and M.H.S.

12. Memories of H. D. Babcock.

13. G.E.H. to Sir W. Huggins, February 24, 1909.

14. A. Schuster to G.E.H., May 4, 1909.

15. G.E.H. to E.C.H., April 20 and April 25, 1909.

16. G.E.H. to E.C.H., May 3, 1909.

17. G.E.H. to E.C.H., May 4 and May 10, 1909.

18. G.E.H. to W. W. Campbell, September 17, 1909.

19. G.E.H. to G. Schiaparelli, January 12, 1910, also April 4, 1910.

20. G.E.H. to R. S. Woodward, February 26, 1910.

21. Carnegie visit, March 17, 1910, described extensively in the local newspapers.

22. H. N. Russell on Hale and H. F. Osborn. Communication sent to W. S. Adams for use by the author, July 23, 1950.

23. F. Schlesinger to G.E.H., June, 1910.

24. J. A. B. Scherer to A. Fleming, September 27, 1910.

25. H. D. Babcock, personal communication to the author.

26. August 31, 1910, newspaper accounts of Solar Union meeting. Also account given by W. S. Adams to the author.

27. G.E.H. on aims of Solar Union. Address at opening of 4th Conference of the International Union for Cooperation in Solar Research, 1910, *Transactions of the International Union for Cooperation in Solar Research*, No. 3 (1911), 11–26.

28. Second 100-inch disk broken in annealing oven, May 10, 1910.

29. Andrew Carnegie to Sir George Reid, June 11, 1910.

30. Memories of G. W. Ritchey, by W. S. Adams and other astronomers at Mount Wilson.

31. G.E.H. to H. F. Newall, October 27, 1910.

32. R. S. Woodward to G.E.H., September 30, 1910.

33. J. D. Hooker to G.E.H., October 6, 1910.

34. G.E.H. to H. M. Goodwin, November 9, 1910.

35. G.E.H. to J. D. Hooker, December 12, 1910.

36. Release from J. D. Hooker, December 15, 1910.

37. G.E.H. to W. S. Adams, December 18, 1910.

38. E.C.H. to W. S. Adams, December 24, 1910.

39. W. S. Adams to G.E.H., December, 1910.

40. J. D. Hooker to Andrew Carnegie, December 15, 1910, in the Library of Congress Archives.

41. Andrew Carnegie to J. D. Hooker, December 21, 1910.

42. G.E.H. to H. M. Goodwin, December 17, 1910.

43. G.E.H. to H. M. Goodwin, March 26, 1911. Also interview of author with Dr. L. Hunnicutt.

44. G.E.H. to Andrew Carnegie, January 23, 1911, also to R. S. Woodward. Also G.E.H., autobiographical notes, and letter to Mrs. Walcott, 1929, *op. cit.*

45. Carnegie's letter of gift, January 19, 1911, in C.I.W. files.

46. G.E.H., diary notation, January 22, 1911.

47. E.C.H. to R. S. Woodward, February 3, 1911.

48. R. Maclaurin to R. S. Woodward, in C.I.W. files, January 24, 1911.

Chapter Thirteen

1. J. H. Breasted to G.E.H., April 25, 1931.

2. Based on material in Charles Breasted, *Pioneer to the Past* (New York: Charles Scribner's Sons, 1947), pp. 134–36.

3. G.E.H. to W. S. Adams, February 8, 1911.

4. G.E.H. to H. M. Goodwin, March 26, 1911.

5. E.C.H. to J. A. B. Scherer, March 1, 1911.

6. Address of T. Roosevelt, published in Pasadena newspapers, March, 1911.

7. C. F. Holder to G.E.H., March 22, 1911.

8. J. A. B. Scherer to G.E.H., February 7, 8, and 11, 1911.

9. Records of B. I. Wheeler in the archives of the president, University of California, notes on Scherer's visit. Also B. I. Wheeler to J. A. B. Scherer, February 15, 1911.

10. Memories of J. A. B. Scherer, and accounts sent to G.E.H.

11. *Pasadena Star News* and other southern California newspapers.

12. J. A. B. Scherer to G.E.H., April 5, 1911.

13. G.E.H. to J. A. B. Scherer, May 7, 1911.

14. G.E.H. to H. M. Goodwin, June 12, 1911. Also G.E.H. to H. F. Newall, 1930.

15. J. H. McBride to G.E.H., July 11, 1911, etc.

16. Accounts of Bethel. G.E.H. to H. M. Goodwin, July 30, August 13, September 7, 1911.

17. G.E.H. to E.C.H., August 13 and September 9, 1911.

18. G.E.H. to M.H.S., August 3, 1911.

19. Elizabeth Connor, personal recollections sent to the author, 1963.

20. Paul Merrill and H. D. Babcock, personal communications to the author.

21. Elmer Grey, personal communication to the author.

22. Margaret Harwood, recollections tape-recorded, 1963.

23. F. H. Seares, "The Scientist Afield," *Isis*, XXX, 2, No. 81 (May, 1939), 260 ff.

24. G.E.H. to W. J. Hussey, November, 1915.

25. G.E.H. to C. Dodge, July 25, 1908, and G.E.H. to R. S. Woodward, September 16, 1908.

26. R. S. Woodward to G.E.H., November 20, 1908.

27. Memories of C. G. Abbot on 150-foot tower telescope, as told to the author via W. S. Adams, January 8, 1950.

28. "The 150′ Tower Telescope of the Mt. Wilson Solar Observatory," *P.A.S.P.*, XXIV (1912), 223–26.

29. G.E.H. to E.C.H., February 18, 1912.

30. G.E.H. to R. S. Woodward, February 12 and February 28, 1913.

31. G.E.H. on polarities in spots: "The Polarity of Sun-spots," with F. Ellerman, read at Evanston meeting, American Astronomical Society, 1914. Abstract in *Pub. A.A.S.*, III (1914), 81–82. See also "Sunspots as Magnets," in *The Depths of the Universe* (New York: Charles Scribner's Sons, 1924), pp. 91–94.

32. G.E.H. to W.B.H., April 29, 1913.

33. G.E.H. to W.B.H., June 18, 1913.

34. G.E.H. to F. R. Moulton, May 26 and June 29, 1914.

35. F. R. Moulton, "Our Twelve Great Scientists," *Technical World*, November, 1914, p. 342.

36. G.E.H. to M.H.S., August 23, 1915. Also memories of H. D. Babcock and W.E.H. II.

37. Memories of W. S. Adams.

38. G.E.H. to H. H. Turner, February 19, 1913.

39. G.E.H. to R. S. Woodward, February 14, 1916.

40. G.E.H. to R. S. Woodward, October 9, 1913.

41. Draft of introduction to *Ten Years Work of a Mountain Observatory*, not used.

42. G.E.H., *Ten Years Work of a Mountain Observatory* pub. No. 235 (Washington, D.C.: Carnegie Institution of Washington, 1915), pp. 5–6.

Chapter Fourteen

1. Background of war, in G.E.H.'s autobiographical notes.

2. G.E.H. to Dr. William H. Welch, July 3, 1915.

3. W. H. Welch to G.E.H., July 14, 1915.

4. Naval Consulting Board formed July 3, 1915.

5. W.B.H. to G.E.H., October 11, 1915.

6. G.E.H. to H. M. Goodwin, February 27, 1916.

7. G.E.H. to R. S. Woodward, March 1, 1916.

8. G.E.H. to E.C.H., April 20, 1916.

9. G.E.H. to E.C.H., April 25, 1916.

10. Resolution sent to W. Wilson by W. Welch, April 23, 1916.

11. Account based on letters and personal recollections of E. G. Conklin, as told to the author at Princeton, N.J. See also W. H. Welch, "Memorandum of a conference with the President of the U.S. at 12:45 noon"; and account sent to E.C.H. by G.E.H., April 26, 1916.

12. G.E.H. to J. A. B. Scherer, May 19, 1916.

13. Title of N.R.C. proposed by A. Noyes, July 6, 1916.

14. G.E.H. to W.B.H., June, 1916.

15. G.E.H. to W. S. Adams, June 22, 1916.

16. G.E.H. to E. M. House, July 21, 1916.

17. E. M. House to G.E.H., July 22, 1916.

18. W. Wilson to W. Welch, July 24, 1916, copy of letter sent by Wilson to G.E.H.

19. G.E.H. to Gano Dunn, July 25, 1916.

20. Michael Pupin, *From Immigrant to Inventor* (New York: Charles Scribner's Sons, 1924). Account of N.R.C. given on pp. 349–87.

21. Account of G.E.H., given to the author in New York by Gano Dunn.

22. W. Welch on G.E.H., records in the Welch Library, Johns Hopkins University. Also Simon and James T. Flexner, *William Henry Welch and the Heroic Age of American Medicine* (New York: The Viking Press, 1941).

23. W. Welch, diary record, September 6 and 7, 1916. See also G.E.H. to Simon Flexner, November, 1934.

24. G.E.H. to Gano Dunn, October 12, 1916.

25. G.E.H. to Woodrow Wilson, October 25, 1916.

26. G.E.H. to E. M. House, October 31, 1916.

27. W. Welch to G.E.H., February 8, 1917.

28. G.E.H. to W. Welch, February 21, 1917.

29. G.E.H. to Woodrow Wilson, February 4, 1917.

30. Telegram W. Welch to G.E.H., February 4, 1917, and Cary Hutchinson to G.E.H., March 1, 1917.

31. G.E.H. to M.H.S., May 12, 1917.

32. Memories of A. Mosgrove, told to the author.

33. G.E.H. to M.H.S., November 16 and November 22, 1917.

34. H. H. Turner to W.S.A., December, 1917.

35. Recollections of Robert M. Yerkes, recorded by the author in meeting in New Haven. Amplified by Dr. Yerkes in letter to the author, June 18, 1952, "My Acquaintance with George Ellery Hale." Also recorded in correspondence between G.E.H. and R. M. Yerkes. Plan for Research Information Committee outlined March 15, 1918.

36. G.E.H. to Woodrow Wilson, March 26, 1918.

37. Woodrow Wilson to G.E.H., April 19, 1918.

38. G.E.H. to E.C.H., April 22, 1918.

39. G.E.H. to E.C.H., May 10, 1918.

40. Executive Order signed by Woodrow Wilson, May 10, 1918.

41. M.H.S., reminiscences of Washington in World War I.

42. G.E.H. to E.C.H., May 30, 1918.

43. G.E.H. diary entry, June 8, 1918.

44. G.E.H. diary entry, July 11, 1918.

45. Annual Report of Mount Wilson Observatory, 1918.

46. Otto Struve, "The Story of an Observatory," *Pop. Astron.* LV (May and June, 1947), 27.

47. Record of Robert Goddard, quoted in Milton Lehman, *This High Man* (New York: Farrar, Straus & Co., 1963), p. 95.

48. R. Goddard, "A Method of Reaching Extreme Altitudes," Smithsonian Miscellaneous Collections, LXXI, No. 2 (1919).

49. G.E.H. on I.R.C., autobiographical notes, and F. H. Seares, "The Scientist Afield," *op. cit.*

50. G.E.H. to A. Schuster, April 18, 1918. See also G.E.H., "Sir Arthur Schuster," *Ap. J.*, LXXXI (1935), 97–106.

51. *Ibid.*

52. R. A. Millikan, personal communication.

53. G.E.H. to E.C.H., undated.

54. G.E.H. to E.C.H., October 12, 1918.

55. *Ibid.*

56. G.E.H. to E.C.H., October 14, 1918.

57. G.E.H. to E.C.H., November 5, 1918.

58. G.E.H. to E.C.H., November 10, 1918. Abstract of paper on "The Nature of Sunspots," given by G.E.H. at Royal Society, pub. in *Proc. R. Soc. Ser. A,* XCV (1919), 234–36.

59. G.E.H. to E.C.H., November 15, 1918.

60. G.E.H. to E.C.H., November 18, 1918.

61. G.E.H. to H. M. Goodwin, November 26, 1918. (See also "International Organization of Science," Report on Conference held at Paris, November 26–29, 1918.)

62. G.E.H. to M.H.S., December 6, 1918.

63. M.H.S. to G.E.H., November 10, 1918.

64. G.E.H. on N.R.C. in peacetime, autobiographical notes.

65. G.E.H. diary note, October 14, 1916. "Maclaurin [at M.I.T.] agreed to establish Research Com. and Research Fellowships."

66. G.E.H. to J.A.B. Scherer at Throop, confirming plan, October 26, 1916.

67. G.E.H. to Julius Stieglitz, January 6, 1917.

68. R. A. Millikan to G.E.H., June 15, 1918.

69. Action of Rockefeller Foundation, April, 1919: $100,000 per year for 5 years voted for National Research Fellowships in chemistry and physics.

Additional background material on the N.R.C. by G.E.H.: Introduction; "Science and War," chap. 1; "War services of the National Research Council," chap. 2; "The Possibilities of Co-operation in Research," chap. 22; "The International Organization of Research," chap. 23. In *The New World of Science,* ed. Robert M. Yerkes (New York: The Century Co., 1920).

Chapter Fifteen

1. G.E.H. autobiographical notes on N.A.S. building. G.E.H. to Simon Newcomb, March 21, 1906, "We really ought to have a separate building of our own, where a library could be maintained and where members could go whenever they visit Washington."

2. G.E.H. to Charles D. Walcott, January 25, 1908. Already (1905) he had advocated the election of ten instead of five members annually, hoping in this way to instill new blood into the mutual admiration society. He suggested a membership limit of 200.

3. G.E.H. to C. D. Walcott on D. H. Burnham, May 17, 1912.

4. G.E.H. to Elihu Root, December 30, 1912, March 3 and May 7, 1913.

5. G.E.H. to Charles Walcott, May 17, 1912.

6. G.E.H. to E. Rutherford, June 1, 1914.

7. "National Academies and the Progress of Research." I. Work of European Academies, *Science,* n.s. XXXVIII (1913), 681–98; II. The first half century of the N.A.S., *Science,* n.s. XXXIX, 189–200; III. The future of the N.A.S., *Science,* n.s. XL (1914), 407–19.

8. G.E.H. to E.C.H., April 18, 1914. Also G.E.H.'s memories of visit to Andrew Carnegie, sent to Burton Hendrick.

9. G.E.H. to E.C.H., April 22, 1914.

10. G.E.H. to Andrew Carnegie, March 3, 1914.

11. H. Pritchett to G.E.H., May 7, 1914.

12. Andrew Carnegie to G.E.H., May 11, 1914.

13. J. Breasted to G.E.H., June 8, 1914.

14. J. A. B. Scherer to G.E.H., June 14, 1914.

15. G.E.H. to J. A. B. Scherer, July 3, 1914.

16. G.E.H., "National Academies and the Progress of Research." IV. The Proceedings of the Academy as a Medium of Publication, *Science,* n.s. XLI (1915), 815–17.

17. G.E.H. to J. A. B. Scherer, May 26, 1914.

18. G.E.H. to M.H.S., February 16, 1919.

19. G.E.H. to E. B. Frost, May 12, 1919.

20. G.E.H., "Some recollections of Bertram Grosvenor Goodhue," in memorial volume, *B. G. Goodhue, Architect and Master of Many Arts* (American Institute of Architects, 1925), 45–50.

21. G.E.H. to E.C.H., April 22, 1920.

22. Memories of Gano Dunn, as told to the author in New York.

23. G.E.H. to W.B.H., February 28, 1923.

24. Wallace K. Harrison, "The Building of the National Academy and the National Research Council," *Architecture,* L, No. 4 (October, 1924).

25. G.E.H., "A National Focus of Science and Research," *Scribner's Magazine,* November, 1922, 515–31 (quote on p. 523).

26. Gano Dunn to G.E.H., April 15, 1924.

27. G.E.H. to H. F. Newall, June 8, 1924.

28. A. A. Michelson to G.E.H. on resolution, April 27, 1927.

Chapter Sixteen

1. Memories of November 1, 1917, recorded in the Prefatory Note to Alfred Noyes, *Watchers of the Sky* (New York: Frederick Stokes, 1922).

2. Memory of night by W. S. Adams, as told to the author. Also memories of G.E.H. in autobiographical notes.

3. Alfred Noyes, personal communication to the author on visit to Hale Solar Laboratory.

4. Alfred Noyes, Prefatory Note to *Watchers of the Sky*, p. v., *op. cit*. See also Alfred Noyes, *Two Worlds for Memory* (Philadelphia: J. B. Lippincott, 1953).

5. Extracts from *Watchers of the Sky, op. cit.*, pp. 4–5.

6. Memories of W. S. Adams.

7. G.E.H. to Sir James Jeans, November 3, 1917.

8. Record in observing log in dome on Mount Wilson, September 22, 1919.

9. G.E.H. to W.B.H., August 10, 1919.

10. G.E.H. to R. S. Woodward, July 2, also September 19, 1919.

11. *The Study of Stellar Evolution, op. cit.*, 1908, pp. 44–45.

12. Heber D. Curtis, *op. cit., P.A.S.P.*, XXV (1913), 15–27.

13. Harlow Shapley, personal communication.

14. Harlow Shapley, *P.A.S.P.*, XXIX (1917), 213.

15. "The Depths of the Universe," *Scribner's Magazine*, LXXI (1922), 689–704.

16. G.E.H. to Harlow Shapley, March 14, 1918, from Washington, D.C.

17. G.E.H., "The New Heavens," *Scribner's Magazine*, LXVIII (1920), 387–402.

18. G.E.H. to A. Lawrence Lowell, March 29, 1920.

19. *Bulletin of the National Research Council*, II, Part 3 (May, 1921).

20. "The Depths of the Universe," *op. cit.*, 1922.

21. G.E.H., "Beyond the Milky Way," *Scribner's Magazine*, LXXX (1926), 276–91.

22. Allan Sandage, *The Hubble Atlas of Galaxies* (Washington, D.C.: Carnegie Institution of Washington (1961), pub. No. 618.

23. Milton Humason, tape recording, 1964.

24. W. S. Adams on G. W. Ritchey's characteristics, as told to the author. Also W. S. Adams to G.E.H.

25. G.E.H. on Carnegie presidency, 1921.

26. G.E.H. to W.B.H., March 23, 1921.

27. Name of Throop changed to California Institute of Technology, February 10, 1920, annual meeting of trustees.

28. E. C. Barrett to G.E.H., January 4, 1920.

29. Warren Weaver, quoted by George W. Gray to the author from his unpublished manuscript.

30. J. A. B. Scherer to Arthur Fleming, March 3, 1920.

31. J. A. B. Scherer autobiography, *op. cit.*

32. G.E.H. to E.C.H., April 15, 1920. Also June 7, 1920.

33. G.E.H. to Arthur Fleming, May 24, 1920.

34. R. A. Millikan to G.E.H., June 19, 1920.

35. G.E.H., diary entry, January 25, 1921.

36. G.E.H. to H. M. Goodwin, February 20, 1921.

37. G.E.H. to Henry M. Robinson, February 25, 1921.

38. G.E.H. to E.C.H., May 17, 1921.

39. G.E.H. to E.C.H., May 18, 1921.

40. G.E.H., diary entry, May 26, 1921.

41. G.E.H. to E.C.H., May 31, 1921.

42. R. A. Millikan, as told to the author.

43. G.E.H. to W. S. Adams, June 6, 1921.

44. H. P. Judson to R. A. Millikan, quoted by R. A. Millikan to the author.

45. Address by G.E.H. at dedication of Norman Bridge Laboratory, "A Joint Investigation of the Constitution of Matter and the Nature of Radiation," *Science*, n.s. LV (1922), 332–34.

46. G.E.H. to H. M. Goodwin, February 5, 1921.

47. J. C. Kapteyn to G.E.H., April 10, 1921.

48. G.E.H., "Observation of Invisible Sunspots," *Proc. N.A.S.*, VIII (1922), 168–70.

49. "Law of the Sun-Spot Polarity" (with S. B. Nicholson), *Ap. J.*, LXII (1925), 270–300.

50. G.E.H. in *The Telescope*, May–June, 1936.

51. G.E.H. to C. G. Abbot, May 3, 1935.

52. H. D. Babcock, personal communication, 1965.

53. G.E.H. to W.B.H., August 22 and September 17, 1922.

54. G.E.H. to M.H.S., October 10, 1922.

55. G.E.H. to M.H.S, December 22, 1922.

56. G.E.H. to H. M. Goodwin, January 26, 1923.

57. "Recent Discoveries in Egypt," *Scribner's Magazine*, LXXIV (1923), 34–49. "The Work of an American Orientalist," *Scribner's Magazine*, LXXIV (1923), 392–404. (For a further account of Hale's role in the carrying on of Breasted's work in Egypt, see Charles Breasted, *Pioneer to the Past, op. cit.*)

58. J. C. Merriam to G.E.H., March 10, 1923.

59. G.E.H. to J. C. Merriam, March 29, 1923.

60. G.E.H. to W. S. Adams, May 17, 1923.

61. J. C. Merriam to G.E.H., May 7 and May 22, 1923. (G.E.H. became honorary director, July 1, 1923.)

62. G.E.H. to H. M. Goodwin, June 1, 1923.

63. G.E.H. to W. S. Adams, May 17, 1923.

64. *Pioneer to the Past, op. cit.*

Chapter Seventeen

1. G.E.H. to H. M. Goodwin, August 23, 1924.

2. G.E.H. to George Sarton, April 29, 1931.

3. G.E.H., unpublished manuscript, "Everyman's London."

4. G.E.H., autobiographical notes. G.E.H. to W.B.H., November 18, 1923.

5. G.E.H., diary entry on spectrohelioscope, September 22, 1922.

6. G.E.H. to H. F. Newall, February 8, 1924.

7. G.E.H., autobiographical notes. Also diary entry, *op. cit.*, describing cost of solar laboratory which he was covering himself. At least $30,000.

8. G.E.H. to Lee Lawrie, November 24, 1926.

9. G.E.H. correspondence with E. B. Frost at Yerkes, January and March, 1924.

10. G.E.H. to Gano Dunn, January 26, 1926.

11. G.E.H., *Signals from the Stars* (New York: Charles Scribner's Sons, 1931), pp. 46, 71–72.

12. G.E.H. to H. M. Goodwin, January 31, 1926. Also described in "The Spectrohelioscope," *P.A.S.P.*, XXXVIII (April, 1926).

13. G.E.H. to G.E.H. II, March 30, 1926.

14. G.E.H. to J. C. Merriam, July 12, 1926. Also "Observations with the Spectrohelioscope," *Nature*, September 18, 1926.

15. G.E.H. to C. G. Abbot, November 1, 1926.

16. G.E.H. to H. H. Turner, November 2, 1926.

17. H. H. Turner to G.E.H., January 2, 1924.

18. "Work for the Amateur Astronomer," *P.A.S.P.*, April, 1916 and October, 1928.

19. Correspondence with Jack Garrison over a period of years, 1919–28.

20. G.E.H. to T. H. Morgan, November 6, 1927.

21. Memories of H. D. Babcock.

22. G.E.H. to J. C. Merriam, February 23, 1926.

23. G.E.H. to J. C. Merriam, July 21, 1925.

24. "Giant Stars," *Scribner's Magazine*, LXX (1921), 3–15; "Cosmic Crucibles," *Scribner's Magazine*, LXX (1921), 387–401.

25. J. H. Breasted to G.E.H., October 20, 1921.

26. G.E.H. to E.C.H., April 29, 1909.

27. G.E.H., unpublished notes on religion.

28. G.E.H. to T. H. Morgan, August 26, 1932.

29. G.E.H. to H. M. Goodwin, August 23, 1924.

30. Medical record of G.E.H. at Johns Hopkins Hospital, examined by the author.

31. G.E.H. to H. M. Goodwin, January 7, 1925.

32. G.E.H. to E.C.H., November 9, 1930.

33. G.E.H. to H. M. Goodwin, October 22, 1925.

34. *Ibid.*

35. *The Memoirs of Herbert Hoover, 1920–1933* (New York: The Macmillan Company, 1952), pp. 73–74.

36. G.E.H. to H. M. Goodwin, January 31, 1926.

37. *Ibid.*

38. *Ibid.*

39. G.E.H. to E.C.H., March 14, 1928.

40. G.E.H., "Science and the Wealth of Nations," *Harper's Magazine*, January, 1928.

41. J. J. Carty to E. A. S. Clarke, June 22, 1929.

42. Pledge by steel industries was made August 19, 1929.

43. G.E.H. to Henry S. Pritchett, October 14, 1929.

44. G.E.H. to Frank B. Jewett, December 17, 1930.

45. G.E.H. to F. B. Jewett, November 14, 1931.

Chapter Eighteen

1. Robert O. Schad, "Henry Edwards Huntington, the Founder and the Library," *Huntington Library Bulletin*, No. 1 (May, 1931).

2. William Hertrich, superintendent of the San Marino Ranch, personal communication. Memories of Huntington, later recorded in *The Huntington Botanical Gardens, 1905–1949, Personal Recollections of William Hertrich* (San Marino, Calif.: The Huntington Library, 1949).

3. G.E.H., diary record, February 4, 1904.

4. Dinner for Huntington, described in local newspapers, 1906.

5. G.E.H., autobiographical notes.

6. Memories of Ernest Batchelder on founding of Pasadena Music and Art Association, personal communication.

7. *Ibid.*, and G.E.H., autobiographical notes.

8. G.E.H. to Arabella Huntington, February 14, 1914.

9. G.E.H., diary record, on meeting with Archer Huntington, April 16, 1914.

10. G.E.H. to E.C.H., April 16, 1914.

11. G.E.H., autobiographical notes.

12. G.E.H. to H. M. Goodwin, May 6, 1914.

13. G.E.H. to Henry E. Huntington, April 17, 1914.

14. H. E. Huntington to G.E.H., April 20, 1914.

15. G.E.H. to H. E. Huntington, May 11, 1914.

16. G.E.H. to J. H. Breasted, May 11, 1914.

17. J. H. Breasted to G.E.H., May 19, 1914.

18. *Ibid.*

19. H. E. Huntington to G.E.H., October 5, 1914.

20. G.E.H. to E.C.H., December 14, 1914.

21. G.E.H. to Henry M. Robinson, January 14, 1916.

22. G.E.H. to A. A. Noyes, March 30, 1916.

23. G.E.H. to H. E. Huntington, March 28, 1916.

24. H. E. Huntington to G.E.H., April 22, 1916.

25. G.E.H. to M.H.S., May 1, 1916.

26. G.E.H. to J. H. Breasted, May 1, 1916.

27. Ernest Batchelder, personal communication.

28. "The Huntington Library," unpublished manuscript.

29. G.E.H. to A. Mosgrove, July 19, 1921.

30. G.E.H. to H. M. Goodwin, February 5, 1922.

31. Ernest Batchelder, *op. cit.*

32. First steps taken to make over center of city, *Pasadena Star-News,* April 6, 1923.

33. Robert Freeman to G.E.H., May 5, 1923.

34. G.E.H. to W.B.H., July 1, 1924.

35. G.E.H., "The Future Development of the Huntington Library and Art Gallery," written August and September, 1925.

36. G.E.H. to J. H. Breasted, August 29, 1925.

37. G.E.H. to Elihu Root, October 15, 1925.

38. G.E.H. to Henry Crew, November 21, 1925.

39. Max Farrand to G.E.H., January 20, 1926.

40. G.E.H. to Max Farrand, July 12 and September 23, 1926.

41. G.E.H. to Elihu Root, January 10, 1927.

42. G.E.H. to Root, January 14, 1927.

43. G.E.H. to J. C. Merriam, January 24, 1927.

44. Farrand offered position as director of research, February 5, 1927. Huntington guaranteed eight millions in additional endowment.

45. W.B.H. to G.E.H., March 15, 1927.

46. G.E.H. to H. M. Goodwin, June 11, 1927, after Huntington's death, May 23, 1927.

Chapter Nineteen

1. G.E.H., "The Possibilities of Large Telescopes," *Harper's Magazine,* CLVI (April, 1928).

2. W. W. Keen to G.E.H., March 22, 1921.

3. G.E.H., "The Possibilities of Instrumental Development," *Pop. Astron.,* No. 31 (1923), 568–74. Published also in Smithsonian Annual Report, 1923, pp. 187–97.

4. G.E.H. to W. W. Keen, September 23, 1926.

5. Visit of H. J. Thorkelson to Solar Laboratory, October 1, 1926. Record in Rockefeller Foundation files.

6. "The Possibilities of Large Telescopes," *op. cit.,* 1928.

7. G.E.H. to Wickliffe Rose, February 14, 1928.

8. W. Rose to G.E.H., February 21, 1928.

9. G.E.H. to W. S. Adams, on meeting with Rose, March 15, 1928. Events also recorded in G.E.H., autobiographical notes, and in diary record.

10. G.E.H. to W. S. Adams, *op. cit.*

11. G.E.H. to R. G. Aitken, June 26, 1928.

12. Raymond Fosdick, *The Story of the Rockefeller Foundation* (New York: Harper & Bros., 1952), p. 13.

13. J. C. Merriam to G.E.H., April 3, 1928.

14. On Rose and Thorkelson's visit to Pasadena, March 26, 1928, Thorkelson's diary record. Also W. S. Adams to G.E.H., March 30, 1928.

15. Rose, diary record, April 10, 1928, in Rockefeller Foundation files.

16. G.E.H. to E.C.H., April 12, 1928.

17. G.E.H. to W. Rose, April 16, 1928.

18. Telegram, A. A. Noyes to E. C. Barrett, April 26, 1928.

19. J. C. Merriam to H. S. Pritchett, April 28, 1928. Record in C.I.W. files.

20. G.E.H. to J. C. Merriam, April 27, 1928.

21. G.E.H. to Elihu Root, April 27, 1928.

22. Telegram, J. J. Carty to G.E.H., April 28, 1928.

23. G.E.H. diary entry, April 29, 1928.

24. G.E.H., notes on train, April 30, 1928.

25. G.E.H. to E.C.H., May 3, 1928.

26. Telegram, A. A. Noyes and R. A. Millikan to G.E.H., May 4, 1928.

27. G.E.H. to E.C.H., May 8, 1928. Also G.E.H. to W.B.H., May 10, 1928.

28. H. S. Pritchett to G.E.H., May 9, 1928.

29. G.E.H., diary entry, May 9, 1928.

30. G.E.H., diary entry, May 10, 1928; also June 19, 1928.

31. G.E.H. to W.B.H., May 10, 1928; and G.E.H., diary entry, May 10, 1928.

32. G.E.H. to E.C.H., May 11, 1928.

33. G.E.H. to J. J. Carty, May 25, 1928.

34. Frederick Keppel to G.E.H., June 5, 1928.

35. G.E.H., diary entry, June 11, 1928.

36. G.E.H. to J. J. Carty, June 12, 1928.

37. G.E.H. to Elihu Root, June 14, 1928.

38. G.E.H. to W.B.H., June 12, 1928.

39. H. M. Goodwin to G.E.H., June 30, 1928.

40. H. M. Goodwin to G.E.H., December 29, 1928.

41. G.E.H. to E.C.H., May 3, 1927.

42. G.E.H. to E.C.H., May 12, 1927.

43. J. H. Breasted to G.E.H., May 18, 1929.

44. G.E.H. to J. H. Breasted, May, 1929.

45. Elihu Thomson to G.E.H., March 27, 1928.

46. Gerard Swope to G.E.H., November 12, 1930.

47. A. L. Ellis to J. A. Anderson, January 6, 1931.

48. G.E.H. to H. M. Goodwin, January 15, 1931.

49. A. L. Ellis to G.E.H., March 23, 1931.

50. Harlow Shapley to W. S. Adams, March 31, 1931.

51. G.E.H. to John Johnston, Director, Research Laboratory, U.S. Steel Corp., April 28, 1931.

52. E. Thomson to G.E.H., July 28, 1931.

53. G.E.H. to Sir Herbert Jackson, September 11, 1931.

54. Ellis to G.E.H., September 3, 1931.

55. G.E.H. to Ellis, September 10, 1931.

56. G.E.H., diary entry, October 3, 1931.

57. Arthur L. Day to G.E.H., April 1, 1928.

58. G.E.H. to Max Mason, February 25, 1931.

59. Milton Humason to G.E.H., March 9, 1931.

60. G.E.H. to Sir Joseph Larmor, June 12, 1931.

61. A. L. Day to G.E.H., September 30, 1932.

62. G.E.H. to H. M. Goodwin, December 13, 1932.

Chapter Twenty

1. G.E.H. to Sir Henry Lyons, March 16, 1932.

2. G.E.H. to E.C.H., quoting from Sir Henry Lyons, May 19, 1932.

3. G.E.H. to H. H. Donaldson, September 7, 1932.

4. G.E.H. to Sir Henry Lyons, September 21, 1934.

5. Memories of W.E.H. II.

6. G.E.H. to M.H.S., December 20, 1932.

7. G.E.H. to H. M. Goodwin, December 13, 1932.

8. G.E.H., autobiographical notes, written February 8, 1933.

9. G.E.H. to M.H.S., June 27, 1934.

10. G.E.H., quoted in *The Telescope,* May–June, 1936.

11. Harold Babcock, personal communication, 1965.

12. G.E.H. to George Sarton, July 24, 1933.

13. G.E.H., "Some of Michelson's Researches," *P.A.S.P.,* XLIII (1931), 175–85.

14. J. A. B. Scherer to G.E.H., July 12, 1932.

15. Extracts from letters of E.C.H., 1934.

16. Theodore Dunham to G.E.H., June 24, 1933.

17. G.E.H. visit to Corning, N.Y., October 26, 1933.

18. Telegram, W. S. Adams to G.E.H., March 26, 1934, and accounts of pouring in *The New York Times,* etc.

19. A. L. Day to G.E.H., June 2, 1934.

20. *The New York Times,* December 3, 1934.

21. *The New York Times,* July 10, 1935, and J. C. Hostetter to G.E.H., July 11, 1935.

22. W. J. Hussey, "Report of Committee on Southern and Solar Observatories," *C.I. Yearbook,* No. 2 (1903), 5–70.

23. The story of the search for sites is told in greater detail in the author's *Palomar, The World's Largest Telescope* (New York: Macmillan, 1952), 73–87.

24. J. A. Anderson to G.E.H., August 19, 1935.

25. Captain Clyde S. McDowell granted leave by Navy, December 11, 1934.

26. January 27, 1936, Westinghouse chosen to build mounting, designed by J. A. Anderson, Mark Serrurier, and R. Edgar. The story of its building is told in *Palomar, The World's Largest Telescope,* 1952, *op. cit.* Additional material on

the site and mounting in G.E.H., "The Astrophysical Observatory of the C.I.T.," *Ap. J.*, CLXXXII (1935), 111–39.

27. 200-inch disk removed from annealer. *The New York Times*, December 8, 1935.

28. G.E.H. to G. McCauley, April 13, 1936.

29. G.E.H. to H. M. Goodwin, April 14, 1936.

30. Robert S. Ball, *The Story of the Heavens* (London: Cassell & Co., 1887), p. 431.

31. Ira S. Bowen, in address given May 15, 1964, at the C.I.W.

32. G.E.H., "The Possibilities of Large Telescopes," *op. cit.*, 1928.

33. Bowen, *op. cit.*

34. *Ibid.*

35. G.E.H. to Harlow Shapley, January 7, 1936.

36. G.E.H. to H. M. Goodwin, April 7, 1936.

37. Speeches given at symposium, reported in newspapers, April, 1936.

38. Editorial in *The New York Times*, April 7, 1936.

39. H. M. Goodwin to G.E.H., April 8, 1936.

40. G.E.H. to H. M. Goodwin, April 14, 1936.

41. Elizabeth Connor, personal communication.

42. Personal recollection of Charles Breasted, as told to the author.

43. G.E.H. to H. M. Goodwin, May 20, 1937.

44. Personal reminiscence of M.H.S.

45. Ruddy Hale to W.B.H., February 22, 1938.

46. *The New York Times* editorial, February 23, 1938.

BIOGRAPHICAL ARTICLES ON GEORGE ELLERY HALE

Abetti, Giorgio. *George Ellery Hale, Notizie Biografiche, Estratto dalle Memorie della Società Astronomica Italiana*, Vol. XI, 3, Pavia, 1938.

Adams, Walter S. "George Ellery Hale," *Astrophysical Journal*, LXXXVII (1938)

————. Biographical Memoir of George Ellery Hale, *National Academy of Sciences, Biographical Memoirs, XXI* (1940).

Aitken, Robert G. "Address of the Retiring President of the Society in Awarding the Bruce Gold Medal to Dr. George Ellery Hale," *Astronomical Society of the Pacific*, No. 162 (February, 1916).

Babcock, Harold D. "George Ellery Hale," *Publications of the Astronomical Society of the Pacific*, L, No. 295 (June, 1938).

Carew, Harold D. "A Man of Many Worlds, George Ellery Hale," *Touring Topics*, October, 1928, p. 28.

Dunham, Theodore, Jr. Obituary Notice of George Ellery Hale, *Monthly Notices of the Royal Astronomical Society*, February, 1939.

Fox, Philip. "George Ellery Hale," *Popular Astronomy*, XLVI, No. 8 (1938).

Moulton, F. R. "Our Twelve Great Scientists," *Technical World*, November, 1914, 342.

Newall, H. F. "Scientific Worthies, XLVII. George Ellery Hale," *Nature*, CXXXII (July 1, 1933), 1.

Seares, Frederick H. "George Ellery Hale, The Scientist Afield," *Isis*, XXX, No. 81 (May, 1939), 241–67.

Stokley, James. "A Tribute to George Ellery Hale, June 29, 1868–Feb. 21, 1938," *The Sky*, April, 1938.

Turner, H. H. Address on George Ellery Hale given at time of award of the Gold Medal of the Royal Astronomical Society, *Monthly Notices of the Royal Astronomical Society*, LXIV (1904), 388.

Obituary Notices of Dr. George Ellery Hale, Foreign Member of the Royal Society of London, *Nature*, March 19, 1938. Articles by F. W. Dyson, J. H. Jeans, H. F. Newall, F. J. M. Stratton.

"The Works of George Ellery Hale," A Survey of the Career of a Great Living Scientist (in three parts). *The Telescope*, 1936.

BOOKS—A SELECTED LIST

Hundreds of books and thousands of articles have been read in the preparation of this biography. Some of these have provided background material; others trace the developments in which Hale played a leading role. For the reader who would like to trace those developments further, a selected list of books is given. (In general, as will be evident from the notes, the writer has depended on primary sources—correspondence, reminiscences, and articles by Hale himself and by those who knew him or participated in the great events of his life.)

Abbot, Charles G. *Adventures in the World of Science*. Washington, D.C.: Public Affairs Press, 1958.

Abetti, Giorgio. *The History of Astronomy*. Translated by Betty B. Abetti. New York: H. Schuman, 1952.

———. *The Sun*. Translated by J. B. Sidgwick. New York: Macmillan, 1947.

———. "Solar Physics," in the *Handbuch der Astrophysik*. Band IV (1929) and VII (1936). Berlin: Julius Springer.

Bailey, Solon I. *The History and Work of the Harvard Observatory*, Harvard Observatory Monograph, No. 4. New York: McGraw-Hill Book Co., 1931.

Ball, Robert. *The Story of the Heavens*. London: Cassell & Co., 1888.

———. *The Story of the Sun*. London: Cassell & Co., 1893.

Ball, W. Valentine. (ed.). *Reminiscences and Letters of Robert Ball*. Boston: Little, Brown & Co., 1915.

Bates, Ralph S. *Scientific Societies in America*. New York: John Wiley & Sons, Inc., 1945. (A Publication of M.I.T.)

Beer, Arthur. *Vistas in Astronomy*, Vol. I. New York: Pergamon Press, 1955.

Bell, Louis. *The Telescope*. New York: McGraw-Hill Book Co., 1922.

Brashear, John A. *An Autobiography of a Man Who Loved the Stars*. ed. Lucien W. Scaife. New York: American Society of Mechanical Engineers, 1924.

Bray, R. J., and Loughhead, R. E. *Sunspots*. New York: John Wiley & Sons, Inc., 1965.

Breasted, Charles. *Pioneer to the Past*. New York: Charles Scribner's Sons, 1947.

Cassell's Book of Sports and Pastimes. London: Cassell, Peter, Galpin & Co., 1881.

Clerke, Agnes M. *A Popular History of Astronomy during the 19th Century*. London: Adam and Charles Black, 1st ed., 1885.

――――. *Problems of Astrophysics*. London: Adam and Charles M. M. Black, 1903.

Dingle, Herbert. *Modern Astrophysics*. London: W. Collins, 1924.

Dreiser, Theodore. *The Titan*. London: John Lane Co., 1914. (The character of Frank Cowperwood in this novel is said to have been suggested by Charles T. Yerkes.)

Dupree, A. Hunter. *Science in the Federal Government*. Cambridge, Mass.: Harvard University Press, 1957.

Eddington, Arthur S. *Stars and Atoms*. New Haven: Yale University Press, 1927.

Flexner, Simon and James T. *William Henry Welch and the Heroic Age of American Medicine*. New York: The Viking Press, 1941.

Fosdick, Raymond. *Adventures in Giving: The Story of the General Education Board*. New York: Harper & Row, 1962.

――――. *The Story of the Rockefeller Foundation*. New York: Harper & Bros., 1952.

Frost, Edwin B. *An Astronomer's Life*. Boston: Houghton, Mifflin Co., 1933.

Ginger, Ray. *Altgeld's America*. New York: Funk & Wagnalls, 1958.

Goodspeed, Thomas W. *The Story of the University of Chicago, 1890–1925*. Chicago: The University of Chicago Press, 1925.

――――. *William Rainey Harper, first President of the University of Chicago*. Chicago: The University of Chicago Press, 1928.

Gray, George W. *The Advancing Front of Science*. New York: McGraw-Hill Book Co., 1937.

――――. *New World Picture*. Boston: Little, Brown & Co., 1936.

Hendrick, Burton. *The Life of Andrew Carnegie*. New York: Doubleday, Doran & Co., 1932.

Holden, Edward S. *Handbook of the Lick Observatory*. San Francisco: The Bancroft Co., 1888.

Holder, Charles Frederick. *All About Pasadena and its Vicinity*. Boston: Lee and Shepard, 1889.

Jaffe, Bernard. *Outposts of Science*. New York: Simon & Schuster, 1935.

Jessup, Philip C. *Elihu Root*. 2 vols. New York: Dodd, Mead & Co., 1938.

Kiepenheuer, Karl. *The Sun*. Ann Arbor: The University of Michigan Press, 1959.

King, Henry C. *The History of the Telescope*. London: Griffin, 1955.

Kuiper, Gerard P. (ed.). *The Sun*. Chicago: The University of Chicago Press, 1953. (See especially the introduction by Leo Goldberg.)

Langley, S. P. *The New Astronomy*. New York: The Century Co., 1884.

Lewis, Lloyd, and Smith, Henry Justin. *Chicago's Great Century*. New York: Harcourt, Brace & Co., Inc., 1929.

Lockyer, J. Norman. *Contributions to Solar Physics*. London, 1874.

Lockyer, T. Mary and Winifred L., with Prof. H. Dingle and contributions of Charles E. St. John and others. *Life and Work of Sir Norman Lockyer*. London: Macmillan, 1928.

Lowell, A. Lawrence. *Biography of Percival Lowell*. New York: Macmillan, 1935.

Maunder, E. W. *William Huggins*. London: T. C. Jack, n.d.

Menzel, Donald. *Our Sun*. Cambridge, Mass.: Harvard University Press, 1959.

Miczaika, G. R., and Sinton, William M. *Tools of the Astronomer*. Cambridge, Mass.: Harvard University Press, 1959.

Millikan, Robert A. *Autobiography*. New York: Prentice-Hall, 1950.

Newcomb, Simon. *Reminiscences of an Astronomer*. New York: Harper & Bros., 1903.

Newton, H. W. *The Face of the Sun*. Harmondsworth, England: Penguin Books Ltd., 1958.

Noyes, Alfred. *Watchers of the Sky*. New York: Frederick A. Stokes, 1923.

Pannekoek, A. *A History of Astronomy*. New York: Interscience Publishers, 1961.

Pendray, G. Edward. *Men, Mirrors and Stars*. New York: Funk & Wagnalls, 1935.

Pupin, Michael. *From Immigrant to Inventor*. New York: Charles Scribner's Sons, 1924.

Secchi, Angelo. *Le Soleil*. 2 vols. Paris: Gauthier-Villars, 1875–77.

Shapley, Harlow. *Source Book in Astronomy, 1900–1950*. Cambridge, Mass.: Harvard University Press, 1956.

Stratton, F. J. M. *Astronomical Physics*. London: Methuen, 1925.

Struve, Otto, and Zebergs, Velta. *Astronomy of the 20th Century*. New York: The Macmillan Co., 1962.

Sykes, Godfrey. *A Westerly Trend*. Tuscon: Arizona Pioneers Historical Society, 1944.

Verne, Jules. *De la terre à la lune*. Paris: J. Hetzel, 1865.

Waterfield, Reginald. *A Hundred Years of Astronomy*. New York: Macmillan, 1940.

Wilson, Carol Green. *California Yankee*. Claremont, Calif.: The Saunders Press, 1946.

Wright, Helen. *Palomar, The World's Largest Telescope*. New York: Macmillan, 1952.

Yerkes, Robert M. *The New World of Science*. New York: The Century Co., 1920.

Young, Charles A. *The Sun*. New York: D. Appleton & Co., 1895.

Zouche (14th baron), Robert Curzon. *Visits to Monasteries in the Levant*. New York: G. P. Putnam, 1849.

ACKNOWLEDGMENTS

In the long years of working on this book hundreds of people and numerous institutions in many different places have helped me in countless ways. I should like to thank them all, but especially the following institutions and individuals who have provided me with recollections, correspondence, or other information, and have given me permission to use these materials in this biography:

Allegheny Observatory of the University of Pittsburgh, and its Director, Nicholas E. Wagman, especially for Brashear, Keeler, and Langley correspondence.

California Institute of Technology, and Robert A. Millikan, for information on the National Research Council and on Dr. Millikan's coming to the institute. Also to E. C. Barrett for material on the early history of Throop.

Carnegie Institution of Washington, and Paul A. Scherer and his wife, Margaret Hale Scherer, for assistance on examining the files of trustees' meetings and of its President, J. C. Merriam. Also Caryl Haskins, now President of the C.I.W.

Corning Glass Works, and especially J. C. Littleton, Director of Research, and Catherine Mack, Librarian, for information on the casting of the 200-inch mirror.

Hartford State Library for material on the history of Connecticut, especially probate and other records relating to members of the Hale family.

Harvard College Observatory, and its former Director, Harlow Shapley, for assistance on tracing the history of G.E.H.'s work at that observatory.

Harvard University and the Harry Elkins Widener Collection for access to the files of the Harvard College Observatory and especially of E. C. Pickering.

Henry E. Huntington Library and Art Gallery, and members of its staff, for use of its manuscript and reference collections.

Johns Hopkins University for access to the H. A. Rowland correspondence in their archives; also the files in the William H. Welch Library for information on Welch, Simon Flexner, and the National Research Council. Also to the Henry Phipps Psychiatric Institute for permission to examine Hale's medical record.

Library of Congress, Manuscript Division, especially for permission to examine the Simon Newcomb, Andrew Carnegie, and Woodrow Wilson files.

Lick Observatory of the University of California for the use of files, especially those dealing with Barnard, Burnham, and Holden, and to Donald and Mary Shane for their help in every aspect of this work.

Los Angeles Public Library for material on local history of southern California.

Massachusetts Institute of Technology for access to its alumni files and to records in the Charles Hayden Library. Also to the following classmates for their memories of college days: Frank Greenlaw, H. M. Goodwin, G. A. Packard, and Willis R. Whitney.

Mount Wilson Observatory and the Hale Solar Laboratory for the chance to go over the thousands of letters, bills, and memorabilia which make up the Hale collection, and to the entire staff for their interest in the biography and for their friendship over the long years of its gestation. I wish especially to thank William C. Miller for his invaluable help in providing most of the photographs in this book.

Mount Wilson Toll Road Company for permission to examine the minutes of their meetings which relate to the early history of the observatory.

National Academy of Sciences and *National Research Council* for access to their files, especially those relating to the history of the academy from 1902 on, and of the council from the time of its birth.

National Archives for examination of files, especially those dealing with the history of the U.S. Naval Observatory, and with Hale's efforts to wrest it from naval control. Also for additional material relating to the National Research Council during World War I.

New York Public Library for material on J. S. Billings and Mrs. Henry Draper in its manuscript division. Also for countless hours spent in the reference and science divisions.

Pasadena Public Library for local history of Pasadena and the surrounding region, and especially for access to its newspaper files.

Princeton University Observatory for material on Charles A. Young and the early history of the observatory. Special thanks to Lyman Spitzer and Martin Schwarzschild for their interest and help.

Rockefeller Foundation for access to its files, especially those dealing with the history of the Hale 200-inch telescope, and to George W. Gray and Thomas B. Appleget, Vice-President of the foundation, for their help.

Royal Astronomical Society for assistance in my search for information on Sir William Huggins and other British astronomers.

Smithsonian Institution for permission to examine the C. D. Walcott correspondence. Also to C. G. Abbot, former Secretary, for his vivid memories of Hale and the early days on Mount Wilson.

University of California at Berkeley for access to the files of President B. I. Wheeler, and for additional information on California history.

University of Chicago, Archives, for material from the files of President Harper and other members of the early university faculty, including that of T. J. J. See.

Vanderbilt University, Nashville, Tennessee, and Carl Seyfert at the Dyer Observatory for information on E. E. Barnard. Also to Mary Calvert for her hospitality and assistance.

Yerkes Observatory of the University of Chicago for weeks spent consulting its files, which provided not only the history of Yerkes but also much concerning the beginnings of Mount Wilson. I am grateful to its staff at that time for their friendship and help, especially to S. Chandrasekhar for his evaluation of

Hale's contribution to science as well as to astronomy, also to Gerard P. Kuiper, Otto Struve and Bengt Strömgren for their encouragement and interest in the biography.

In addition to these institutions and to those individuals already mentioned, I should like to thank the following: Giorgio Abetti of Florence, Italy, for his memories of Hale on Mount Wilson; Charles Breasted, for his understanding and interest in this biography during the early days of its writing; Ernest Batchelder for his accounts of meetings of the Pasadena Music and Art Association and of the Pasadena Planning Commission; Edwin G. Conklin at Princeton, for his memories of the beginnings of the National Research Council and of the development of the National Academy of Sciences; Leander Conklin for his account of Evelina Hale's childhood in Brooklyn and Madison; Elizabeth Connor, Librarian at Mount Wilson, for her recollections of Hale; Henry Crew at Northwestern, for his memories of the beginnings of the *Astrophysical Journal,* and of Hale and other astronomers in the early days in Chicago; Margaret Harwood of the Maria Mitchell Observatory in Nantucket for her memories of meetings with Hale; William Hertrich for his account of Henry Huntington; Milton and Helen Humason, for their hospitality and for vivid accounts of the early days on Mount Wilson; Dr. Leland Hunnicutt, for the account of Hale's illness which Hale confided to him; George Jones, the Superintendent of Construction on Mount Wilson for his early memories; Winchester Jones of California Institute of Technology for his account of Hale and the invention of the spectroheliograph; Paul Merrill, for his tape-recorded account of G.E.H. at Mount Wilson; Cora Keeler Moore, daughter of James E. Keeler, whom I visited in San Antonio, for her memories of her father and the early days at Lick Observatory; Alfred Noyes, the poet, who visited the Solar Laboratory and there recounted the origin of his epic poem, *Watchers of the Sky;* R. O. Redman, Director of the Observatories in Cambridge, England, for his assistance on my visit there; Henry Norris Russell for his reminiscences; George Sarton at Harvard for his memories of G.E.H. and of his help on the history of science project; Frederick H. Seares for his recollections of his friendship with Hale, and for his knowledge of the early history of Pasadena; Seward Simons for his reminiscences recounted to me in Oakland, Calif.; Dorothy Michelson Stevens, daughter of Albert Michelson, for extracts from her father's letters; F. J. M. Stratton at Cambridge University, England, for his accounts of Hale's part in the founding of the International Astronomical Union, as well as of his work on the sun; Seymour Thomas, portrait painter, for his memories of Hale as a subject for his brush; George Van Biesbroeck for his account of Hale's return to Yerkes; W. H. Wright for his account of the early days at Lick as told to Mary Shane; Fred E. Wright, my father, for his accounts of work on Mount Wilson and the development of the 200-inch, which I shared before I ever entered the field of astronomy; also for his knowledge of the history of the National Academy of Sciences and the National Research Council, which were part of my growing up in Washington; Robert M. Yerkes for his vivid accounts of Hale's influence on his life before World War I; and finally, for secretarial and research assistance, to Pamela King Bartle and Mary Beecher.

INDEX